BIRDS OF TEXAS

A TIMBER PRESS

FIELD GUIDE

Photo and illustration credits appear on page 603.

Timber Press
Workman Publishing
Hachette Book Group, Inc.
1290 Avenue of the Americas
New York, New York 10104
timberpress.com

Timber Press is an imprint of Workman Publishing, a division of Hachette Book Group, Inc. The Timber Press name and logo are registered trademarks of Hachette Book Group, Inc.

Printed in China on responsibly sourced paper
Cover design by Vincent James based on a series design by Adrianna Sutton
Text design by Will Brown based on a series design by Adrianna Sutton
The publisher is not responsible for websites (or their content) that are not owned by the publisher.
The Hachette Speakers Bureau provides a wide range of authors for speaking events. To find out more, go to hachettespeakersbureau.com or email hachettespeakers@hbgusa.com.

ISBN 978-1-64326-199-7
A catalog record for this book is available from the Library of Congress.

OPPOSITE Great Blue Heron

David Sarkozi

BIRDS OF TEXAS

CONTENTS

SPECIES ACCOUNTS

INTRODUCTION:
BIRDWATCHING IN TEXAS

TEXAS IS THE CROSSROADS of the bird world in North America. Here, east meets west and north meets south. The Texas Rio Grande Valley in the south is decidedly tropical, with colorful Green Jays and Hooded Orioles. The Texas Panhandle is the end of the Great Plains, with Lesser Prairie-Chickens and Western Meadowlarks. North America's Eastern Deciduous Forest's western edge is the mixed forest of the Texas Pineywoods, with Pileated Woodpeckers and White-breasted Nuthatches. The Chihuahuan Desert includes the Texas Trans-Pecos region in Big Bend country, with Chihuahuan Raven and Crissal Thrasher. The Texas "sky islands" of the Chisos, Davis, and Guadalupe Mountains give Texas a taste of the Rocky Mountains, with Steller's Jays and Townsend's Solitaires.

Such varied ecological regions ensure that there will be rich and varied birdlife. More than 660 species have been accepted to the official Texas list maintained by the Texas Ornithological Society's Texas Bird Records Committee. Approximately 360 species are known to breed in Texas.

Birdwatching in Texas is as varied as the birds. Many birdwatchers are casual, interested in knowing the birds they see in their yard and that come to their feeder or birdbath. Some are laser-focused on a single species like Eastern Bluebirds or the swarms of Ruby-throated Hummingbirds that migrate along the Texas coast in September. Some are hybrids, combining birdwatching with photography. Many are hunters, enjoying the birds while pursuing their quarry. (Indeed, are birdwatchers not just a specialized type of hunter?)

Birdwatching is an obsession for many in Texas. They keep a life list and check daily for reports of new species to pursue and add to their lists. Some may have a go bag ready to take with them on a moment's notice to chase a report of the next new species for Texas. "Big Years" are popular—seeing how many species you can record in a single year in a county, the state of Texas, or even the United States. The current record for a Texas Big Year stands at 522 species.

Birdwatching tourists flock to Texas, pun intended. Some come for spring migration on the Texas coast hoping to see a "fallout" of thousands of colorful warblers, tanagers, and buntings. Some come for the tropical species at the southern tip of Texas with exotic names like Great Kiskadee, Common Pauraque, and Morelet's Seedeater. Some come for the diversity of hummingbirds in West Texas, or to see one of the rarest birds in the world, the Whooping Crane.

I've lived my whole life in Texas and have been pursuing birds for more than forty years, ever since I saw a Painted Bunting in my backyard in central Texas. "How could any bird that amazing be in my backyard?" was soon followed by the question "What other amazing birds are in my backyard?"

My hope is that *Birds of Texas* will serve you no matter what your level of interest when you pick up this book. Keep it handy as a quick reference. Study the accounts of those species you hope to find and use it to learn about the new ones you have just seen.

There are many ways to enjoy the birds of Texas, but the only wrong way is to not have fun while doing it.

Painted Bunting

Morelet's Seedeater

BINOCULAR BASICS

BIRDING IS FUNDAMENTALLY a
low-equipment endeavor. In its most basic
form, you only need suitable binoculars and
knowledge from a reliable reference, like this
field guide. Binoculars come, however, in a be-
wildering number of configurations, and not
all those configurations are suitable for birding.
You can find low-magnification, very compact
binoculars more appropriate for use at the opera
and high-power giants used by astronomers that
must be mounted on a tripod. Binoculars for
birding come somewhere in between.

When shopping for binoculars you will
see nearly every pair with a number like 10x25,
15x42, or 10x50. These two numbers tell you
two things about binoculars. The first number
is magnification, how many times larger the
image appears. So 10x means it's 10 times larger,
15x seven-and-a-half times larger, and so on.
The second number is the diameter of the objective lens, how big the end is that you point
at the bird you want to see. A diameter of 25
means 25 mm, or about the size of a US quarter;
42 would be 42 mm or about 1.6 inches, about
the diameter of a golf ball; and 50 would be
50 mm, or about 2 inches in diameter.

The larger the objective lens, the more light it
gathers and the brighter the image. This can be
very important early and late in the day and in
dense woodlands. Subtle details and even color
will be more visible with a larger objective. The
downside of larger objective lenses is . . . well,
they're bigger and heavier. Carrying around a
pair of binoculars with 50 mm lenses can be fa-
tiguing after a few hours. Most birders find that
40 to 45 mm objective lenses are a good compro-
mise between weight and light gathering

You might think the more magnification
the better, but that's not always true. While an
8x binocular magnifies the image 8 times, it
also magnifies vibrations or "shakes" 8 times.
You can find 12x and 15x binoculars. Without
a way to steady the binoculars, wind, caffeine,
and car vibrations may make them unusable
for many people. I used to use 10x binoculars
but tried a pair of 8x and found I could see
so much more detail because the binoculars
didn't magnify the imperceptible shake from
too much coffee consumption. If you have exceptionally steady hands, you might find 10x
binoculars usable, but most birders find they
prefer 7x to 8x binoculars as a balance between
steadiness and magnification.

A third number you need to know about
binoculars is eye relief. Binoculars project
the image out behind the exit pupils, or back
lenses. The distance at which your eye sees
this image full-size is known as eye relief. The
smaller the eye relief, the closer your eyes
need to be to the exit pupils to see the image
well. Too close and you won't see the full field
of view, or maybe even a distorted or "blacked
out" center of the image. Too far back and
you only see part of the image as a dot in the
center of a black field. This is why binoculars
have folding or twist-up-and-down eye cups
for getting them just right. If you wear glasses, you won't be able to get as close to the exit
pupils and will need a pair of binoculars with
more eye relief. If you don't wear glasses, you

will be able to tolerate smaller eye relief. If you have the option of getting glasses specifically for birding, you should ask your optometrist for glasses that sit closer to your eyes. If you will be birding with glasses, look for binoculars that have eye relief of at least 15 mm, and preferably 20 mm.

Look for binoculars that say they are "fully multi-coated." When light passes from the air to the glass, or even between two pieces of optical glass, some of the light is lost. Modern optical glass has special coatings that reduce the amount of light lost. Some of the most advanced coatings now transmit an amazing 95% of the light through the glass. When a manufacturer states binoculars are fully multi-coated, it means all the optical

surfaces have these coatings and you are getting as much light to your eye as the manufacturer can deliver.

Finally, birding is usually an outdoor activity. You should have "fully waterproof" binoculars. Even if you can't see yourself birding in the rain, we all get accidentally caught in the rain from time to time and you can save yourself a lot of stress. It's not just rain either—fog or even high humidity can affect non-waterproof binoculars.

You may find hanging even a light pair of binoculars around your neck for a few hours makes your neck sore. Many birders find relief by using a binocular harness not unlike a backpack in reverse that keeps the weight off the neck.

RARE BIRDS AND HOW
TO DOCUMENT THEM

NEW BIRDERS ARE OFTEN dismayed
when they are asked to submit documentation on a notable sighting. While birding is
often on the honor system, the Texas Bird
Records Committee tries to create a record of
Texas birdlife worthy of scientific scrutiny.
The committee is made up of professional
ornithologists and knowledgeable peers in
the birding world. When the TBRC votes to
accept a record, they are saying the record
supports the sighting. If they fail to accept a
record, except in a few cases where they can
positively show the bird was misidentified,
they are only saying the submitted documentation does not support the identification of
the rare bird. They are not calling the report-
ing birder a liar, just stating that the documentation was insufficient.

So how does one go about documenting
a rare bird? While physical evidence like a
photo, a recording, or a specimen are ideal, it's
not a requirement. Many records are accepted
on just a detailed written description of what
was observed.

It's a terrific idea to practice making the
kind of detailed notes that make a good TBRC
report. In a perfect world you would make
these notes in writing or a recording while ob-
serving the bird, but the world isn't perfect.
You can practice by making note of every detail you can in your head.

Start with the big things: Is the bird dark
above? Is it light below? Does it have streaks or
spots? How heavy are the streaks or spots? Are
there wingbars? What color is the bill? What is
the shape of the bill? Is the upper bill a different color than the lower bill? Is there a supercilium, or eyebrow stripe?

Then move on to the fine details: Is there
a malar, or mustache-like stripe on the face?
Is there an orbital ring around the eye? What
color is the eye? What about the feathers—do
they have light or dark edges? Are the center of
the feathers light or dark?

Don't forget the proportions of the bird.
How long is the bill? Describe it in terms of
how deep the head is. If you laid the bill back
against the head would it go to the back of the
head? Are the wings longer or shorter than
the tail (a factor known as wing extension)? If
the bird is perched or standing, how far down
does the tail extend? Is the posture upright or
horizontal?

As Sherlock Holmes says, "When you
have eliminated the impossible, whatever
remains, however improbable, must be the
truth." Holmes was on to something. Most
important when documenting a rare bird is
to prove it is not a common bird! Once you
show why it isn't an expected bird, you can
show what you observed that makes you believe it is a rare bird.

OPPOSITE Golden-cheeked Warbler

GLOSSARY OF PLUMAGE AND ANATOMY TERMINOLOGY

ALTERNATE PLUMAGE: Breeding plumage

AURICULAR: The area around the ear on the side of the head, behind the eyes

AXILLARIES: Feathers in the space under the wing, corresponding to "wingpits" or armpits in a human

BASIC PLUMAGE: Nonbreeding plumage

CAP: The top of the head, often referred to as the crown

CERE: The non-feathered area at the bill base, enclosing the nostrils

CHIN: The area immediately under the bill

CREST: The tuft of feathers on top of a bird's head that either stick up or can be raised

CROWN: The top of the head, often referred to as the cap

CULMEN: The center ridge of the upper mandible of a bird's bill

DECURVED: Curved downward

EYEBROW STRIPE: A stripe of contrasting shade running laterally above the eye, also referred to as a supercilium

EYE RING: A feathered ring of contrasting shade encircling or partially encircling the eye

EYE STRIPE OR EYE LINE: A stripe of contrasting shade running laterally through the eyes on the sides of the face

FLANKS: The area along the bird's sides, beneath the wings and above the base of the legs

FOREHEAD: The front of the head, above the bill

GORGET: A patch of iridescent feathers on the throat of mainly male hummingbirds

LORE: The space between the eye and the base of the bill

MALAR: The cheek area

MALAR STRIPE: Cheek stripe

MANDIBLE, LOWER: The lower half of the bill

MANDIBLE, UPPER: The upper bill, also known as the maxilla

MANTLE: The upper back, often including the back, tertials, and scapulars taken together

MAXILLA: The upper bill, also referred to as the upper mandible

NAPE: Back of the neck

NECKLACE: A band of contrasting shade running across the upper breast

ORBITAL RING: A ring of bare skin surrounding the eye

PRIMARIES: The outer flight feathers, projecting from the outer half of the wing

PRIMARY COVERTS: The row of feathers that veils the base portion of the primary flight feathers on the dorsal or upper surface of the wing

RECTRICES: The long, stiff feathers of the tail (singular, rectrix)

RECURVED: Curved upward

SCAPULARS: Feathers at the base of the upper wing surface along each side of the bird's back

SECONDARIES: The inner flight feathers, projecting from the inner half of the wing

SECONDARY COVERTS: Feathers covering the upper surface of the wing exclusive of the secondary flight feathers. There are three rows, starting at the leading edge of the wing: lesser secondary coverts, median secondary coverts, and greater secondary coverts

SEMIPALMATED: partially webbed toes

SPECTACLES: contrasting colored lore and eye ring combination

SPECULUM: the row of colorful and often iridescent feathers on the upper wing surface of many ducks

SUBTERMINAL BAND: a band of contrasting shade across the tail, just above the tip

SUPERCILIUM: A band of contrasting shade running laterally above the eye, also referred to as an eyebrow stripe

SUPRALORAL: The area on the face above the lores

TAIL COVERTS AND UPPER TAIL COVERTS: Small feathers on the upper side covering the base of the tail

TERTIALS: The inner flight feathers at the base of the wings

THROAT: The area below the bill, between the chin and the upper breast

UNDERTAIL COVERTS: The feathers covering the underside of the base of the tail

WINGBARS: Stripes in the wing surface usually produced by differently colored wing covert tips

WINGTIPS: The outer ends of the primary flight feathers

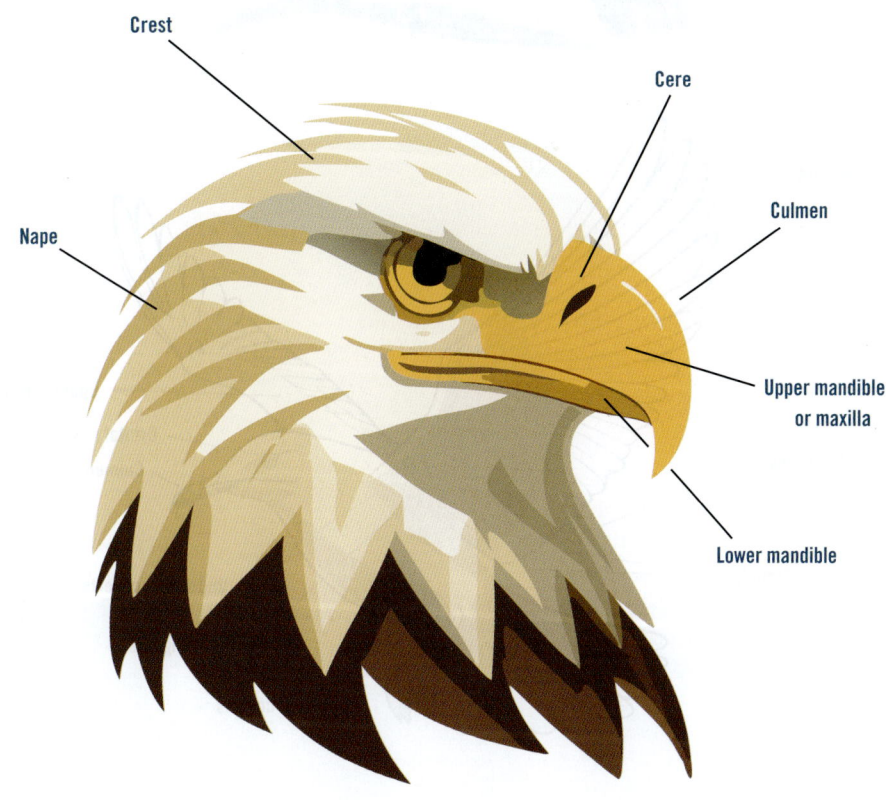

Crest

Cere

Culmen

Nape

Upper mandible or maxilla

Lower mandible

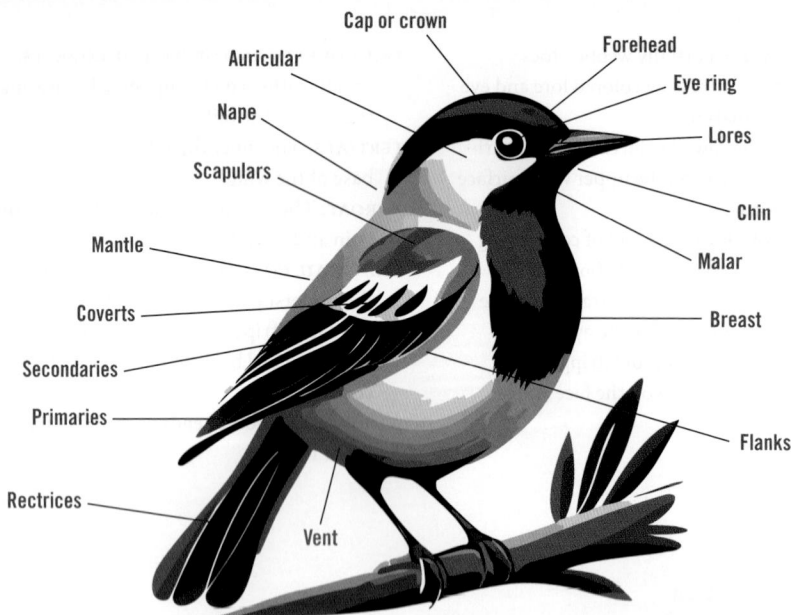

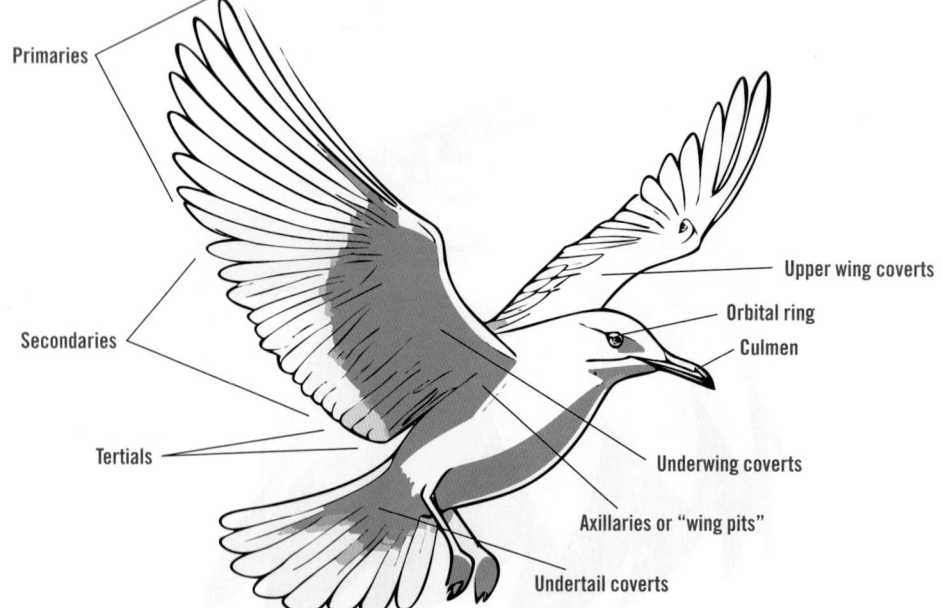

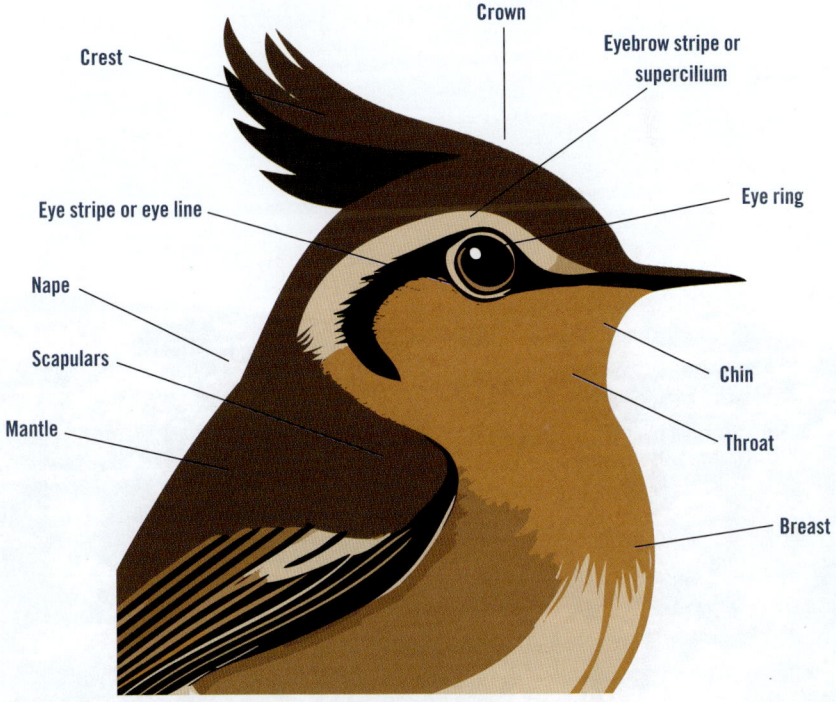

Crown

Crest

Eyebrow stripe or supercilium

Eye stripe or eye line

Eye ring

Nape

Scapulars

Chin

Mantle

Throat

Breast

ECOREGIONS OF TEXAS

TEXAS IS USUALLY DIVIDED into ten natural regions, or ecoregions. These ecoregions are why the birds in Texas are so diverse.

OPPOSITE Cenizo, or Purple Sage, in bloom in Val Verde County, at the edge of the Edwards Plateau

PINEYWOODS

The Texas Pineywoods begin just north of Houston and Beaumont and extend north to Texarkana. The terrain is rolling and covered with pines and oaks. Soils are generally acidic and sandy. There are scattered areas of croplands, native pastures, and planted pastures. Typical birds of the Pineywoods are Pine Warblers, Brown-headed Nuthatches, and Red-bellied Woodpeckers.

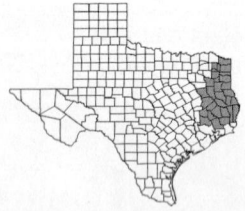

GULF PRAIRIES AND MARSHES

The Texas Gulf Prairies begin at Sabine Pass and extend along the coast to South Padre Island. On the Upper Texas Coast, the prairies extend inland about 100 miles west of Houston to just a thin strip of coastline and barrier islands in the south. The land is low, almost always less than 150 feet in elevation. It includes bays and estuaries, salt grass marshes, oak mottes, and tall woodlands in the river bottoms. The region is home to the now-endangered Attwater's subspecies of Greater Prairie-Chicken, and the extinct Eskimo Curlew once migrated through by the millions. Millions of waterfowl and shorebirds winter on the coast. Less than 1% of the native tallgrass prairie still exists.

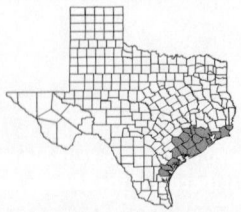

POST OAK SAVANAH

This is a transitional area of oak forests crossing strips of grasslands. The region starts just south of I-10 in Guadalupe and Gonzales counties and extends northeast to the Red River in Red River and Lamar counties. The terrain is rolling or hilly from 300 to 800 feet in elevation. Typical breeding birds are Painted Buntings and Eastern Bluebirds. Up to 12 species of sparrow winter in the region and waterfowl can be abundant in large stock ponds and small lakes.

BLACKLAND PRAIRIE

The Blackland Prairie is named for the fertile black soils that make up the region from northern Bexar County around San Antonio through Dallas to the Red River. Once a tallgrass prairie, its fertile soils were converted to agriculture.

OPPOSITE TOP Trinity River National Wildlife Refuge in the Pineywoods

OPPOSITE BOTTOM Bosque County in the Cross Timbers

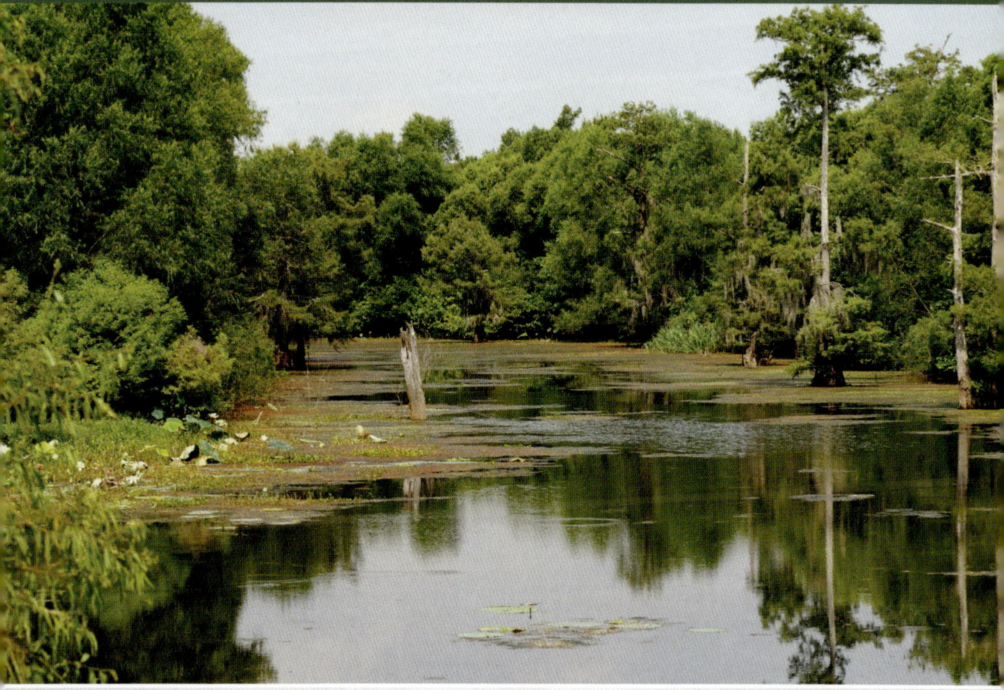

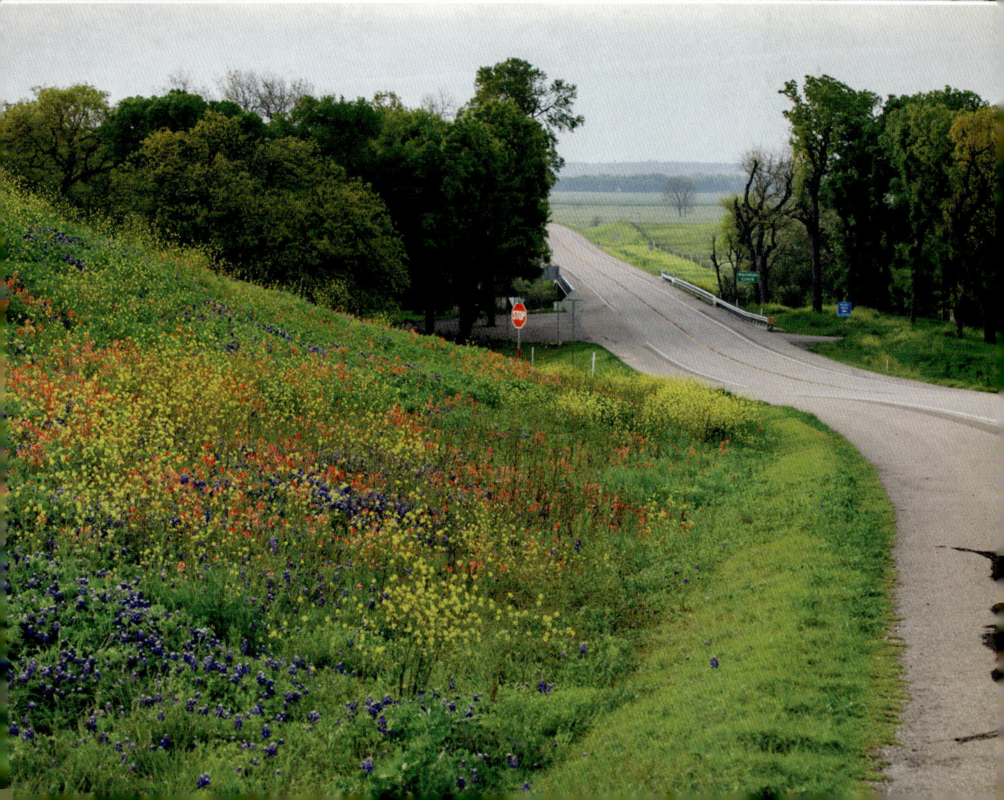

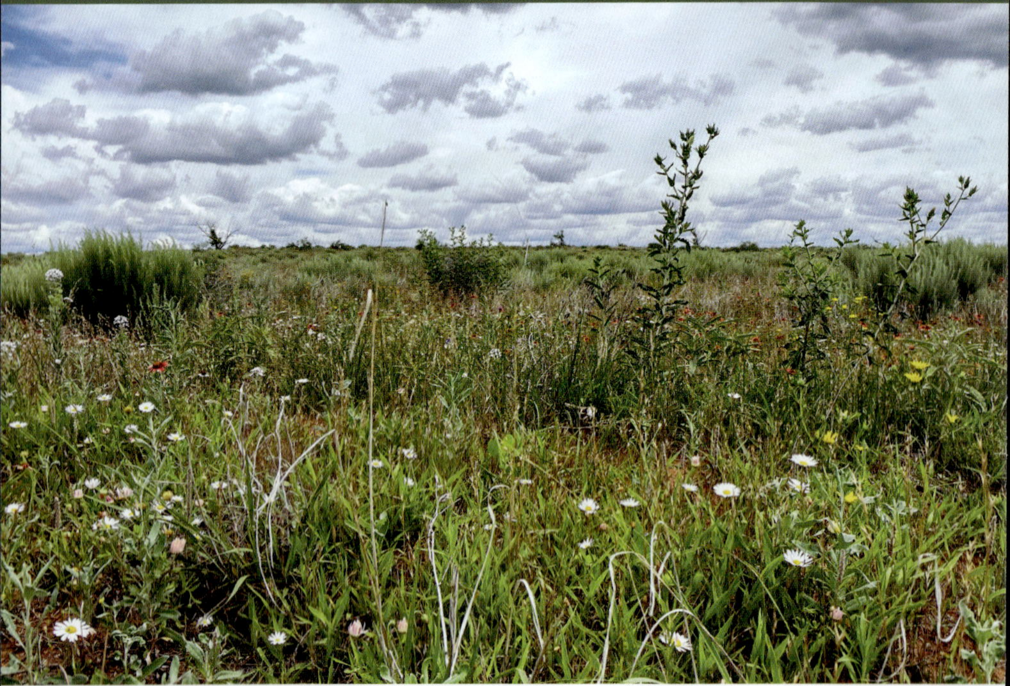

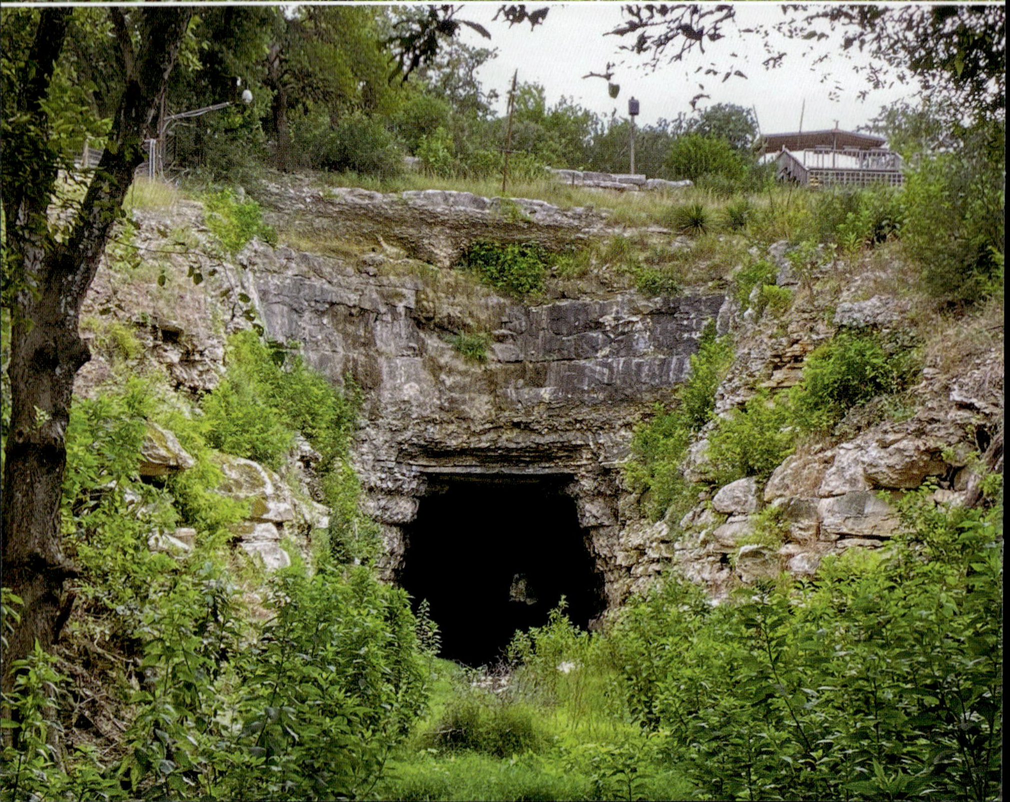

CROSSTIMBERS

From just north of Austin to Fort Worth and Wichita Falls, the timbered areas were a barrier that had to be crossed while traveling between the easier-to-traverse prairies to both the east and west. The open areas between the timbered areas make up the most southern tallgrass prairie.

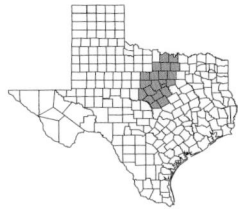

SOUTH TEXAS PLAINS

These plains of thorny scrubs and trees extend from Hwy. 90 between San Antonio and Del Rio to the Rio Grande River and the scattered patches of palms and subtropical woodlands in the Rio Grande Valley. Once open grasslands and more extensive woodland, today it is made up of mesquite, acacia, and prickly pear with open grasslands. Typical birds are tropical and subtropical species like Long-billed Thrasher, Olive Sparrow, and western species like Bullock's Oriole and Lesser Goldfinch.

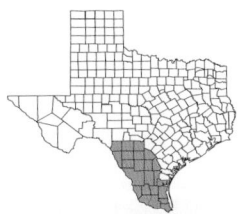

EDWARDS PLATEAU

Also known as the Texas Hill Country, this region extends from the Balcones Escarpment just west of I–35 between San Antonio and Austin to the Pecos River in the west. It has many springs, rocky hills, and steep canyons lined with juniper and live oak trees. Elevations range from 100 feet to 3000 feet. The endangered Golden-cheeked Warbler breeds nowhere else, and it is the core breeding range of the equally endangered Black-capped Vireo.

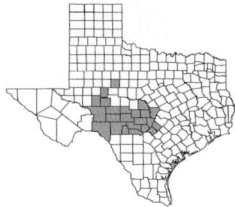

ROLLING PLAINS

These gently rolling hills run from San Angelo and Abilene to the eastern part of the Texas Panhandle. This is the "Big Sky Country" of Texas. Once mostly tall and mid-grasses, uncontrolled grazing has turned it into a mesquite shortgrass savannah.

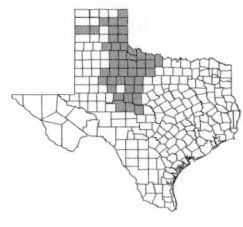

OPPOSITE TOP Matador Wildlife Management Area in the Rolling Plains

OPPOSITE BOTTOM Old Tunnel State Park on the Edwards Plateau

HIGH PLAINS

The Texas High Plains extend from about Seminole in the south, north through the Panhandle and west of Lubbock and Amarillo. Historically shortgrass prairie, mesquite and yucca are spreading in the region. Historically Lesser Prairie-Chickens were common. Thousands of waterfowl and Sandhill Cranes winter in the playa lakes of the region.

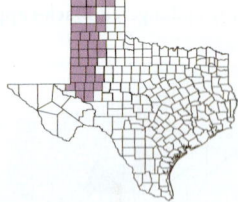

TRANS-PECOS

This region is west of the Pecos River and ranges from about 2500 feet to the high point of 8749 feet at Guadalupe Peak. The three major "sky islands" are here: the Chisos Mountains of Big Bend National Park, the Davis Mountains, and Guadalupe Mountains National Park. The lower-elevation grasslands are home to western species like Scaled Quail and Greater Roadrunners and the mountains have species like the Colima Warbler, Hepatic Tanager, and Black-chinned Sparrow.

OPPOSITE TOP Palo Duro Canyon in the High Plains

OPPOSITE BOTTOM Le Petit Teton, a Trans-Pecos landmark

MAJOR BIRDING SITES IN TEXAS

TEXAS IS BIG AND BOASTS many birding sites. The website eBird.org lists more than 6000 hot spots for Texas. Birds can be found everywhere in Texas, no doubt, but some sites stand out for their diversity and/or the uniqueness of the birds there. The sites highlighted below were chosen based on the number of species reported compared to other sites in the same region.

OPPOSITE Lark Bunting

THE UPPER TEXAS COAST

High Island (379 species)

High Island is a small unincorporated community on the eastern edge of Galveston County and the Bolivar Peninsula. It is located on top of an ancient underground salt dome and from the distance appears like an island. If you're a neotropical songbird at the end of a 17-hour crossing of the Gulf of Mexico it is a lifesaving refuge. The Houston Audubon Society and the Texas Ornithological Society have a total of six sanctuaries in High Island. From late March through mid-May millions of songbirds migrate through and often stop for a day or two to refuel after burning sometimes as much as 50% of their body weight making the perilous trans-Gulf crossing. When the winds are favorable only a small percentage stop here and birding is pleasant, but five or six times a year a rapidly moving cold front with north winds slams into the coast, forcing the birds to literally seem to fall out of the skies. In what seems like a moment, the quiet woods can come alive with birds of all colors. Birders greedily make lists of 20, 25, even 30 species of warblers seen in a day.

Fall can also be an excellent time to visit High Island, as the diversity of migrants is high from mid-September to mid-October.

Recommended sanctuaries at High Island are the Houston Audubon Society's Boy Scout Woods and Smith Oaks Sanctuary, and the Texas Ornithological Society's Hooks Woods.

Texas Ornithological Society's Sabine Woods Sanctuary (328 species)

Located on a feature known as a chenier, an ancient beach dune system now covered with large live oaks, Sabine Woods is located 30 miles east of High Island just outside of Sabine Pass. Like High Island, Sabine Woods is a "migrant trap," and spring and fall are the times to visit. Just a few acres in size, Sabine Woods often produces as many species in a day as the hundreds of birding friendly acres in High Island. Even though it's in close proximity to High Island as the bird flies, it does seem to produce a few more sightings of eastern migrants—such as Cape May Warblers and Black-throated Blue Warblers—than High Island does.

Anahuac National Wildlife Refuge (346 species)

Located just west of High Island, the 40,000-acre Anahuac National Wildlife Refuge stretches from the Gulf of Mexico 10 miles north through a mosaic of saltgrass prairie; salt, brackish, and freshwater marshes; and upland prairies. Established as a waterfowl refuge, the wintering waterfowl are a star here. November through February, large flocks of Snow and Greater White-fronted Geese use the refuge. The Western Hemisphere Shorebird Reserve Network has designated Anahuac National Wildlife Refuge a site of international importance for the numbers of shorebirds using the refuge. In spring and fall, migrating songbirds can be found in the small migrant traps on the refuge. The secretive Yellow and Black Rails are also found here. The Shoveler Pond loop trail is one of the best places to see Least Bitterns and Purple Gallinules in the breeding season.

OPPOSITE Anahuac National Wildlife Refuge

Bolivar Flats Shorebird Sanctuary (324 species)

The Bolivar Flats are sand and mudflats created by the century-old Galveston Jetties on the Bolivar Peninsula. On any given day, thousands of gulls, terns, and pelicans use the flats. From fall through spring, thousands of shorebirds congregate, including Piping, Wilson's, and Snowy Plovers. Least Terns breed here. In winter, as many as 10,000 American Avocets may form one large flock on the flats.

Lafitte's Cove Nature Preserve (329 species)

This 24-acre sanctuary is located inside a well-to-do subdivision on Galveston Island. It has become one of the premier sites to see migrating songbirds on the island. The attraction is a large grouping of mature trees that provide a much-needed sanctuary for songbirds once they complete their marathon Gulf of Mexico crossing. The sanctuary has been enhanced with water features that provide fresh water for the birds. The best times to visit are mid-March to mid-May and September through mid-October.

THE RIO GRANDE VALLEY

Bentsen–Rio Grande Valley State Park (370 species)

Bentsen–Rio Grande Valley State Park maintains a network of feeding stations November through March that attract species you can see nowhere else in the United States but in the Texas Rio Grande Valley. Plain Chachalacas forage under feeders looking like small dinosaurs, brightly colored Green Jays try in vain to defend the feeders from each other, orange Altamira Orioles streak by. In spring and fall there is a hawk watch from a spectacular tower close to the Rio Grande River.

Estero Llano Grande State Park (367 species)

Estero Llano Grande State Park is built around wetlands alongside a linear lake known as the Llano Grande. It's home to tiny Least Grebes and giant Ringed Kingfishers. Emerald-colored Green Kingfishers dart along the water. Buff-bellied Hummingbirds breed here and frequent the hummingbird feeders. Skilled or lucky observers look under the bushes for perfectly camouflaged Common Pauraques. Photographers fill their cameras with spectacular shots from the headquarters observation deck overlooking the wetlands.

BELOW The headquarters building and wetlands at Estero Llano Grande State Park

South Padre Island (420 species)

This busy beach town has some very fine migration birding. There are two major sites on the island. The first site is the Valley Land Fund lots of West Sheepshead Street at Laguna Boulevard. These seven undeveloped lots are a spectacular migrant trap, and when there is a north wind in spring you may find as many as 20 species without walking a single step.

The second site, at the north end of town, is the South Padre Island Convention Center. The landscaping for this public facility has become another spectacular migrant trap. With views of adjacent flats and a boardwalk through mangrove trees one can make a half day of birding here, or even a whole day.

EDWARDS PLATEAU

South Llano River State Park (290 species)

Located about 90 minutes west of San Antonio in Junction, this park is one of the best places to see two of Texas's special birds, the Golden-cheeked Warbler and the Black-capped Vireo. Both species are widespread on Edwards Plateau but there is a dearth of easily accessible places where you can see both. This park has great camping facilities and good facilities for day use. November through March, several feeder-blinds are stocked and often very active.

WEST TEXAS

Big Bend National Park (356 species)

Big Bend National Park is one of the largest national parks in the lower 48 states. Within the park are the Chisos Mountains, one of the so-called "sky islands" of West Texas. This small mountain range is the only location for two species in Texas, the Colima Warbler and the Mexican Jay. Mexican Jays are resident in the Chisos while the Colima Warbler breeds in the high elevations and is present from late March through August. Many birders have made the trek to Boot Springs to see this special warbler.

Besides the high country of the Chisos Mountains there are many riparian areas along the Rio Grande River and thousands of acres of Chihuahuan Desert dotted with small oases in the form of springs. Most visit the park in spring for the Colima Warblers and in winter for cooler weather and abundant wintering species.

BELOW The Boot at Big Bend National Park

The Nature Conservancy's Davis Mountains Preserve (233 species)

The Nature Conservancy's Davis Mountains Preserve was established to preserve the unique ecosystem of one of Texas's three sky islands. This is a cooler, wetter landscape surrounded by an arid lowland desert. Combined with more recent land acquisitions and conservation easements on adjoining properties, the preserve now protects almost 110,000 acres.

The preserve includes the 8300-foot Mount Livermore. The high country of the preserve is one of the few places to find montane species like Steller's Jay, Dusky-capped Flycatcher, and Painted Redstart in Texas

Access to the preserve is limited to a few open weekends every year. Check the preserve's website for upcoming dates.

Balmorhea Lake (323 species)

Balmorhea Lake is a privately owned lake open to the public in West Texas. At 556 acres, it is one of the few large lakes easily accessible by birders in West Texas. Large bodies of water are a magnet for many species in this arid landscape and have a concentrating effect for any water-dependent species. Clark's and Western Grebes are resident here, large numbers of shorebirds use the shoreline in migration, and large numbers of waterfowl are often present in winter and during migrations.

BELOW Mount Livermore in Davis Mountains Preserve

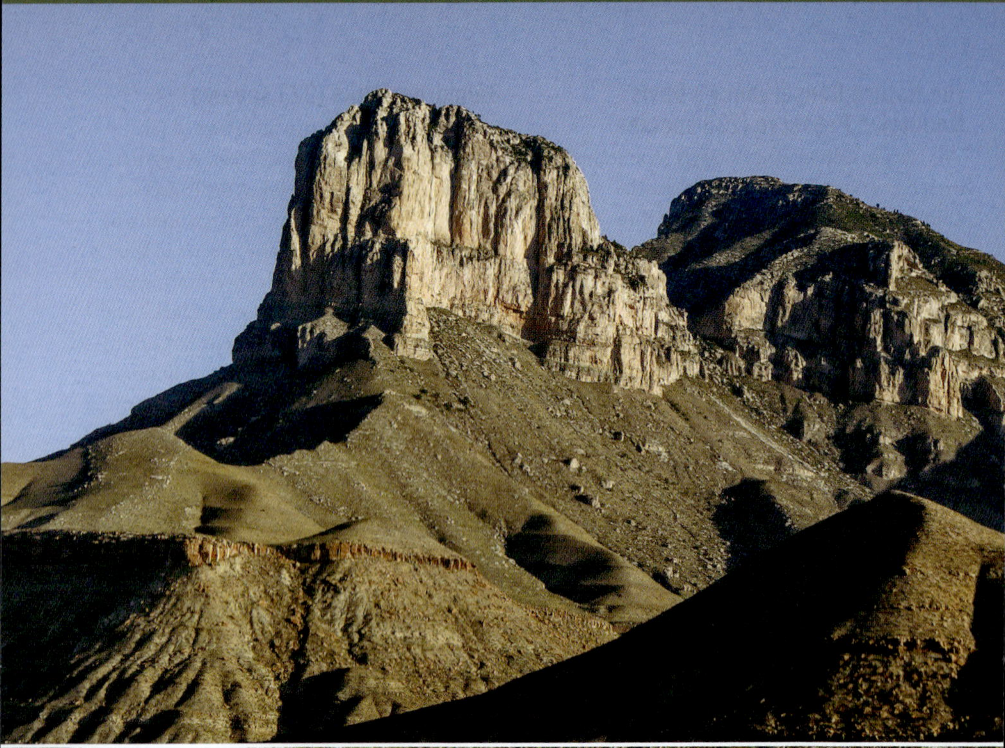

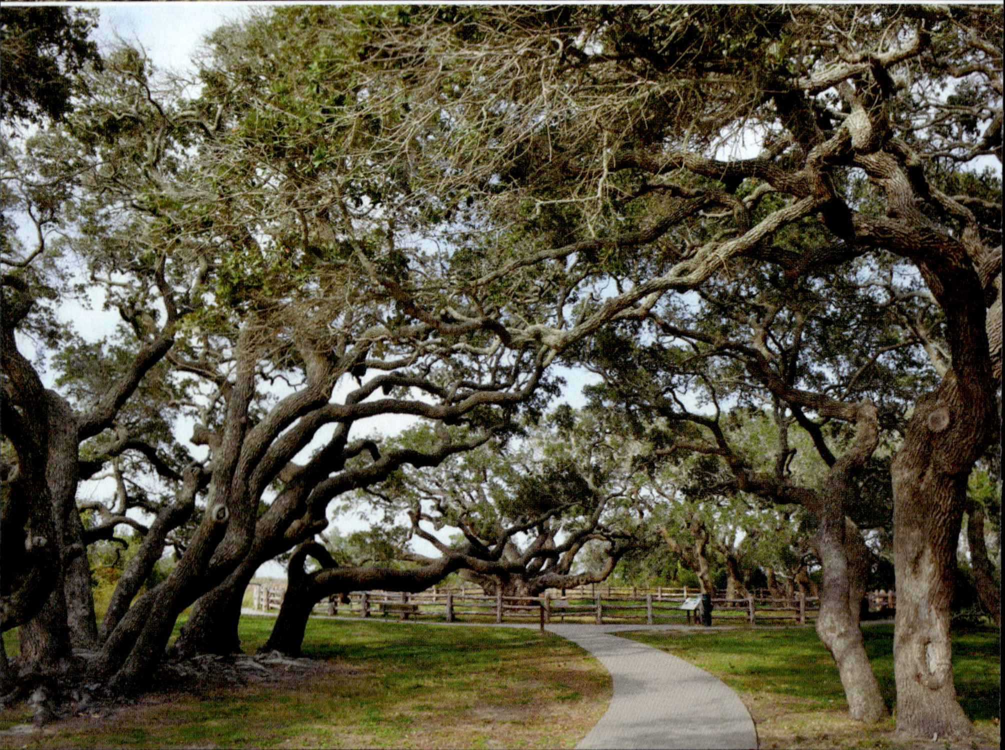

Guadalupe Mountains National Park (282 species)

This national park protects the third of the West Texas sky islands. The Guadalupe Mountains are the "roof of Texas." Nine of the top ten highest peaks in Texas are here, including Guadalupe Peak, the highest at 8751 feet. Several species are found exclusively in Texas here or are easiest to find here. Specialty birds for the Guadalupe Mountains are Juniper Titmouse, Steller's Jay, Pygmy Nuthatch, Sagebrush Sparrow, Gray Vireo, Western Bluebird, and Flammulated Owl.

CENTRAL TEXAS COAST

Hazel Bazemore Park (354 species)

Located on the far northwestern end of Corpus Christi, Hazel Bazemore Park has become known as one of the best hawk watches in North America. Every fall more than a million hawks stream by the counters on the dedicated hawk watch platform. The flights peak in the third week of September, and at times the kettles of hawks can look like smoke.

Many of the birds in the area are more closely associated with the Rio Grande Valley at the southern tip of Texas. Green Jays, Great Kiskadees, Olive Sparrows, and Green Kingfisher can be found in this park.

OPPOSITE TOP El Capitan in Guadalupe Mountains National Park

OPPOSITE BOTTOM The Big Tree in Goose Island State Park

Goose Island State Park (367 species)

Located at the end of the Lamar Peninsula in Aransas County, Goose Island State Park is both a migrant trap and an excellent shorebird location. At low tide, conditions are perfect for shorebirds. In spring and fall, the oak trees on the camping loops can teem with migrants when conditions are good.

Goose Island State Park is also home to the Big Tree. This ancient tree was the state champion live oak until 2003, when a larger one was found. It is still one of the largest live oaks in the world. The fields around the Big Tree often have Whooping Cranes in winter.

Port Aransas (385 species)

The small beach community of Port Aransas has become a Texas birding destination in recent years. Several city parks offer a variety of birding venues. The oldest site to make a name for itself was the Leonabelle Turnbull Birding Center. Before Hurricane Harvey devastated the area in 2017 there was a boardwalk and tower extending out into the shallow lagoon formed by the outflow from the water treatment plant. The boardwalk did not survive. In the years since the hurricane, this site and several others have been combined into one master plan and the sites are being linked together. The birding center now has a new, better boardwalk offering close access to the lagoon. The Port Aransas Nature Preserve at Charlie's Pasture was also wrecked by Hurricane Harvey, but a new boardwalk will link to the birding center boardwalk. The nearby Joan and Scott Holt Paradise Pond is a hot spot for migrating songbirds. The Port Aransas Wetland Park offers excellent shorebirding. And finally, the Port Aransas South Jetty offers access a kilometer into the Gulf of Mexico.

Choke Canyon State Park (353 species)

Choke Canyon State Park consists of two separate park units on the south shore of Choke Canyon Reservoir. The reservoir, one of the newest in Texas, was only filled in 1987. Located near the centerline of Texas, the birdlife here has an east-meets-west flavor to it. Species most often associated with West Texas can be found here, like Vermilion Flycatcher, Bell's Vireo, and Bullock's Oriole. Eastern Species like Blue Jay and Carolina Chickadee are found here, and some South Texas specialties like Long-billed Thrasher, Green Jay, and Great Kiskadee are residents in the park.

CENTRAL TEXAS

Mitchell Lake Audubon Center (358 species)

The Mitchell Lake Audubon Center is a wildlife refuge operated by Audubon in San Antonio. The property is owned by the San Antonio Water System and was once the city's sewage treatment facility. It is primarily a wetlands facility and is the best place in the region to see waterfowl and shorebirds. It is revegetating to native South Texas Brushlands habitat and is the northernmost outpost of many species representing South Texas, like Long-billed Thrasher, Lesser Goldfinch, and Audubon's Oriole.

Hornsby Bend Bird Observatory (368 species)

The Hornsby Bend Bird Observatory is a program of the Austin Water Center for Environmental Research and is an active sewage or biosolids management plant. The site consists of 1200 acres of ponds, woods, agricultural fields, pastures, and 3½ miles of Colorado River bottom. The Austin area does not have many wetlands, and the artificial wetlands created to process the biosolids is a magnet for waterfowl and shorebird species. The varieties of habitat contained within the site allow birders to grow their large list of birds almost year-round in a half day of birding. Numerous rarities have been found here over the years. The ponds and the settling basins are often the star, but the trails in the river bottom along the Colorado River can offer excellent birding during migrations.

Granger Lake (352 species)

Granger Lake is a flood control reservoir on the San Gabriel River in Williamson County. The US Army Corps of Engineers operates three parks on the lake: Taylor Park, Willis Creek Park, and Wilson H. Fox Park. There are four wildlife areas around the lake: Pecan Grove, San Gabriel, Sore Finger, and Willis Creek. These wildlife areas and parks, and the surrounding agricultural fields, have made this area a hot spot for Austin-area birders.

BELOW Hagerman National Wildlife Refuge

NORTH-CENTRAL TEXAS

Village Creek Drying Beds (323 species)

The Village Creek Drying Beds are a working wastewater treatment facility. The site is large and has more than 4 miles of trails. Some are on elevated levees and some are lower alongside the marshy settling ponds. The site encompasses a variety of habitats.

Hagerman National Wildlife Refuge (352 species)

Hagerman National Wildlife Refuge was established in 1946 to protect migratory birds and native wildlife. The refuge is a complex of more than 11,000 acres of shallow marshes, creeks, bottomland hardwood forests, upland forests, grasslands, and managed farm fields on Lake Texoma. This great variety of habitats in a compact area makes for a very high diversity of birds.

THE PANHANDLE

Kiowa and Rita Blanca National Grasslands (117 species)

The Rita Blanca National Grassland and the Kiowa National Grassland in New Mexico are administered together by the National Forest Service. The national grasslands were created after the famous Dust Bowl in the 1930 and '40s to restore damaged lands, and they now consist of about 230,000 acres in Texas, New Mexico, and Oklahoma. While the species count in the grasslands isn't high, birders who venture to the top of the Texas Panhandle can be treated to some species rarely seen elsewhere in Texas, like Northern Shrike, American Tree Sparrow, Prairie Falcon, and Rough-legged Hawk.

The restored grasslands represent the southernmost part of the Great Plains. A drive down High Lonesome Road will take you through the best of the grasslands and give you a taste of what the original Great Plains were like.

Lake Meredith (275 species)

The Lake Meredith National Recreation Area is located on Lake Meredith. Created in the 1960s on the Canadian River, it was the primary drinking water source for Amarillo and Lubbock. Drought has led to dramatic decreases in the lake level, and much of the former lake bottom is becoming grasslands on the shore of the lake. The lake is still a large reservoir and a magnet for birds and wildlife in a relatively arid part of Texas. Many wetlands-associated species of birds that are scarce in the Panhandle can be found here.

OPPOSITE **Rita Blanca National Grassland**

USING THIS BOOK TO IDENTIFY BIRDS

BIRDS OF TEXAS **PROVIDES** the essential information to help you identify the birds most likely to be encountered while birding in Texas. As of December 31, 2023, the Texas Bird Records Committee has accepted records of 666 species of birds in Texas, more than any other state in the United States. This book covers 481 species of *regularly occurring* birds in Texas. Those species that the TBRC considers review species—species that occur four or fewer times per year anywhere in the state over a ten-year average—are not covered in this book. Additionally, those species that have a limited and restricted access range are not covered—for example, pelagic species such as Band-rumped Storm-Petrel and Audubon's Shearwater. The full list of known species for Texas is contained at the end of this book.

The species in *Birds of Texas* are presented in order of the 64th supplement of the American Ornithological Society's *Checklist of North American Birds.* This order groups families and closely related species together, which is handy for identification because it places birds that are most alike together for comparison. All the hawks, kites, and eagles are presented together as a family, for example. Families are further grouped together in genera (plural for genus). Red-tailed Hawk and Rough-legged Hawk are both in the genus *Buteo* and are presented next to each

other. Once you have determined what kind of bird you have—for example, a duck—you need only explore the family of ducks grouped together.

Besides the pictures, a lot of details are given in the text about each species. Don't overlook this resource. Reading the description of a species complements the pictures, drawing your eye to details that may not have jumped out at you when you looked at the bird as a whole in the picture. These details or field marks are often the key to telling similar species apart. Great-tailed Grackle and Boat-tailed Grackle are presented together, and at first glance you might just see two glossy black birds with a purplish sheen. You say to yourself, "How can I tell these apart?" The text points out that Great-tailed Grackles have yellow eyes and Boat-tailed Grackles have dark brown eyes, and presto! It's easy to tell the two species apart.

The text often gives a mnemonic of what the bird sounds like. These mnemonics are not meant to be an accurate representation of the call or song, but a guide to the rhythm and pace of a bird's call. The song of a White-throated Sparrow isn't literally "Old Sam Peabody Peabody," but if you pronounce it as written, it very closely matches the rhythm and meter of the song. Saying these mnemonics out loud is a huge help in making sense of them.

Range maps are presented for all species in this book. Range maps were developed based on the county where the bird species is expected to occur. This works well in most cases because most Texas county boundaries are based on physical features like rivers, and very often the physical features define the boundaries of ecoregions. There are cases where the appropriate habitat only occurs in a small part of the county. In that instance the whole county may be shaded for the species, but the species only occurs where the habitat is appropriate for it. While out-of-range species do occur and are one of the great thrills of birding in general, they don't occur that often. The range maps answer two really important questions of bird identification: should I expect this bird here and should I expect to see this bird at this time of year? The range maps show the breeding range, winter range, year-round range, and migration range. If it's January and the range map shows the bird you've just identified as only expected in that location during spring and fall migration, you should consider other species expected at that time and place.

Finally, if you stick with birding you will eventually find birds that are out of range or season, or potentially not in *Birds of Texas*. Every year or so a bird never before seen in North America shows up to delight birders in Texas. Remember, though, they are called rarities for a reason—they are rare. Follow the steps outlined to identify your bird, and if you've eliminated all the expected species, congratulations, you've found a rarity!

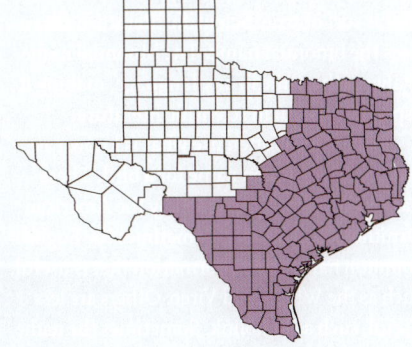

BIRD NAMES

ALL LIVING THINGS ARE GIVEN a name known as a binomial. As you can guess from the name "binomial" it consists of two parts, the genus name and the species name. For example, the Eastern Kingbird's binomial is *Tyrannus tyrannus*. In this case the genus name and the species name are the same because Eastern Kingbird was the first species named in the genus *Tyrannus*. Other species in the same genus share the same genus name in their binomial. The closely related Western Kingbird's binomial is *Tyrannus verticalis*. Here the name shows that close relationship by indicating they are in the same genus, but they are distinct by having a unique species name—in this case *verticalis*.

Binomial names are often called "Latin names", but the truth is they are often not based on Latin but are Latinized words taken from many languages. The binomial is also sometimes called the "scientific name," and that is sort of true. Scientific literature always uses the binomial name. The binomial name is always the same in any language, making it especially useful in scientific literature.

Birders most often use what is known as the English common name for a bird. These names are officially assigned by the American Ornithological Society. Sometimes the English common name adds information that is useful, such as the White-eyed Vireo. Others are less useful, such as Bobolink. Sometimes the name is even misleading; one doesn't really expect to find Connecticut Warblers in Connecticut.

It's not actually all that uncommon for the English common name of a bird to change.

There are lots of reasons why this happens. In some cases, it's because two different jurisdictions used different names for the same bird and reconciliation was needed. Sometime a species is split into two species, or two species were lumped into one. For example, Bullock's and Baltimore Orioles were lumped into one species for a time and called Northern Orioles, but further research showed that was an error and they were split back up. Depending on when a field guide was published, the bird's name would be different.

Some birds have been named after people. These are known as eponyms. Sometimes it was meant to convey an honor—for example, Baird's Sparrow was named for Spencer Baird, a prominent ornithologist of the time, by one of his peers. Other names were a bit more frivolous. Lucy's Warbler was named by Spencer Baird for Lucy Baird, his daughter, because . . . well, because he could.

History has not treated all the people who have had birds named for them well. In November 2023 the American Ornithological Society committed to changing the name of all the eponyms to something more useful. This will be done in batches, with the most offensive names addressed first. The binomial will stay the same, but the English common name will be changed to something more descriptive. Meriwether Lewis's namesake Lewis's Woodpecker will lose that moniker for something like "Pink-bellied Woodpecker," but the binomial *Melanerpes lewis* will stay the same.

OPPOSITE White-faced Ibis (breeding)

SPECIES ACCOUNTS

ANATIDAE
(DUCKS, GEESE, AND SWANS)

The family Anatidae, commonly known as waterfowl, is the family of ducks, geese, and swans. Waterfowl are one of the most diverse families of birds. The giant Trumpeter Swan can weigh 30 pounds and the diminutive Green-winged Teal weighs less than a pound. Mergansers have long serrated bills for catching fish, scoters have heavy, thick bills for crushing mussels and snails, and the familiar Mallard have classic duck bills for their mostly vegetarian diet.

Waterfowl are one of the most cosmopolitan families of birds too. They can be found from the edge of the arctic circle to the southern tip of South America. They inhabit every habitat except the deep ocean waters. Many species are long-distance migrants and can be found migrating in large flocks. Blue-winged Teal breed in North America and winter as far south as Brazil and Argentina, with a large portion of the population wintering in the northern parts of South America.

Waterfowl have been valuable to humans from prehistory to the present day. Long popular as a food source, industrial scale market hunting devastated North American waterfowl populations. But their recovery has been paid for, at least in part, by hunters. Funds generated by the federally required duck stamp that each hunter must buy annually have funded most of the national wildlife refuges in Texas. Founded by waterfowl hunters in 1937, the habitat conservation organization Ducks Unlimited has conserved more than 15 milion acres of waterfowl habitat in North America, roughly equal to the size of West Virginia.

Adults

Ducklings

Black-bellied Whistling-Duck

(Dendrocygna autumnalis)

IUCN RED LIST STATUS (2021): Least Concern
POPULATION TREND: Increasing

Social and often forming very large flocks, these neotropical ducks have been expanding for several decades in Texas. Arboreal and often seen perched in the cavities of large dead trees. They frequent nest boxes and urban and suburban environments, including retention ponds and golf courses.

LENGTH: 21 inches. **WINGSPAN:** 30 inches.
ADULT: Sexes are identical. In flight, the bold white wing patch extending from the base of the primaries to the base of the tertials stands out. The trailing edge of the wings are black. Underwings and belly are black. Bill is red. Face is gray with a pale eye ring. Neck, breast, and back are a rich chestnut. Tail and undertail are dark gray to black. Legs are bright pink. **JUVENILE:** Lack the rich chestnut color and have gray bills and legs. Dark parts are a light charcoal color and not dark black.

Ducklings are straw colored with dark zebra stripes. **VOICE:** Multi-syllable *yip yip yip yip yeeeee* in flight. Often makes a single-syllable *yip* when not in flight. **BEHAVIORS:** Forms permanent pairs. Flies in loose lines and flocks while calling nearly continuously. **HABITAT:** Usually near shallow water no deeper than the legs. **STATUS:** Abundant near the coast in appropriate habitat. The range is expanding inland and locally common east of a line from Del Rio to Fort Worth. Rare to the southern Panhandle.

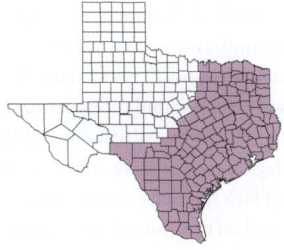

Fulvous Whistling-Duck

(Dendrocygna bicolor)

IUCN RED LIST STATUS (2016): Least Concern
POPULATION TREND: Decreasing

This species is one of the least studied of the common waterfowl in North America. Fulvous Whistling-Ducks have a cosmopolitan distribution, mostly in the southern hemisphere with resident populations along the Gulf of Mexico, South America, sub-Saharan Africa, and the southern Indian subcontinent. Birds in Texas are migratory, with the majority migrating south and only a small portion of the species wintering over.
LENGTH: 19 inches. **WINGSPAN:** 26 inches. **ADULT:** Sexes are identical. Distinctive in flight, dark upper and underwings contrast with a rich brown, almost cinnamon head, neck, and belly. Back is dark as well. Back scapular feathers are edged in cinnamon. Flanks have distinctive white stripes. Bill and legs are a steel gray. Immatures are like adults but lack the rich cinnamon tones. **VOICE:** In flight and on the ground, often gives a distinctive high-pitched *pee-chee* or *wee-whoo*. Sometimes gives the first syllable multiple times.
BEHAVIORS: Flies in unorganized flocks. Does not perch in trees like Black-bellied Whistling-Ducks do. **HABITAT:** Makes heavy use of artificial wetlands like rice fields. Little breeding documented in natural wetlands.
STATUS: Breeding restricted to wetlands, primarily in rice production areas near the coast. Wandering birds have been documented as far north as Hagerman National Wildlife Refuge, Grayson County. Most records are east of I–35, although there are records as far west as El Paso County.

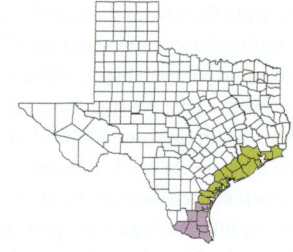

Blue morph in flight with three white morphs

White morph

Snow Goose

(Anser caerulescens)

IUCN RED LIST STATUS (2021): Least Concern
POPULATION TREND: Increasing

Few sights are more spectacular than a large flock of wintering Snow Geese in flight. Snow Geese are dimorphic, coming in a light morph (white) and a dark morph (blue). Expect to find both in most flocks.

LENGTH: 28–31 inches. **WINGSPAN:** 53–56 inches. **ADULT:** Sexes are identical. White morph essentially all white with dark primaries. Dark or blue morph has a white head, gray to blue-gray neck and body. Primaries and secondaries are black. Upper wing is lighter gray. Underwing is white. Vent is white, with gray tail with a pale terminal band. Bill is pinkish with a black cutting edge, often referred to as a grin patch, and a white nail. Legs are bright pinkish. **IMMATURE:** White morph has a pale gray head, neck, and back. Chest and belly are dirty white. Dark or blue morphs are essentially gray to blue-gray all over. Bill and legs are dark gray. **VOICE:** Call is a harsh, honking *whouk* or *heenk*. **BEHAVIORS:** Flies in loose skeins, calling frequently. On wintering grounds forms large flocks. **HABITAT:** Winters in marshes and coastal prairies. Frequents agricultural fields on the coast and in central Texas and the Panhandle. **STATUS:** Wintering range is expanding from a narrow band on the upper and central coast to central and north-central Texas. Wintering birds begin to arrive in late October.

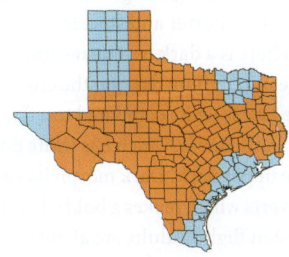

Ross's Goose

(Anser rossii)

IUCN RED LIST STATUS (2021): Least Concern
POPULATION TREND: Increasing

Reducing market hunting pressure and warming in the arctic where they breed have allowed this once-rare species to rebound to between 1.7 and 2 million individuals.
LENGTH: 23 inches. **WINGSPAN:** 45 inches.
ADULT: Clean white over the whole body with black primaries and dark primary coverts. Compared to Snow Geese, bill is shorter, smaller, and pink with a bluish base without the dark cutting edge, or grin patch. Legs are pink. Neck is shorter and stockier than Snow Geese. There is a dark or blue morph, about 1% of the population: dark on the crown, but only the face is white compared to a dark morph Snow Goose, which has white down onto the upper neck. Dark morph has white wing coverts which makes a bold white bar when not in flight. Adults are about 60%–65% of the size of adult Snow Geese. **JUVENILE:** (Aug.– Jan.) Similar to adult but washed in gray. Bill and legs are a neutral gray. **VOICE:** Less vocal than Snow Geese, but in flight and resting gives a *keek-keek* higher in pitch and more rapid. **BEHAVIORS:** Often mixed in Snow Geese flocks. **HABITAT:** Feeds in agricultural fields and shallow wetlands. Roosts on artificial and natural wetlands. **STATUS:** Winters on the High Plains of the Panhandle and the coastal prairies. Migrants found in the eastern two-thirds of the state.

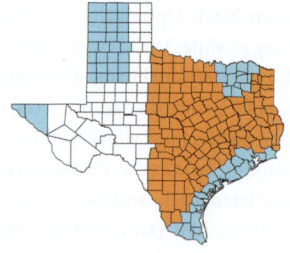

Greater White-fronted Goose

(Anser albifrons)

IUCN RED LIST STATUS (2021): Least Concern
POPULATION TREND: Stable

Almost circumpolar in distribution and the only one of the five gray goose species found in the Americas, this is a popular species with sport hunters, often known by its nickname of "speckle-belly" or simply "specks." Texas birds are almost exclusively of the Tundra subspecies, *A. albifrons frontalis.*

LENGTH: 28 inches. **WINGSPAN:** 53 inches.

ADULT BREEDING: The only gray-brown goose with pink bill and orange feet. White forehead, or front. In flight, the white under-tail, white-tipped tail, and white rump are obvious. Underwings are same gray-brown as the body. Variable dark barring, or speckles, on the belly. At rest shows a white flank-line. **JUVENILE:** (Aug.–Dec.) Much like an adult but the bill is dull yellow. Belly barring less pronounced or absent. The white flank-line is absent. **VOICE:** A high-pitched laughing *kyow-lyow* or *kilik.* **BEHAVIORS:** Strong flier in single or multiple skeins or loose V formation. Strong walker when foraging. **HABITAT:** In migration uses a variety of wetlands, both shallow and deep water. Overwintering birds are usually in or near agricultural land. Uses open water and unvegetated shores for roosts. **STATUS:** Appears in September ahead of other geese. Some birds linger into early April. Uncommon to abundant wintering bird of the coastal prairies and agricultural areas from south of Edwards Plateau. Common migrant through the central state, uncommon in the western third and Pineywoods.

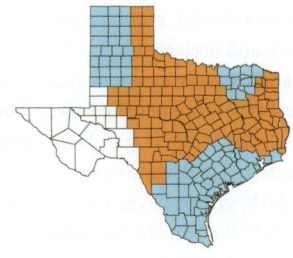

Cackling Goose (foreground) and Canada Goose. Note the shorter neck on the Cackling Goose.

Cackling Goose

(Branta hutchinsii)

IUCN RED LIST STATUS (2021): Least Concern
POPULATION TREND: Increasing

Until 2004, the four subspecies of Cackling Goose were considered subspecies of Canada Goose. Subspecies vary in color from darkest in the west to lightest in the east (Texas). Wintering Cackling Geese in Texas are almost exclusively from the subspecies *B. hutchinsii hutchinsii*, Richardson's Goose. **LENGTH:** 24.8 inches. **WINGSPAN:** 43 inches. **ADULT:** Very similar to Canada Goose. Size is variable but shorter-necked than Canada Goose with a shorter, more steeply sloped bill. Head, bill, and neck are black with a white cheek patch, usually indented at about the eye. Chest is pale but usually darker than Canada Goose. In flight, gray below with a white undertail and white rump. Tail and legs are dark. **VOICE:** Gives a traditional honking call like Canada Goose but slightly higher-pitched. **BEHAVIORS:** Strong fliers,

forming well-organized V formation like Canada Goose. In winter usually roosts on water and flies to feed in agricultural fields. Wintering birds are almost exclusively grazers on grass shoots. **HABITAT:** Winters almost exclusively in agricultural fields. Usually congregates around water for loafing and roosting. In the Panhandle will use small ponds in towns and urban areas. **STATUS:** Formerly common wintering on the coastal prairie but rare in recent decades. Most birds now winter on the Texas High Plains, where they can be common in urban areas around water and golf courses. Uses playas and agricultural areas.

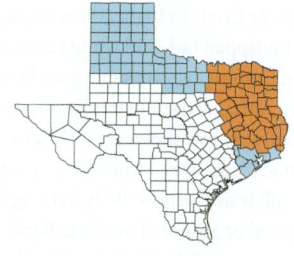

Canada Goose

(Branta canadensis)

IUCN RED LIST STATUS (2021): Least Concern
POPULATION TREND: Decreasing

Before the split of this species into Cackling and Canada Geese it had a whopping eleven subspecies, ranging from the tiny Mallard-sized Aleutian subspecies to the Giant Canada Goose or *B. canadensis maxima,* which can weigh in excess of 14 pounds. Since 2004 there are seven recognized subspecies. **LENGTH:** 25–45 inches. **WINGSPAN:** 43–60 inches. **ADULT:** Head and neck are black except for the white cheek patch, which unlike a Cackling Goose's is not indented at the eye. Dark neck is sharply demarcated from the light-colored breast. Upper parts of the body are grayish brown. Belly and undertail are very light to white. The soft parts—bill, legs, and feet—are black. Some birds are much darker than others. In flight, the appearance is all dark with no contrast on the wings either above or below. **JUVENILE:** (Aug.–Jan.)

Similar to adults, but feather edges much less crisp looking, making barring on the back and breast much less pronounced. **VOICE:** Call is the familiar *h-ronk.* **BEHAVIORS:** Flies in a well-defined V. **HABITAT:** Wintering birds use shallow wetlands, usually near agricultural fields. Nonmigratory resident birds favor park-like settings and golf courses, almost always with a body of water. **STATUS:** Abundant winter resident on the High Plains in the Panhandle, less common on coastal prairies. Canada Goose has become a localized breeder south to I–10.

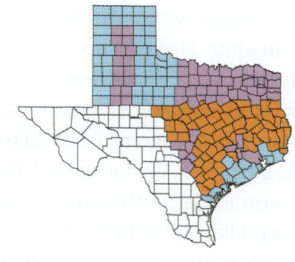

Mute Swan

Tundra Swan

Tundra Swan

(Cygnus columbianus)

IUCN RED LIST STATUS (2016): Least Concern
POPULATION TREND: Unknown

Formerly known as Whistling Swans, Tundra Swans breed in the arctic of both the New and Old Worlds and form lifetime monogamous pairs. They are long-lived species; one wild bird recorded was twenty-one years old. In Texas one may encounter a feral or introduced species of white swan, the Mute Swan of Europe, easily distinguished by its orange bill. Records of these birds are about as common as Tundra Swans in Texas.
LENGTH: 52 inches. **WINGSPAN:** 66 inches.
ADULT: All white. First summer adults have grayish heads and necks. Bill, legs, and feet are black. Most have yellow lores. **JUVENILE:** (Aug.–Mar.) Grayish overall. Legs and feet are black. Bill is pink, becoming black from the base. **VOICE:** Clear mellow *kloo* or *kwoo*. **BEHAVIORS:** Flies mainly between roosting and foraging sites in Texas. Takes off easily,

often with running steps, glides in, and lands feet forward. In migration can fly very high, and groups fly in V formation. Juveniles stay with their parents until they return to the breeding grounds. **HABITAT:** Freshwater lakes and ponds, agricultural fields used for foraging. **STATUS:** Most records are from late October through mid-March. Since 2010 most records in the state are from the Red River drainage or the eastern Panhandle. There are isolated records further south to Chambers and Matagorda counties, and west to Balmorhea Lake, Reeves County.

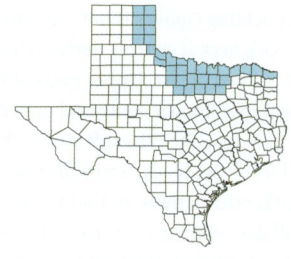

Muscovy Duck

(Cairina moschata)

IUCN RED LIST STATUS (2018): Least Concern
POPULATION TREND: Decreasing

There are two populations of Muscovy Ducks in Texas. In deep southern Texas one can find birds from the wild population in Mexico almost exclusively along the Rio Grande River. These birds are a distinctive black with iridescent green. The second, presumably feral, population of domesticated ducks are common in parks across the state. They are larger and have much more variability in plumage. **LENGTH:** 25–31 inches. **WINGSPAN:** 38–48 inches. **MALE:** Males are dark, almost black iridescent green. Upper and underwing covert feathers are bright white. Males have a black and yellow bill, warty face, and prominent crest. Tail is long and rounded. Feet are short and dark. Most of the feral population has considerable white in the plumage, large facial warts, and various amounts of green. **FEMALE AND JUVENILE:** Like adult males but lack the bright white wing coverts, crest, and face warts. Juvenile bills are duller and have less contrast than adults. **VOICE:** Almost completely silent. Frightened females may quack or croak. Females make soft, shrill calls to ducklings. **BEHAVIORS:** Roosts in trees and nests in tree cavities and nest boxes. **HABITAT:** Texas birds considered of wild origin are almost exclusively found along the Rio Grande River north to the Laredo area. **STATUS:** Records away from the Rio Grande are not uncommon but are regarded as feral or domestic birds.

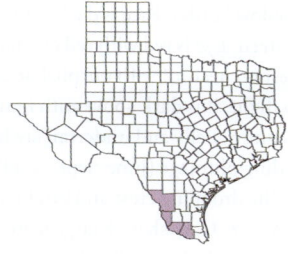

Male

Female

Wood Duck

(Aix sponsa)

IUCN RED LIST STATUS (2021): Least Concern
POPULATION TREND: Increasing

Wood Duck males are almost certainly the most striking ducks native to Texas and the most successful of the cavity-nesting ducks in North America. Some use nest boxes and nests excavated by woodpeckers, but most nest in tree cavities caused by a branch breaking at the trunk.

LENGTH: 18.5 inches. **WINGSPAN:** 30 inches. **MALE:** Head is rounded with a drooping green iridescent crest. Bill is red with black tip and yellow border. Face is dark with white bridle pattern. Eye is red with red orbital ring. Chest is chestnut with purplish iridescence. In flight, it is dark above with indigo blue speculum. Belly and undertail are light. Nonbreeding plumage (June–Sept.) similar but lacks the drooping crest and bright iridescence. **FEMALE:** Grayish with large teardrop patches around dark eyes. Flanks are grayish

and spotted. **JUVENILE:** Similar to female but lacks teardrop patch around eye. **VOICE:** Thin squeal *jeweeep* or *sweeoooo*. **BEHAVIORS:** Flies quickly through wooded habitat, calling as it does. **HABITAT:** Prefers to nest in mature hardwoods near water. Flooded shrubs, button bush, and other wetlands with lots of cover used for feeding. **STATUS:** Uncommon to common and local in the eastern three-quarters of Texas. Summer resident in the western Panhandle. A rare resident along the Rio Grande River in West Texas. More common in East Texas in winter, when migrants join the resident birds.

Male (left) and female (right)

Blue-winged Teal

(Spatula discors)

IUCN RED LIST STATUS (2021): Least Concern
POPULATION TREND: Increasing

Blue-winged Teal are one of the earliest migrating species of ducks in fall. Migrants begin arriving in Texas by the end of August, and by mid-September numbers are great enough that there is an early teal season for Texas duck hunters.
LENGTH: 15.5 inches. **WINGSPAN:** 23 inches.
ADULT BREEDING: (Nov.–June) Males have dark bill, slate-colored head, and bold white crescent on face. Brown with dark speckles on body. Dark rump with white hip patch. In flight, light underwings with dark leading and trailing edge below. Above, light blue wing patch with greenish secondaries near the body. Females are grayish-brown overall with vague white crescent on face. In flight, wings like breeding males but lacking greenish secondaries. **NONBREEDING:** (July–Oct.) Males, females, and juveniles (Aug.–Oct.) all like breeding female. **VOICE:** Female call is a coarse quack. Males, especially when courting, give a thin, whistled *pwis* or *peew*. **BEHAVIORS:** Rests on logs and rocks in water. Flies in fast, erratic flight in a compact group. **HABITAT:** Nests in shallow ponds with grassy vegetation or cover. **STATUS:** Common migrant statewide. Uncommon summer resident in the Panhandle and on the upper Texas coast. Uncommon to common winter resident on the Texas coast and South Texas, becoming less common north. Spring migration mid-February–April. Fall migration late July–late October.

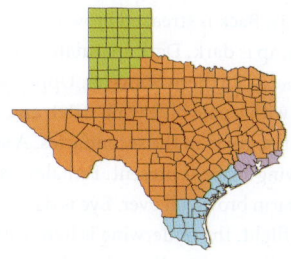

Male (background) and female (foreground)

Cinnamon Teal

(Spatula cyanoptera)

IUCN RED LIST STATUS (2022): Least Concern
POPULATION TREND: Decreasing

The male Cinnamon Teal is unmistakable
with its rich cinnamon-red color, but the females are very similar to female Blue-winged
Teals, hinting at their close relationship.
Cinnamon Teal is one of three closely related
"blue-winged" ducks, along with Blue-winged
Teal and Northern Shoveler.
LENGTH: 16 inches. **WINGSPAN:** 22 inches.
ADULT BREEDING: (Oct.–June) Males are
rich reddish cinnamon color on the head
and body. Back is streaked brown. Eye is
red. Rump is dark. Dark spatulate bill almost like a Northern Shoveler. Upper wing
is dark with a pale blue wing patch and small
amount of green in the secondaries. A small
white wing stripe is present. Females are
plain warm brown all over. Eye is dark, not
red. In flight, the underwing is light with
dark leading and trailing edges. Upper wing
is dark with dull blue wing patch and small
amount of green in the secondaries. A small
white wing stripe is present. **NONBREEDING:**
(July–Sept.) Males are reddish brown overall with red eye. Females and juveniles are
like breeding females. **VOICE:** Females quack.
Males make a dry, rattling *gredek*. **BEHAVIORS:**
Mostly aquatic. Takes flight by leaping near
vertically off the water. **HABITAT:** Mostly
freshwater. **STATUS:** Common migrant in the
western half of the state. Spring migration late
February–April. Fall migration mid-August–
early October. Local summer resident in the
Panhandle.

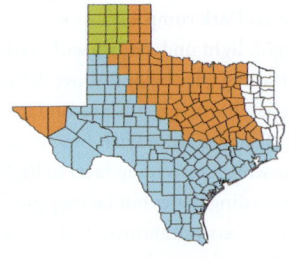

Male (above) and female (below)

Northern Shoveler

(Spatula clypeata)

IUCN RED LIST STATUS (2019): Least Concern
POPULATION TREND: Decreasing

The Northern Shoveler is aptly named. Its broad shovel-like bill is perhaps its most distinguishing feature. Hunters know this bird as a "spoonbill" or a "spoony."
LENGTH: 19 inches. **WINGSPAN:** 30 inches.
ADULT BREEDING: (Dec.–May) Males have large dark spatulate, or shovel-shaped bill. Head is green with yellow eyes. Breast is white with rufous flanks. Rump is dark. Back is slate gray with white on the sides. In flight, underwings are light with dark leading and trailing edges and light blue wing patch with a white bar above. Tail has dark center with white outer edges. Females are brownish with broad, pale-edged feathers. Bill is orange with dark mottling. Underwings are like the male's. Upper wings are dark gray with a thin white bar. **NON-BREEDING:** Males (May–Aug.) have dark, dusky heads and bills more like a female's.

Body is a pale rufous color. Fall males (Sept.–Nov.) similar to nonbreeding males but some white crescent on the face and white chest molting in. **VOICE:** Females give a deep, hoarse *kwarsh.* Males give a nasal *paay.* **BEHAVIORS:** Feeds by continuously dabbling and straining with its bill. Leaps into flight from the water. **HABITAT:** Wide range of wetland habitats. Prefers shallow, open waters. **STATUS:** Common to abundant migrant and winter resident statewide. Fall migration late August–October. Spring migration February–May.

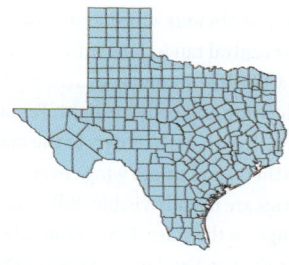

Female (foreground) and male
(with crest raised)

Gadwall

(Mareca strepera)

IUCN RED LIST STATUS (2016): Least Concern
POPULATION TREND: Increasing

Gadwalls are not flashy ducks. Their somber
gray and tan colors are intricately patterned,
and many are impressed with their sub-
dued elegance.
LENGTH: 20 inches. **WINGSPAN:** 33 inches.
ADULT BREEDING: (Sept.–May) Males are
mostly gray with dark gray bill. Head is
puffy with a blocky appearance. Darker on
the crown and chest. Upper wing coverts are
brown/rufous, visible at rest and in flight.
Dark rump is obvious at rest and in flight. In
flight, the central tail is dark with light gray
corners, and white secondaries are easy to
see. Underwing and belly are light. Females
are brownish with dark centers to the mantle
(back feathers) and upper wing coverts. White
secondaries are usually visible. Bill is dark
with orange on the sides. Fine vermiculated
streaks in the head and neck. **NONBREEDING:**
(June–Aug.) Males are like breeding females
but with silvery gray visible on the tertials.
JUVENILE: Similar to females but face is plainer. **VOICE:** Female quack is like a Mallard's.
Courting males give a low burping *mepp*.
BEHAVIORS: Takes off straight from water.
Rides high in water picking food from the
surface. **HABITAT:** Occasionally found on the
coast but avoids marine and turbulent water.
Found in fresh and brackish wetlands with
leafy vegetation. **STATUS:** Common to abundant migrant and winter resident throughout
the state, arriving in October and present
until mid-May.

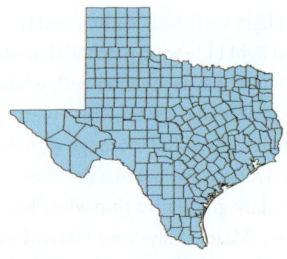

Female (left) and male (right)

American Wigeon

(Mareca americana)

IUCN RED LIST STATUS (2021): Least Concern
POPULATION TREND: Decreasing

The American Wigeon is distinctive, especially the male with its white crown and forehead, which earned it the nickname "baldpate." **LENGTH:** 20 inches. **WINGSPAN:** 32 inches. **ADULT BREEDING:** (Oct.–June) Light blue-gray bill with a dark tip. Forehead is white or pale buffy. Cheeks are gray with dark green face mask. Body and upper wings are pinkish-brown. Large white upper wing patch obvious in flight and usually visible in resting birds. Dark secondaries visible in flight and at rest. In flight, the primaries are gray. Rump is dark. Belly and center of the underwings are white. Tail is long and pointed. Females have light gray heads, dusky around the eye. Body is like males, with dark centers to the back feathers. White wing patch is attenuated to only a white bar. **NONBREEDING:** (July–Sept.) Head is dusky gray with a little green around the eye. White forehead is reduced. **JUVENILE:** (Aug.–Nov.) Like females but more mottled in appearance. **VOICE:** Male whistles *wi-WIW-weew*. Female growls *warr warr warr*. **BEHAVIORS:** Takes off almost vertically from water. Flies erratically like a teal. Swims more between tip-ups than other species. **HABITAT:** Found in salt, brackish, and freshwater habitats with emergent vegetation. **STATUS:** Common to abundant migrant and winter resident statewide. Fall migrants arrive in early August, and spring migrants depart in March but a few linger into May.

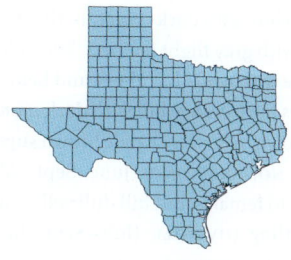

Female (left) and male (right)

Mallard

(Anas platyrhynchos)

IUCN RED LIST STATUS (2019): Least Concern
POPULATION TREND: Increasing

The Mallard is the most familiar of the dabbling ducks. If something is decorated with a duck motif, it is most likely a male Mallard's distinctive head, so iconic that hunters often refer to Mallards as "green heads." **LENGTH:** 23 inches. **WINGSPAN:** 35 inches. **ADULT BREEDING:** (Oct.–May) Males have dark green head with a bright yellow bill, white neck ring above a chestnut chest, and orange feet. Body is pale with a dark rump. Tail is white with dark center. In flight, pale below with gray flight feathers. Secondaries are blue with white bar above and below. Females have orange bill with dark mottling. Pale brown overall with pale brown supercilium. **NONBREEDING:** (June–Sept.) Males similar to females, but bill dull yellow with no mottling. **JUVENILE:** (July–Sept.) Much like females. **VOICE:** Females give the familiar and archetypal duck quack. Males give a similar but short and raspy *quehp*. **BEHAVIORS:** Can take off near vertical from water when alarmed. Can maneuver with rapid turns in flight. **HABITAT:** Can occur on almost any wetland that is relatively shallow. **STATUS:** Common to abundant winter resident in the northern half of the state. Less common southward along the coast. Mostly absent from South Texas Brushlands. Uncommon summer resident in the Panhandle and a rare and local breeder in many other parts of the state.

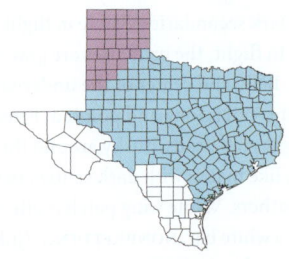

Mexican Duck

(Anas diazi)

IUCN RED LIST STATUS: Has not been assessed

Mexican Duck is one of the least studied North American ducks. Mostly found in Mexico, its range barely extends into Texas, New Mexico, and Arizona. It has perhaps the smallest total population of any of the North American waterfowl, with an estimated population of just 55,000 birds prior to the breeding season.

Mexican Ducks most closely resemble Mottled Ducks, and genetic studies show that Mottled Ducks are their closest relatives. Mexican Ducks are part of the "black duck trio," comprising Mexican, Mottled, and American Black Ducks. This trio is closely related to the common Mallard, and where ranges overlap Mexican Duck hybridizes with both Mallard and Mottled Ducks. Their range may be expanding into the Lower Rio Grande Valley, meaning more hybrids with Mottled Ducks are likely to be found.

LENGTH: 21 inches. **WINGSPAN:** 31.5 inches. **ADULT:** Males are very similar to nonbreeding male Mallards but a warmer brown overall with narrow white bars around the speculum. Crown is dark. Females are nearly identical except for an olive-drab to orange bill. **VOICE:** Nearly identical to Mallards. **HABITAT:** Small wetlands, slow rivers, and artificial wetlands. **STATUS:** A resident in a narrow band about one county deep along the Texas Rio Grande River. Uses wetlands where available in Brewster, Jeff Davis, and Presidio counties. Scattered records north in the western part of the state to Lubbock.

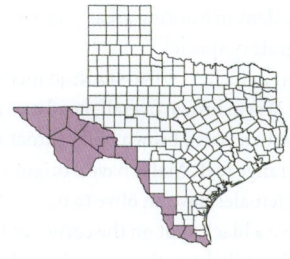

Male

Female

Mottled Duck

(Anas fulvigula)

IUCN RED LIST STATUS (2016): Least Concern
POPULATION TREND: Decreasing

Mottled Duck is closely related to the North American Black Duck and the ubiquitous Mallard. In the Florida population, hybridization with released Mallards is a big problem, but while hybrids aren't unknown in Texas, hybridization hasn't been a threat to the population. Mottled Ducks are especially susceptible to lead contamination from the ingestion of lead pellets from shotgun shells. They are exposed more than other species because they are a resident in hunting areas year-round, unlike migratory species.

LENGTH: 22 inches. **WINGSPAN:** 30 inches. **ADULT:** Warm brown overall, head and neck lighter with a warm brown buffy throat and bright orange feet. Males have a bright yellow bill and females have an olive to orange bill. Both have a black spot on the corner of the bill. In flight, light underwings. Speculum is dark blue with black borders. Very narrow white bars are possible. Upper wing and tail are all dark. **VOICE:** Very Mallard-like. **BEHAVIORS:** Typical of dabbling ducks, explodes into the air from water. Walks well on land. **HABITAT:** Found in non-tidal, fresh water to brackish ponds on coastal prairies. Makes frequent use of urban retention ponds in range that hold water. **STATUS:** Locally common resident along the coastal prairies. Vagrants have been found as far north as Travis County. The Texas population has declined dramatically over the last twenty years.

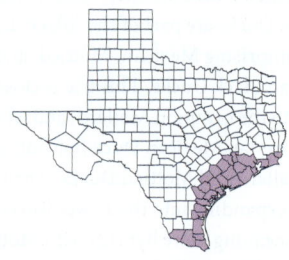

Female

Male

Northern Pintail

(Anas acuta)

IUCN RED LIST STATUS (2019): Least Concern
POPULATION TREND: Decreasing

The Northern Pintail is circumpolar in distribution and breeds all across northern North America, Europe, and Asia. Named for the distinctive long, black central tail feathers that jut out from the shorter tail, it has the longest tail of any freshwater duck in North America. A flock of Northern Pintails is one of the most distinctive species of ducks on the wing. **LENGTH:** 21 inches. **WINGSPAN:** 34 inches. **ADULT BREEDING:** (Nov.–June) Males are unmistakable with brown head, blue-gray bill with dark marks, white breast, and neck stripe to the nape. Gray body, back, and upper wings. Speculum is greenish with a broad white trailing edge. Distinct long central tail feathers. Light belly. Females are light buffy-brown overall, with a plain head and gray bill. **NONBREEDING:** (July–Oct.). Males have buffy-brown head and blue-gray bill with dark marks. Body grayish overall. The signature long central tail feathers are absent. **VOICE:** Females quack quieter and hoarser than a Mallard. Males give a wheezy *whee* call year-round. **BEHAVIORS:** Swift acrobatic flier. Explodes from the water like other dabbling ducks. **HABITAT:** Variety of shallow, inland, fresh water wetlands and intertidal wetlands. Makes use of flooded rice fields. Will forage in dry agricultural fields. **STATUS:** Locally common to abundant migrant statewide August–late February, lingering birds present into May. Winter resident statewide. Uncommon summer resident in the Panhandle.

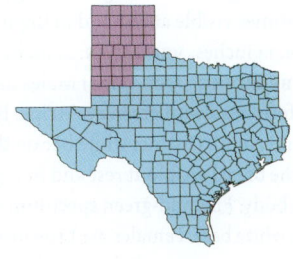

Male   Female

Green-winged Teal

(Anas crecca)

IUCN RED LIST STATUS (2020): Least Concern
POPULATION TREND: Increasing

The Green-winged Teal is one of the most abundant of the dabbling ducks and one of the most popular game species. Green-winged Teal are the second most common ducks taken by hunters. Outside of the breeding season they often fly in tight, synchronized flocks, making them easy targets for hunters.

Green-winged Teal are the smallest of the dabbling ducks in North America. Both sexes have bright iridescent green wing patches, sometimes visible at rest and in flight. **LENGTH:** 14 inches. **WINGSPAN:** 23 inches. **ADULT BREEDING:** (Oct.–June) Males have dark rufous head with green eye stripe. Buffy breast and rump. Distinct white bar on the side of the breast visible at rest and in flight. Grayish body. Emerald-green speculum with buffy or white bars. Females are brownish overall. **NONBREEDING:** (July–Sept.) Males

look much like females. **JUVENILE:** (Aug.–Sept.) Similar to females. **VOICE:** Females quack usually four to seven times in repetition, decreasing in volume. Males give a melodious whistling *prip-prip*. **BEHAVIORS:** May walk or hop from shore to water. Leaves the water to loaf. Flights are rapid and agile. **HABITAT:** Typically, shallow and coastal wetlands. Avoids open salt water. **STATUS:** Common to abundant migrant. Locally common winter resident statewide. Abundant winter resident on the upper Texas coast. Migrants begin arriving in late August and most depart by late April. Lingering birds seen until June.

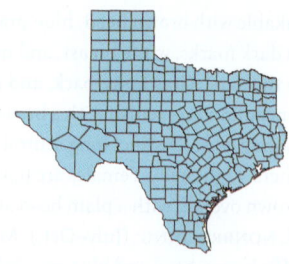

Male

Female

Canvasback

(Aythya valisineria)

IUCN RED LIST STATUS (2021): Least Concern
POPULATION TREND: Increasing

The Canvasback has been called "the aristocrat of ducks." The large head and bill smoothly slope together and can be recognized at great distance. Canvasbacks are one of the most sought-after ducks by hunters because of their large size and reputation for being tasty. Their fine flavor is attributed to their favorite food outside of the breeding season, wild celery. Canvasbacks are so associated with the vegetable that the specific name of Canvasbacks, *valisineria*, is taken from the genus name for wild celery. Canvasbacks have one of the smallest populations of any of the ducks in North America.
LENGTH: 21 inches. **WINGSPAN:** 29 inches.
ADULT BREEDING: (Oct.–June) Males have dark red head, red eye, and black bill that slopes smoothly into the forehead. Black chest and white body, dark rump. In flight, very light above and below. Females have brownish head and chest, grayish body. Bill is black.
NONBREEDING: (July–Sept.) Males similar to breeding males but body is grayish with light mottling. **JUVENILE:** (Oct.–Apr.) Brownish overall with pale belly. **VOICE:** Females give a low growling *grrt grrt*. **BEHAVIORS:** Takes flight by running on the water. Rarely on dry land. **HABITAT:** Variety of habitats from deep fresh water to coastal brackish and salty estuaries. Abundance of food determines use.
STATUS: Uncommon to locally common migrant and winter resident statewide. Migration mid-October–mid-March.

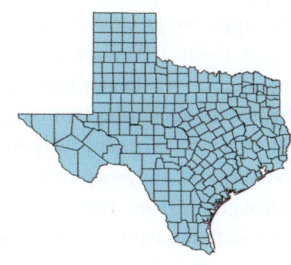

Male (foreground) and female (background)

Redhead

(Aythya americana)

IUCN RED LIST STATUS (2021): Least Concern
POPULATION TREND: Increasing

Redheads are the most notorious of the brood parasites among waterfowl. A brood parasite lays eggs in other birds' nests and leaves their young to be raised by involuntry foster parents.
LENGTH: 19 inches. **WINGSPAN:** 29 inches.
ADULT BREEDING: (Oct.–June) Males have bright rufous-red head with dark-tipped blue bill. Dark breast with gray back and upper wings. In flight, pale flight feathers contrast with darker upper wings. Underwings and belly are light. Rump and tail are dark. Females are plain brown with gray dark-tipped bill. In flight, dark upper wings with light flight feathers. Underwings and belly are pale. **NONBREEDING:** (July–Sept.) Male are browner than gray with less contrast. **JUVENILE:** (Aug.–Oct.) Similar to females. **VOICE:** Mostly silent in Texas. Males at night may call a faint *zoom-zoom*. **BEHAVIORS:** Seldom seen on dry land; may rest on unvegetated shorelines. Takes off by running on the water. **HABITAT:** Most of the winter population uses the shallow seagrass beds of the Laguna Madre in South Texas and other bays that host good beds of seagrass. **STATUS:** Uncommon to common migrant throughout the state. Rare to local common winter resident throughout the rest of the state. Uncommon to locally common summer resident in the Panhandle. Arrives in Texas in late September and most have departed by the end of March, though a few linger into April.

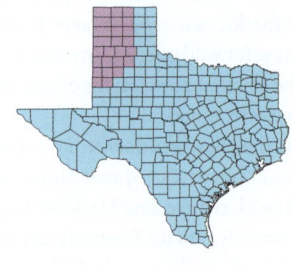

Female

Male

Ring-necked Duck

(Aythya collaris)

IUCN RED LIST STATUS (2021): Least Concern
POPULATION TREND: Increasing

Ring-necked Duck is named for the hard-to-see cinnamon ring around its neck. Many birders have commented that a more appropriate name for this duck would be "Ring-billed Duck" because of the striking white ring around its bill. Females are brown instead of the bold black-and-white pattern of males, but they share the white ring around the bill.

LENGTH: 17 inches. **WINGSPAN:** 25 inches.
ADULT BREEDING: (Oct.–June) Males have dark head and neck with sharp peak to the rear of the crown. Head shows a purple iridescence. Gray bill is lined in white with a dark tip. Eye is bright yellow. Dark back and rump. Light sides with white spur on shoulder. In flight, lighter flight feathers form a light gray wing stripe. Female are brown overall with dark cap, white eye ring, and usually

a pale facial crescent. Bill is gray with dark tip. **NONBREEDING:** (July–Sept.) Males are similar to breeding males but sides are dingy gray-brown. **JUVENILE:** (Aug.–Oct.) Similar to female but facial crescent reduced to a spot. **VOICE:** Mostly silent in Texas. **BEHAVIORS:** Rarely leaves water. Runs on water to take flight. Often slides on the water when landing. **HABITAT:** Variety of freshwater wetlands. Very often found in small ponds, rarely in salt or brackish water. **STATUS:** Uncommon to locally common migrant and winter resident statewide.

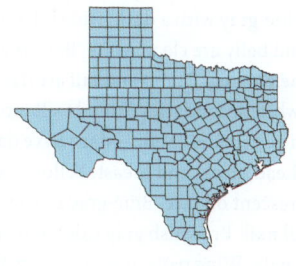

Greater Scaup

(Aythya marila)

IUCN RED LIST STATUS (2018): Least Concern
POPULATION TREND: Decreasing

Separating Greater and Lesser Scaup is a non-trivial problem for birders. The differences are subtle: The white wing-stripe of a Greater Scaup extends well into the primaries, while Lesser Scaup's is restricted to the secondaries. **LENGTH:** 18 inches. **WINGSPAN:** 28 inches. **ADULT BREEDING:** Males (Oct.–June) have dark head, neck, and breast and yellow eye. Compared to a Lesser Scaup, the crown slopes back with a small peak to the rear. Bill is blue-gray with a dark, wide bill nail. Sides and belly are clean white. Back is gray with fine barring. Rump and tail are dark. In flight, white secondaries extend well into the primaries. Females (Sept.–Mar.) have dark brown head, neck, and breast. Yellow eye with white crescent on face. Blue-gray bill with wide bill nail. Brownish gray sides and back. Belly is pale. Wing pattern is same as males.

NONBREEDING: Male (July–Sept.) head is darker brown but lacks the female's crescent. Sides and back brownish gray. Females (Mar.–Sept.) are similar to breeding females but often show distinct cheek patch. **VOICE:** In Texas, females sometimes give a series of raspy *karr karr karr.* **BEHAVIORS:** Feeds primarily by diving and taking food from soft sediments. **HABITAT:** Winters on bays and sheltered shorelines on the coast or large reservoirs in eastern Texas. **STATUS:** Rare to uncommon migrant statewide.

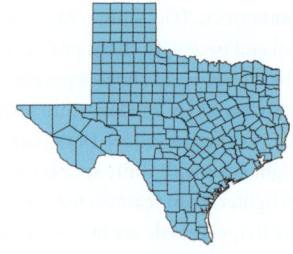

Male

Female

Lesser Scaup

(Aythya affinis)

IUCN RED LIST STATUS (2021): Least Concern
POPULATION TREND: Stable

Lesser Scaup is the more common of the two scaup species in Texas. Large rafts of Lesser Scaup winter in the state's bays and larger lakes.

LENGTH: 16.5 inches. **WINGSPAN:** 25 inches.
ADULT BREEDING: Males (Nov.–June) have dark head, neck, and breast. White sides and belly, gray back. Yellow eye. Crown is flat, giving the head a slight peak to the front. Bill is blue-gray with a small dark bill nail. Rump and tail are dark. White in the secondaries does not extend into the primaries. Females (July–Oct.) have brown head, neck, and breast and yellow eye. Bill is gray with a small bill nail. White crescent on face. Gray-brown sides.
NONBREEDING: Males (July–Oct.) are duller, sides brownish. Females (Mar.–Sept.) are brown overall with white face crescent. **JUVENILE:** (Aug.–Nov.) Like nonbreeding females.

VOICE: Generally silent in Texas. **BEHAVIORS:** In winter, usually takes flight running across water. Often lands feet forward, "skiing" across the water to stop. Walks on land more easily than Greater Scaup. **HABITAT:** Found mainly on lakes and reservoirs, fresh to brackish. Can use hypersaline waters near fresh water. **STATUS:** Common to abundant migrant and winter resident east of Trans-Pecos. Common to uncommon migrants in Trans-Pecos. Rare summer visitor statewide. Fall birds arrive in October, most birds have departed by early April. Some linger into mid-May.

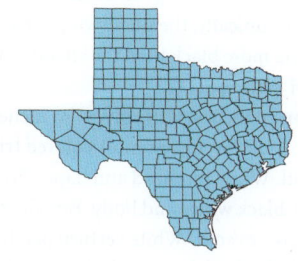

Male

Female

Surf Scoter

(Melanitta perspicillata)

IUCN RED LIST STATUS (2018): Least Concern
POPULATION TREND: Decreasing

Surf Scoters are odd-looking ducks—medium-size with a very heavy bill adapted to crushing the bivalves that are its preferred food outside of breeding season. Like other scoters, the Surf Scoter is mostly black. It differs from the similar Black Scoter by having a white nape and forehead (males). Females are mostly black, unlike the dun-colored Black Scoter females. A Surf Scoter is mostly black in flight, but orange legs and feet may be visible, and ironically, the underwings of a Surf Scoter are more black than the silvery underwings of a Black Scoter.

LENGTH: 20 inches. **WINGSPAN:** 30 inches.
ADULT: Male has heavy, multicolored triangular bill. White forehead and nape. Orange legs. All-black wings and body. Female is dark brown-gray overall. White vertical patch at base of bill. Cheek has small white patch.

JUVENILE: (Aug.–Mar.) Very much like female but white patches have more contrast and are larger. **VOICE:** Usually silent. **BEHAVIORS:** Rarely observed on land. Unlike other scoters, rarely flies in line. Flies low over water. **HABITAT:** Winters in shallow coastal waters. Power plant cooling ponds are a hot spot for scoters in general. **STATUS:** Uncommon and local winter resident on the coast north of Baffin Bay. Fall migrants begin arriving late October–December. Spring migrants are present inland late February–early April. Some birds linger well into June.

Male

Female

White-winged Scoter

(Melanitta deglandi)

IUCN RED LIST STATUS (2018): Least Concern
POPULATION TREND: Decreasing

White-winged Scoters are the least common of the three scoter species that occur in Texas. They nest on freshwater lakes and wetlands in boreal forests of the northwestern interior of North America. Nests can be far from water. Twelve to twenty-four hours after hatching, the female leads the ducklings to water, and ducklings may have to swim a long distance (3–8 kilometers) before they reach the brood area. Ducklings may have to dive repeatedly during this swim to evade gulls. White-winged Scoters are the easiest to identify of Texas scoters, due to their bold white secondaries and greater secondary coverts.
LENGTH: 21 inches. **WINGSPAN:** 34 inches.
ADULT: Male is all black with bold white speculum wing patch and distinctive white comma below the eye. Thick triangular bill with orange tip. Female is all dark with white oval patch behind the bill and over the ear. **JUVENILE:** (Aug.–Mar.) Like female with pale belly. White face patches more prominent. **VOICE:** Mostly silent. **BEHAVIORS:** Takes off running on water. Flies in strings low over water. Sits low in water. In deeper water arches downward. **HABITAT:** Coastal estuaries and bays with shallow water over shellfish beds. **STATUS:** Very rare winter coastal resident, most common in the upper and central Texas coast. Very rare migrant statewide. Inland fall migrants early November–mid-December, inland spring migrants mid-February–mid-April.

Male

Female

Black Scoter

(Melanitta americana)

IUCN RED LIST STATUS (2021): Near Threatened
POPULATION TREND: Decreasing

Black Scoter is one of the least known of North American waterfowl, mostly due to its remote and split breeding range in northern Quebec and northwestern Alaska. Another factor in the lack of research is the perceived lack of interest by hunters. Most of the hunting of this species is in the Atlantic flyway, and only about 0.6% of the Atlantic flyway bag of ducks consists of Black Scoter. Sea duck hunting is on the rise, though, and there is concern that hunting at the present level isn't sustainable. Sea ducks in general have lower reproductive potential than other ducks, and populations take much longer to recover from the loss of adult birds.

LENGTH: 19 inches. **WINGSPAN:** 28 inches.
ADULT: Male is all black. Head is more rounded than other scoters. Bright yellow knob on the bill. Female is dark gray with a dark cap. Pale cheeks and throat. Lacks the yellow knob on the bill. **JUVENILE:** (Aug.–Mar.) Like females but the belly is pale. **VOICE:** More vocal than other scoters, but near silent in Texas. **BEHAVIORS:** Flies in lines, sometimes just above the water. Sometimes forms bunches then spaces back out into lines. **HABITAT:** Mostly bays and near shore waters. **STATUS:** Very rare winter resident, mostly on the upper and central coast. Migrants arrive in late October and depart by early April.

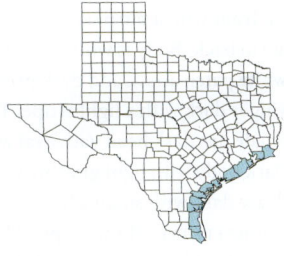

Long-tailed Duck

(Clangula hyemalis)

IUCN RED LIST STATUS (2018): Least Concern
POPULATION TREND: Vulnerable

Long-tailed Duck is an arctic species, breeding as far north as 80°N in tundra and taiga. Most Long-tailed Ducks winter in the Great Lakes, western Greenland, and off the north Atlantic and Pacific coasts of North America, but small numbers occur annually in Texas. They are one of the deepest-diving ducks, often diving as deep as 500 feet.

LENGTH: 16.5 inches. **WINGSPAN:** 28 inches.
ADULT WINTER: (Nov.–Apr.) Male has white crown and neck with dark cheek. Small-billed. White rump and belly with dark chest. White back. Distinctive long central tail feathers are rarely observed in Texas. Female has dark crown, white face, and dark cheek. Dark back and chest. White rump and belly. **ADULT SPRING:** (May–June) Spring plumage is almost never observed in Texas. Male is very dark with white face, rump, and belly. Long central tail feathers. Female is all dark with bold white eye ring tailing into a strip down the neck. White rump. **VOICE:** Males make a mellow *haw a-haw-ra.* The call is *hap-hap-hap.* **BEHAVIORS:** Not adept at walking but able to. Flight is erratic and low over water. **HABITAT:** Coastal bays, sometimes inland lakes. **STATUS:** Very rare to rare winter visitor along the coast, very rare to casual winter visitor in all other parts of the state. Fall birds have been found from early November and spring birds until April.

Female

Male

Bufflehead

(Bucephala albeola)

IUCN RED LIST STATUS (2022): Least Concern
POPULATION TREND: Increasing

A male Bufflehead is a study in black and white, with the front of the head being black and the back being a large puffy white patch. This puffy head is where the Bufflehead gets its name (short for "Buffalo-Head"). The female is duller, with a small elongated ear patch. Buffleheads are one of the few duck species that form long-term pairs.
LENGTH: 13.5 inches. **WINGSPAN:** 21 inches.
ADULT BREEDING: (Oct.–May) Male head is dark in front with a purplish iridescence, and white in back. Body is white. Wings have a large white patch above and dark primaries. Underwing is dark with a light stripe. Tail is dark. Female has dark head with oval white patch. Dark body with light gray breast. Wings and back are dark with small white secondary patch. **NONBREEDING:** (June–Sept.) Male has dark head with large circular white patch.

Body is gray. **JUVENILE:** (Aug.–Oct.) Like female but white patch is smaller, less distinct.
VOICE: Usually silent. **BEHAVIORS:** Seldom walks on dry land. Takes flight by running on water but needs only short distance to take off. Flies high over land, low over water. **HABITAT:** Open water where they can dive for aquatic invertebrates. Uses bays, coves, inland waters, slow-moving rivers. **STATUS:** Common migrant and winter resident statewide. Birds begin arriving in late September; most leave by early April.

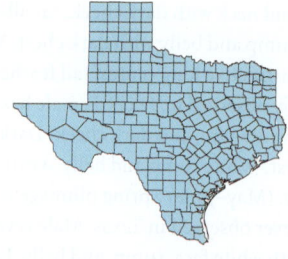

Immature male

Female

Common Goldeneye

(Bucephala clangula)

IUCN RED LIST STATUS (2018): Least Concern
POPULATION TREND: Stable

Common Goldeneye is a cold-hardy duck that breeds worldwide in northern boreal forests. In flight, its wings make a distinctive whistling sound, and hunters often refer to them as "whistlers."
LENGTH: 18.5 inches. **WINGSPAN:** 26 inches.
ADULT BREEDING: (Oct.–June) Male has dark head and bill, with round patch behind the bill. Bright golden-colored eyes. White on the breast, sides, and belly. Back is dark with a white border. In flight, upper wing shows a large white patch and dark primaries with dark leading edge. Female head is dark brown with bright golden-colored eye. Bill is dark with orange tip. Body is mostly slate-colored. Upper wing has large white wing patch. **NON-BREEDING:** (Nov.–June) Males have dark head with small crescent behind the bill. White neck ring. Body and back are gray. **JUVENILE:** (Aug.–Oct.) Dark brown head and dark bill. Eye is dark. Body is dark gray with white belly. **VOICE:** Mostly silent. **BEHAVIORS:** Spends very little time on land. Needs to run only a short distance on water to get airborne. Flight is swift in small compact groups. **HABITAT:** Primarily found on quiet bays and large reservoirs. **STATUS:** Common to uncommon winter resident on upper and central coasts. Uncommon to rare migrant and winter resident in the rest of the state. Birds begin to arrive in early November and depart in early March.

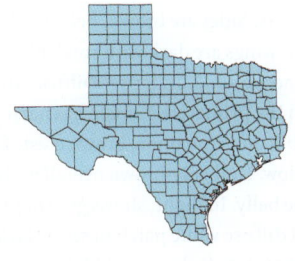

Female

Male

Hooded Merganser

(Lophodytes cucullatus)

IUCN RED LIST STATUS (2021): Least Concern
POPULATION TREND: Increasing

Hooded Merganser is the smallest of the three
North American mergansers and the only one
that is endemic to North America.
LENGTH: 18 inches. **WINGSPAN:** 24 inches.
ADULT BREEDING: (Sept.–June) Male has
dark head with dark, narrow bill and bright
yellow eye. When the large crest is lowered,
shows a broad white stripe. When the crest
is raised, shows a large white patch on the
back half of the head. Breast is white with two
black spurs. Sides are brown, belly is white.
In flight, wings are dark with small blue-gray
patch and white secondaries. Diffuse white
patch in the primaries. Female has narrow yellowish bill and cinnamon/brown crest. Eye is
dull yellow. Dull gray-brown overall with long
tail. Pale belly. In flight, shows gray in primaries and diffuse white patch in secondaries.
NONBREEDING: (July–Sept.) Male very similar
to female but crest not as bright. **JUVENILE:**
(Aug.–Sept.) Dark version of the female, with
dark eyes. **VOICE:** Mostly silent. Males make a
croaking *craaa-crrroooo* in courtship displays.
In flight, the wings make a high-pitched trill.
BEHAVIORS: Takes flight by running across
water. **HABITAT:** In winter, occupies shallow
freshwater and brackish bays, estuaries, and
ponds. Can be found in small stock tanks.
STATUS: Uncommon to common migrant and
winter resident statewide. Rare but increasing
resident in northeast Texas south to about
Nacogdoches County.

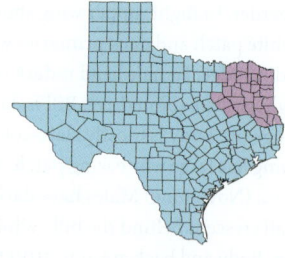

Male

Female

Common Merganser

(Mergus merganser)

IUCN RED LIST STATUS (2018): Least Concern
POPULATION TREND: Unknown

The Common Merganser nests worldwide in northern forests near large lakes and rivers—in North America primarily in Canada. It winters across the continent in large lakes, rivers, and reservoirs. In Texas its footprint in winter is small, but it can be abundant on large reservoirs, particularly on Lake Meredith in the Panhandle.

LENGTH: 25 inches. **WINGSPAN:** 34 inches.
ADULT BREEDING: (Nov.–July) Male has dark green head, dark eyes, and thin orange bill. Mostly white below with a faint rusty wash on the belly. Dark back and upper wings. White on the lower half of the wings in flight. Female has brown head, orange bill and contrasting white throat patch. Grayish overall. Dark primaries and white secondaries. **NONBREEDING:** (July–Oct.) Males similar to females but whiter in the wings. **JUVENILE:** (Aug.–Sept.)

Very much like a female, but belly is mostly white. **VOICE:** Typically silent. Gives a harsh *gruk* or *grrr* when disturbed. **BEHAVIORS:** Spends little time on land. Runs across water for several yards to take flight. Flies in pairs or small groups with a broad front. **HABITAT:** Large lakes and reservoirs. **STATUS:** Locally common and sometimes abundant winter resident in the Panhandle, El Paso, and western Trans-Pecos. Uncommon to rare winter resident on the South Plains to San Angelo. Very rare eastward. Fall migrants arrive in mid-November and depart by early April.

Male

Female

Red-breasted Merganser

(Mergus serrator)

IUCN RED LIST STATUS (2018): Least Concern
POPULATION TREND: Stable

The Red-breasted Merganser is the medium-size merganser in North America. Similar in distribution and ecology to the Common Merganser, it differs by occuring more frequently in salt water and typically nesting on the ground, unlike the Common Merganser which is a cavity-nesting species. It is a late breeder, often not fledging until late September. It is among the fastest flying ducks, having been clocked at speeds up to 80 miles per hour.

LENGTH: 23 inches. **WINGSPAN:** 30 inches.
ADULT BREEDING: (Nov.–May). Male has dark head with wispy crest, orange eye, and thin orange bill. Reddish brown breast with white neck ring. Light belly and dark back. Underwing is light with dark primaries. White secondaries and wing patch. Orange feet. Female has orange bill and eye. Head is brown. Body is gray with paler belly. White secondaries in flight. Primaries are dark. **NON-BREEDING:** (May–Oct.) Male is like female. **JUVENILE** (Aug.–Sept.) Like female with white lores and eye ring. **VOICE:** Generally silent. **BEHAVIORS:** Walks awkwardly and infrequently on land. Strong flier. Usually runs on water to take off. **HABITAT:** Primarily winters in coastal bays and estuaries, rare in inland reservoirs. **STATUS:** Common winter resident on the coast. Uncommon to rare migrant and winter resident on inland reservoirs in the eastern half of Texas. Uncommon locally in El Paso area.

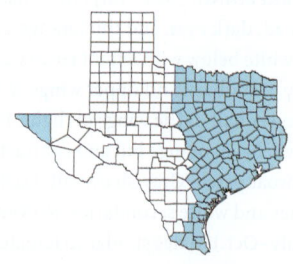

Female

Male

Ruddy Duck

(Oxyura jamaicensis)

IUCN RED LIST STATUS (2018): Least Concern
POPULATION TREND: Decreasing

The brightly colored male Ruddy Duck is a treat for Texas birders to see. Ruddy Duck is only an occasional breeder in Texas, but it is an abundant winter resident, both in saltwater estuaries and on large inland reservoirs. Most breeding takes place in the prairie pothole region of North America. Both adults and ducklings are mostly carnivorous, and they are almost specialists on midge larvae. **LENGTH:** 15 inches. **WINGSPAN:** 18.5 inches. **ADULT BREEDING:** (Mar.–Aug) Male has black cap and neck, bright white cheeks, and bright sky-blue bill. Body is warm brown with light belly. Female has dark cap and pale cheek with dark facial stripe. Bill is dark gray. Body is brownish-gray with light barring. Light belly. Males and females have distinctive stiff upright tails. **NONBREEDING:** (Sept.–Mar.) Male has dark cap and gray bill. White cheeks. Body like females. **VOICE:** Mostly silent. **BEHAVIORS:** One of the least mobile on land of all ducks; walks only with great difficulty. Runs on water to take off. Swift and direct flight. Rarely flies during the day, mostly at twilight or night. **HABITAT:** Coastal brackish and freshwater open-water habitats. **STATUS:** Common migrant and winter resident statewide in Texas. They begin arriving in early September and most have departed by mid-May. An occasional breeder in the state, with records from most of the state for breeding.

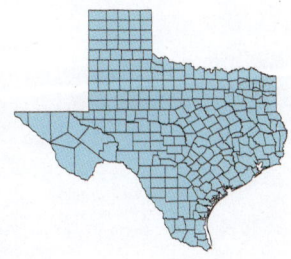

CRACIDAE (GUANS, CHACHALACAS, AND CURASSOWS)

The cracids are a tropical family that includes curassows, guans, and chachalacas. Only the Plain Chachalaca's range includes North America and Texas. Like the other galliformes, or chicken-like birds, they generally have heavy bodies, strong legs, long necks, and small heads. They have rounded wings for quick bursts of speed. They are adapted to life on the ground but are also arboreal. They are just as at home in the trees, running along branches, as they are on the ground.

Most of the cracids are known for their booming voices, from the Plain Chachalaca's raucous dawn call to the strange nighttime booms and rattles of guans in tropical forests. In many of the cracids, the Plain Chachalaca included, the male's voice is much lower than the female's. This is because of the male's unusual syrinx anatomy. The syrinx extends down to the belly of the bird in a long loop. Much like extending the slide on a trombone, this lowers the vocal pitch.

Cracids are a popular game species where they are found. Plain Chachalaca even has a hunting season in Texas. Hunting pressure has had a negative effect on populations, but habitat destruction remains the greater threat.

Plain Chachalaca

(Ortalis vetula)

IUCN RED LIST STATUS (2021): Least Concern
POPULATION TREND: Stable

Plain Chachalaca is a tropical species that occurs widely along the Atlantic slope of Central America, from Texas to Nicaragua, but only occurs in the deep south of Texas, primarily in the Lower Rio Grande Valley. It can be a reasonably urban species, sometimes common in residential areas if there are alleys and vacant lots for nesting and foraging in. They can be amazingly tame and quickly grow accustomed to bird feeders.

LENGTH: 22 inches. **WINGSPAN:** 26 inches.
ADULT: Long-necked and uniform drab olive-brown color. Long wide tail with white corners. Bare red skin on the sides of the throat is not always visible. **VOICE:** A loud raucous *KLOK-a-TOK KLOK-a-TOK KLOK-a-TOK*, usually given in the morning. In groups, it gives a quiet, almost purring *grrr grrr grrr.* **BEHAVIORS:** Heavy wing beats and short glides in flight. **HABITAT:** Dense scrubby thickets. **STATUS:** Uncommon to common resident in the Lower Rio Grande Valley, north to southern Kenedy and Webb counties.

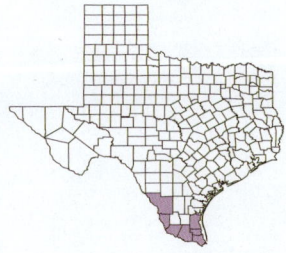

ODONTOPHORIDAE (NEW WORLD QUAIL)

The New World quail's family name, Odontophoridae, refers to the toothlike serated edge of the lower beak. This is a unique trait and one of the morphological characteristics that led to scientists removing New World quail from the pheasants and grouse family to their own group. The family includes thirty-three species of quail, including two species in Africa that genetic studies show are much closer genetically to the New World quail than to the Old World grouse they were previously placed with.

The whistled and throaty calls of quail are often the only way one knows they are present. All six North American species of quail use dense, brushy habitats, so getting a good view of any of these species is a rare treat. Most of the neotropical species in this family occupy forested habitats.

Most species of the quail family are easily found in some part of their range. Many, like the Northern Bobwhite, have adapted to living with agriculture if a brushy edge is left. Changing farming practices are eliminating these brushy refuges and have negatively impacted the population. Montezuma Quail is the most specialized of the North American quail, and its mid-level to high-level grass and pine-oak scrub habitat is very sensitive to overgrazing.

Male (foreground)
and female (background)

Northern Bobwhite

(Colinus virginianus)

IUCN RED LIST STATUS (2021): Near
Threatened POPULATION TREND: Decreasing

The clear-whistled call of the Northern
Bobwhite is familiar to many as this most popular of game birds in eastern North America
seems to say its name. Its popularity with
hunters has made it one of the most studied
of North America's birds starting in the 1930s.
Northern Bobwhite ranges through most of
Texas, except for the Trans-Pecos region. It
has likely benefited from habitat changes that
humans brought to Texas, including the opening of the eastern forest into more parklike
conditions.
LENGTH: 9.75 inches. WINGSPAN: 13 inches.
ADULT: Male has chestnut head with broad
white eye stripe and white throat. Dark
neck ring. Rufous on the body with thin
white streaks on the sides and pale belly. In
flight, wings and tail are gray. Female is similar to male but with buffy eye stripe and
buffy throat. Head is more rounded. VOICE:
Onomatopoeic call is a whistling *whip-WAAYK*
or *bob-WHITE*. BEHAVIORS: Walks quickly on
the ground. Only flies when forced; explodes
into the air, usually very close to the disturbance. Flight is low and short. HABITAT: Can
use a variety of rangelands with a mosaic of
small patches of fields and forest. STATUS:
Occurs throughout the state except for western Trans-Pecos. Has been declining, likely
due to habitat changes, particularly the removal of hedgerows in agricultural fields.
Populations are becoming very localized.

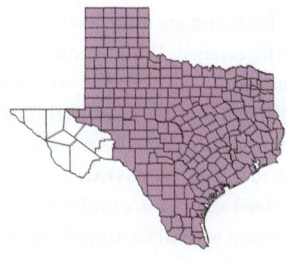

Scaled Quail

(Callipepla squamata)

IUCN RED LIST STATUS (2021): Least Concern
POPULATION TREND: Decreasing

Scaled Quail get their name from the dark edges on the blue-gray feathers of their breasts and napes, giving them a scaly appearance. They also have distinctive white-tipped crests. The bluish feathers on their breasts have given them the popular name "blue quail" with hunters, and their crests give them their other colloquial name of "cottontop." Their loud *kuk-kyuur* call seems to carry for miles. In winter, they form large coveys of up to thirty birds and are more likely to run than fly when threatened. While a popular game bird, it has not been studied as extensively as the Northern Bobwhite, most likely due to its limited range within North America. **LENGTH:** 10 inches. **WINGSPAN:** 14 inches. **ADULT:** Head and face are uniformly dull grayish-brown with white-tipped crown feathers. Mostly gray background color. Neck, breast, and belly are intricately scale-patterned. Males differ chiefly from females with slightly longer crown feathers, and females have more fine streaking on the throat and side of the face. Birds in South Texas are more rufous on the belly, especially males. **VOICE:** A loud hoarse *QUER-esh* and a rhythmic *kuk-kyuur kuk-kyuur*. **BEHAVIORS:** Runs on the ground, especially if cover is sparse. When flushed, flies only a short distance. **HABITAT:** Grasslands with scattered shrubs. **STATUS:** Uncommon to locally common resident west of a line from Starr to Childress counties.

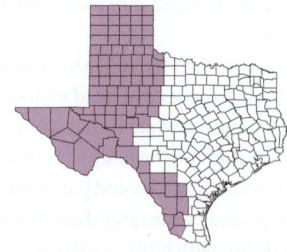

Male (left) and female (right)

Gambel's Quail

(Callipepla gambelii)

IUCN RED LIST STATUS (2021): Least Concern
POPULATION TREND: Stable

Gambel's Quail was named after Philadelphia naturalist William Gambel, who collected the first specimens along the Sante Fe Trail in 1841. It is one of the four species in the scaled quail complex, along with Scaled, California, and Elegant Quail. Its jaunty, drooping topknot plume makes it an instantly recognizable species of the interior desert.

Gambel's Quail are gregarious and form stable family groups. Several groups may join together to form a large covey. Winter aggregations of several dozen birds may form. Their diet is 90% plant matter, and they are subject to boom-and-bust cycles with rainfall and habitat quality. Dry years mean fewer young birds. **LENGTH:** 10 inches. **WINGSPAN:** 14 inches. **ADULT:** Male has black face and throat bordered in white. Chestnut cap above a white supercilium. The dark forehead plume is unique among quail in Texas. Belly is creamy and unmarked. Sides are rufous with white streaks. Uniform gray above. Female differs from male by a plain gray head and face. **VOICE:** Hoarse drawn-out *where* and nasal *whup-way-wawa whup-way-wawa*. **BEHAVIORS:** Runs, even in dense cover. Seeks escape by running ahead of pursuit. Explodes into the air when forced, but flight is not particularly swift. **HABITAT:** Brushy and thorny vegetation, often concentrated near a water source. **STATUS:** Common resident in El Paso, Hudspeth, and Culberson counties and south into Presidio County.

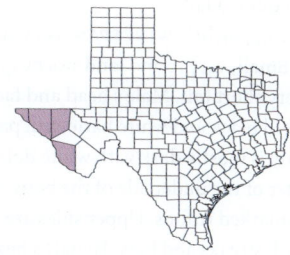

Male

Female (center) with juveniles

Montezuma Quail

(Cyrtonyx montezumae)

IUCN RED LIST STATUS (2021): Least Concern
POPULATION TREND: Decreasing

Montezuma Quail was described in 1830 from a specimen collected in Mexico and named for the Aztec Emperor Moctezuma Xocoyotzin, known in Europe as Montezuma II. The Montezuma Quail is shy and one of the least studied of the oak woodland birds. Its habit is to crouch and hide if possible, and detection is difficult. It's been said that one can hike for days through its habitat and walk by dozens of birds unknowingly. It has also been known as the Harlequin Quail.

LENGTH: 8.75 inches. **WINGSPAN:** 15 inches.
ADULT: Small, very round, and stocky quail with a very short tail. Male's head and face have a bold black and white harlequin pattern. Body is mostly dark with white dots. The center of the underside of the body is dark, unmarked rufous. Upper sides are buffy, with streaks and bars. Female's head has an indistinct pattern like males. Body is brownish, marked with indistinct barring and streaks. **VOICE:** Male's song is a long, drawn-out whistling trill: *vwirrrrrr.* **BEHAVIORS:** Generally walks and runs. When flushed, explodes into air then glides to a landing and typically runs. **HABITAT:** Usually found on steep hillsides with grasses. **STATUS:** Uncommon resident of the Davis Mountains and northern Brewster County. A small population still exists in Edwards and northern Kinney counties. Reintroduced birds are present in the Guadalupe Mountains.

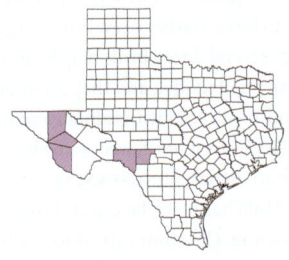

PHASIANIDAE (PHEASANTS, GROUSE, AND ALLIES)

This large worldwide family currently contains 186 species, including the genus *Gallus*, the jungle fowl that are the wild form of our common chicken. Relatives of the humble chicken include some spectacular birds such as the showy pheasants and peacocks. Several species in this family have been domesticated for food: chickens, guinea fowl, and turkeys.

Texas is home to four species in this family. The Wild Turkey is found in much of the state. The Old World pheasant has been introduced to the Panhandle. Two species of grouse, the Lesser Prairie-Chicken of the Panhandle and the endangered subspecies of Greater Prairie-Chicken, are not included in this guide because they are very range-restricted, cryptic, and rarely encountered even though they are native species.

Wild Turkey

(Meleagris gallopavo)

IUCN RED LIST STATUS (2018): Least Concern
POPULATION TREND: Increasing

The Wild Turkey was introduced to Europe from domesticated birds brought back to Spain by the conquistadores. The American turkey was apparently confused with African guinea fowl, which had come to Europe via the Turkish empire. Thus, this endemic species of North America came to be known by the name of a faraway country. Domestic turkeys and the Wild Turkey are the same species, but domesticated varieties have been bred to be all white because the pin feathers on a dressed bird are less noticeable.

A displaying male Wild Turkey is a sight to behold. It fans its tail in the familiar Thanksgiving pose, spreads its wings, and rattles them loudly.
LENGTH: Male 46 inches, female 37 inches.
WINGSPAN: Male 64 inches, female 50 inches.
ADULT: Huge size, dark colored, almost appears black. Long legs and tiny head make it distinctive. Males have bare unfeathered head. **VOICE:** When displaying, males give the familiar gobble. Females give a repeated *tuk* or *yike* call. **BEHAVIORS:** Mostly travels on foot, running to escape threats. Roosts in trees. Takes flight by running and hopping. **HABITAT:** Mostly in wooded open habitat and riparian habitats in the southwestern part of the state. **STATUS.** Common to uncommon resident from the eastern Panhandle, south to the southern coastal prairie, north to Jackson County, west to Trans-Pecos. Has been reintroduced to the Davis Mountains.

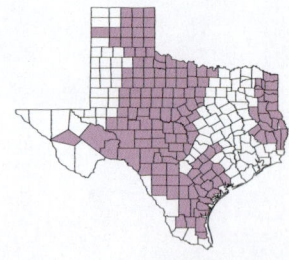

Ring-necked Pheasant

(Phasianus colchicus)

IUCN RED LIST STATUS (2016): Least Concern
POPULATION TREND: Decreasing

Ring-necked Pheasant was introduced to Texas in the 1930s for hunting. There were once small populations in Chambers County and the Trans-Pecos. The original native range of Ring-necked Pheasant is fragmented from Eastern Europe (Bulgaria) to Central China and Korea. It has been introduced in many places over thousands of years and is now Holarctic in distribution.

Ring-necked Pheasant has as many as thirty recognized subspecies and geographic variations. Because introduced birds came from many different populations, birds in Texas are not identifiable to subspecies.
LENGTH: 21 inches. **WINGSPAN:** 31 inches.
ADULT: Male has dark bluish head and red facial skin. Broad white neck-ring. Bright rufous flanks with dark spots and dark belly. Back is brown with white spots. Wings and rump are gray. Signature long brown tail with dark bars. Female is pale-drab tan with dark spots and bars. White teardrop shaped eye ring.
VOICE: When displaying, males make a hoarse *urrk-iik*. When alarmed, they give a hoarse, cackling *kutt-it kutt-it kutt-it.* **BEHAVIORS:** Spends most of the time on the ground. Often prefers to run to cover. Capable of powerful, quick flight for short distances. **HABITAT:** Grasslands with cover and access to water. **STATUS:** Common resident in the northern and western Panhandle, becoming less common southward and eastward.

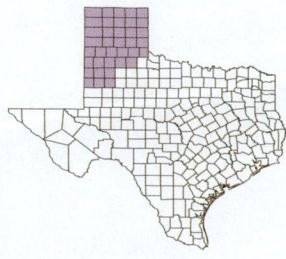

PODICIPEDIDAE (GREBES)

The grebes have traditionally been placed close to the loons taxonomically. This makes superficial sense. Loons and grebes share aquatic habitats, have similar body shapes, and both have their feet placed far back on the body. Recent studies, though, suggest a counterintuitive relationship between flamingos and grebes. Some experts disagree over this relationship, but there is growing evidence that grebes and flamingos are sister families.

The grebe family is an ancient one. The genus *Podiceps* extends back 30 million years to the Oligocene epoch. In modern Texas this genus is represented by Red-necked, Eared, and Horned Grebes.

The name Podicipedidae comes from the Latin *podiceps*, for "rump-footed," referring to the placement of their legs far back on the body. This makes them awkward at best on land but powerful swimmers. Their toes are widely lobed with pads that expand on the power stroke to propel the bird forward, and fold up bladelike on the return stroke for maximum efficiency.

Least Grebe

(Tachybaptus dominicus)

IUCN RED LIST STATUS (2021): Least Concern
POPULATION TREND: Stable

The Least Grebe is the smallest of the New World grebes. Widely distributed in the American tropics and subtropics, it is expected in North America only in South Texas, but increasingly is regular and breeding occasionally as far north as the Houston area. Rarely seen flying, they have low wing loading and can use small temporary ponds. Much smaller than the more common Pied-billed Grebe, the Least Grebe is best distinguished by its thin dark bill and bright golden eye.
LENGTH: 9.5 inches. **WINGSPAN:** 11 inches.
ADULT BREEDING: (Feb.–Oct.) Black face and bill with bright yellow eye. Dark slate-gray overall. In flight, secondaries and lower primaries appear white, belly is pale. **NONBREEDING:** (Sept.–Mar.) Similar to breeding, but cheeks and throat lighter colored. Bill is more gray than black.

JUVENILE: (June–Oct.) Paler than adults with low-contrast pattern on face. **VOICE:** Rapid trill: *chi-chi-chi-chi-chi*. Alarm call is a loud and high-pitched *EAP*. **BEHAVIORS:** Glides swiftly across water without bobbing. Often flutters across water when fleeing or aggressive. **HABITAT:** Found in small ponds to lakes, usually in slow or still water. **STATUS:** Uncommon to common in the Lower Rio Grande Valley, north through the South Texas Brush Country to about I–10. Increasingly regular on the upper Texas coast to Jefferson County and west along the Rio Grande River to about Del Rio.

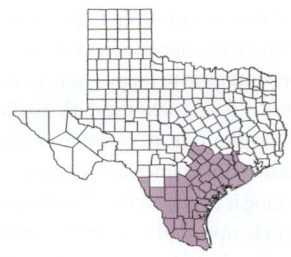

Winter

Pied-billed Grebe

(Podilymbus podiceps)

IUCN RED LIST STATUS (2021): Least Concern
POPULATION TREND: Increasing

Pied-billed Grebe is the most widely distributed of any of the New World grebes, from the boreal forests of Canada to southern Chile and Argentina.
LENGTH: 13 inches. **WINGSPAN:** 16 inches. **ADULT BREEDING:** (Feb.–Sept.) Stocky brownish-gray. Bill is thick and stout, mostly white with a black band, thicker than any other Texas grebe. Thin white eye ring. Dark throat. Wings often held high so fluffy pale undertail coverts visible. In flight, wings show a faint pale edge and light below, pale belly. **NONBREEDING:** (Sept.–Mar.) Thick bill is mostly pale. Lacks the dark marking of breeding plumage. **JUVENILE:** Similar to nonbreeding. Face patterned gray and brown. **VOICE:** Song is a long series of *wook-wook-wook* slowing in tempo. Call is a rapid chattering *heh-heh-heh-heh*. **BEHAVIORS:** Awkward on land because feet are well back on the body. Prefers to flee by diving. Will often observe threats by raising only the eyes and top of the head from the water. Needs to run a long distance on the water to take flight. **HABITAT:** Wetlands with dense stands of emergent vegetation or aquatic vegetation that can support its nest. **STATUS:** Uncommon to common migrant and winter resident statewide. Uncommon and local breeder in the eastern half of the state and a rare breeder in Trans-Pecos. Migrants begin arriving in mid-August and most have left by early May.

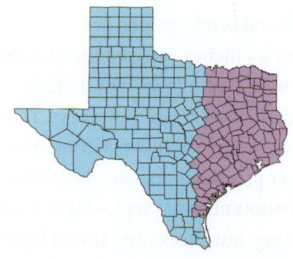

Horned Grebe

(Podiceps auritus)

IUCN RED LIST STATUS (2018): Vulnerable
POPULATION TREND: Decreasing

Horned Grebes are Holarctic breeders. They get their name from the "horns" they develop during the breeding season, yellowish feather patches behind the eyes that can be raised at will. During nesting season, they select small ponds and lake inlets with emergent vegetation and open water for breeding. Their approachability during the breeding season has made them one of the most studied species in the Old and New Worlds.
LENGTH: 14 inches. **WINGSPAN:** 18 inches.
ADULT BREEDING: (Feb.–Aug.) Dark head with solid yellow wedge from the eye to the back of the crown. Bright red eye. Neck is rufous, back is dark gray with a scaled pattern. Rufous sides. In flight, shows white secondaries, belly, and undertail. **NONBREEDING:** (Sept–Mar.) Dark crown and red eye inside the dark area. White cheek and neck, dark nape. Back gray. **JUVENILE:** (Aug.–Oct.) Similar to nonbreeding plumage but lower contrast, cheeks more smudgy. **VOICE:** Usually silent in Texas. On breeding grounds, a repeating creaky *creek creek creek.* **BEHAVIORS:** Awkward on land. Generally runs on the water to take off. **HABITAT:** Large to moderate bodies of water, saltwater and freshwater. **STATUS:** Uncommon to locally common winter resident on reservoirs in the northeast corner of Texas. Uncommon to rare on central and upper Texas coast. Winter residents begin arriving in mid-October and most leave by mid-March. Some linger into mid-May.

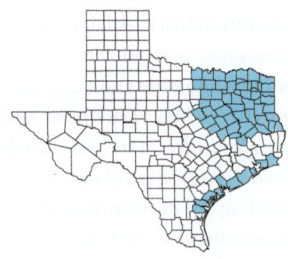

Eared Grebe

(Podiceps nigricollis)

IUCN RED LIST STATUS (2018): Least Concern
POPULATION TREND: Unknown

Holarctic in distribution and North America's most abundant grebe, the Eared Grebe is usually seen in its drab winter plumage. Only during the short breeding period in the Pacific Northwest does it turn its signature black with rufous sides and golden tufts on the sides of its head.
LENGTH: 13 inches. **WINGSPAN:** 16 inches.
ADULT BREEDING: (Apr.–Sept.) Dark head with wispy yellow tufts on the rear of the head. Eye is red. Neck is black. All dark back and dark rufous sides. In flight shows white secondaries and white belly and undertail. **NON-BREEDING:** (Oct.–Mar.) Dark head and pale throat, usually dusky on the neck. Dark back and pale sides. **JUVENILE:** (Aug.–Nov.) Similar to nonbreeding but duskier chest and neck. **VOICE:** Generally silent in Texas. **BEHAVIORS:** Walks awkwardly in an upright position. Flies with head and neck stretched low and forward, feet trailing behind. **HABITAT:** Prefers deep unvegetated waters, both salt and fresh. **STATUS:** Common to uncommon migrant in the western three quarters of the state. Rare in the Pineywoods. Common to locally abundant winter resident on the coast. Uncommon to rare summer resident in the Panhandle and El Paso County. Rare breeder in other parts of the state. Fall migrants begin to arrive in the state in late July, with most having departed by the end of April. Some linger into June.

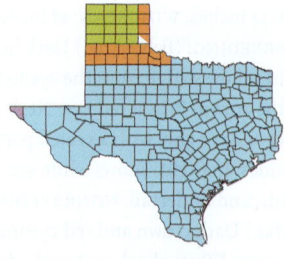

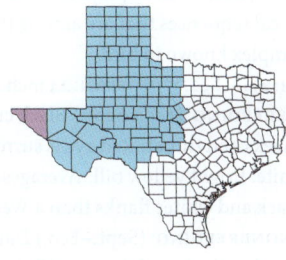

Western Grebe

(Aechmophorus occidentalis)

IUCN RED LIST STATUS (2019): Least Concern
POPULATION TREND: Stable

For more than 100 years, the Western Grebe was considered cospecific with Clark's Grebe. Western was the dark morph and Clark's was the white morph. This changed when it was discovered that even when both forms were present together, they mated assortatively, or like with like.

The two species are unique among grebes in that they have a mechanism in their necks that allows them to thrust forward like a spear. **LENGTH:** 25 inches. **WINGSPAN:** 24 inches. **ADULT BREEDING:** (Feb.–Sept.). Black head with red eyes, surrounded by the dark on the crown. Bill yellowish to olive in color. Throat and breast are white. Back of the neck and the back are dark. In flight, white below. Dark above with an indistinct pale wing stripe. **NONBREEDING:** (Sept.–Feb.) Similar to breeding with less contrast on the face, more gray than black around the eye. **VOICE:** Harsh drawn-out *creeeet creeet*. **BEHAVIORS:** Seldom ever comes on land. Diving average about 30 seconds; has been recorded diving for more than 60 seconds. **HABITAT:** Western lakes and along the coast. **STATUS:** Uncommon to locally common resident in the western Trans-Pecos. Rare to casual winter resident farther east to the Pineywoods. Occasional winter resident on the coast in sheltered bays. Breeds in Balmorhea Lake and a few other lakes in Hudspeth and El Paso counties.

Clark's Grebe

(Aechmophorus clarkii)

IUCN RED LIST STATUS (2016): Least Concern
POPULATION TREND: Decreasing

From 1886 to 1985, Clark's and Western Grebes were considered cospecific, with Clark's being the white morph of Western Grebe. However, it was found that they mate assortatively and DNA-DNA hybridization studies showed that their genetic differences were comparable to other closely related species.

Clark's and Western Grebes are best known for their elaborate and energetic courtship rituals, a series of displays in ritualized mechanical sequences that are among the most complex known in birds. **LENGTH:** 25 inches. **WINGSPAN:** 24 inches. **ADULT BREEDING:** (Feb.–Sept.) Black crown, white cheeks and throat. Red eye is surrounded by white. Bright yellow bill. Averages a grayer back and whiter flanks than a Western Grebe. **NONBREEDING:** (Sept.–Feb.) Dark crown touches the top of the eye. Whiter in the lores than on a Western Grebe. **VOICE:** Call is similar to Western Grebe but more a single-syllable *kreeeed*. **BEHAVIORS:** Seldom on land. Uses springing dive more often than Western Grebes and dives less frequently. **HABITAT:** Freshwater lakes and marshes with extensive areas of open water bordered by emergent vegetation. Breeding requires open water of at least several square kilometers. **STATUS:** Common to locally abundant resident in the western Trans-Pecos. Rare migrant and winter resident in the rest of Trans-Pecos and the western Panhandle. Like Western Grebe, the population in Trans-Pecos is increasing and is now more common than Western Grebe.

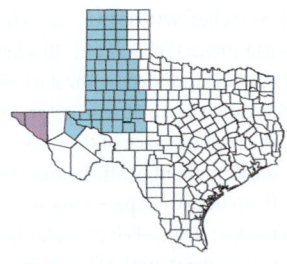

COLUMBIDAE (PIGEONS AND DOVES)

The family of pigeons and doves is a large grouping consisting of 353 current species worldwide. They occupy almost every terrestrial habitat. While the species found in North America are mostly gray and brown, many species elsewhere in the world are quite showy and colorful.

The family Columbidae is not thought to be closely related to any other family and is placed in its own order, Columbiformes. The order includes the now-extinct family Raphidae, commonly known as the dodos.

There is no exact definition of what is a pigeon and what is a dove. The distinction is largely based on size, with the larger species usually known as pigeons and the smaller known as doves. In the middle of the size range, though, you can find doves that are larger than pigeons.

The now-extinct Passenger Pigeon was native to North America and Texas. It was once perhaps the most numerous bird in the world, certainly the most numerous in North America with an estimated population between 3 and 5 billion birds. There were so many that a professional hunter in Washington County in Texas reported a flock took three days to pass in 1853. By 1900 there were none left in the wild, and the last capitve Passenger Pigeon died in 1914.

Thirteen species of pigeons and doves have been recorded in Texas, including Passenger Pigeon. Nine species are regularly occuring in Texas, and three—White-crowned Pigeon, Ruddy Ground Dove, and Ruddy Quail-Dove—are rare visitors to Texas.

Rock Pigeon

(Columba livia)

IUCN RED LIST STATUS (2019): Least Concern
POPULATION TREND: Decreasing

Rock Pigeon, formerly known as Rock Dove, was introduced to the New World in 1606. Domesticated descendants of the European wild Rock Dove have been around for thousands of years. They have been selectively bred to almost any variation imaginable, and the variations of wild birds today are stunning. Most birds have a form that closely approximates the wild form of a dark gray head and neck, lighter gray body, and two dark bars on the wings, but a cursory look through any North American flock will show almost-all-black birds, brown birds, tan birds, and even sometime snow-white ones.

The wild Rock Pigeon is declining, but the feral domestic variety thrives wherever humans provide artificial cliffs for nesting in the form of buildings, bridges, and grain silos. **LENGTH:** 12.5 inches. **WINGSPAN:** 28 inches.

ADULT: Extremely variable. Natural ancestral form most common. Dark slate-colored head, white base on the bill. Body and underside light gray. In flight, underwings light with dark trailing edge. Rump is pale. Tail gray with dark band. **VOICE:** A low *whoo, hoo-witooo-hoo*. **BEHAVIORS:** Walks or runs with head-bobbing motion. Powerful flier, often at great speed. **HABITAT:** Urbanized areas that mimic cliffs of the ancestral population—tall buildings, overpasses, grain elevators. **STATUS:** Common and abundant statewide, where they are usually in population centers. Becoming almost rare where there are no artificial cliff substitutes.

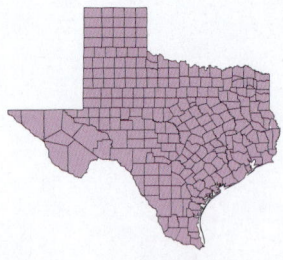

Red-billed Pigeon

(Patagioenas flavirostris)

IUCN RED LIST STATUS (2020): Least Concern
POPULATION TREND: Decreasing

The Red-billed Pigeon is a large, dark-colored, fruit-eating pigeon of the lowlands and foothills of Mexico and Central America. In North America, it has a small presence in the Lower Rio Grande Valley. The bill is mostly light or yellow-colored but is red on the base and the cere. The life history of the Red-billed Pigeon is poorly known, as there has been little research done on this species. However, the Band-tailed Pigeon provides a model for the life history and approximates the behavior and breeding of Red-billed Pigeons.

The Texas population is in decline, primarily observed in the riparian woodlands of the Rio Grande River.

LENGTH: 15 inches. **WINGSPAN:** 24 inches.
ADULT: Reddish-gray neck and head. Bill is short and red with pale tip. Overall, body is dark gray. **JUVENILE:** (May–Sept.) Similar to adults, but neck and head are the same gray as the body. **VOICE:** Hoarse cooing *whoah wha-wha-whoooo.* **BEHAVIORS:** Mostly arboreal. Rarely found on the ground except to drink. Their flight is swift and strong, similar to a Rock Pigeon's. **HABITAT:** In Texas, limited to the riparian forest along the Rio Grande River. Most of the remaining tall timber is on islands in the river. **STATUS:** Locally uncommon to rare summer resident in the western or upper Lower Rio Grande Valley, from Starr to Maverick counties.

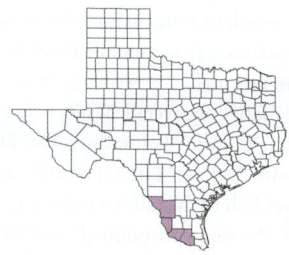

Band-tailed Pigeon

(Patagioenas fasciata)

IUCN RED LIST STATUS (2016): Least Concern
POPULATION TREND: Decreasing

Band-tailed Pigeons at first glance resemble the common Rock Pigeon but are distinguished by yellow bills and feet and distinct white collars on their napes. There is a long history of sport and market hunting of these birds. Under the pretense of crop protection, they were heavily hunted in the past century. There was a worry among naturalists they could go the way of the extinct Passenger Pigeon.

Today, though low numbers are actually taken through hunting in the US, the population is declining at an annual rate of 2.8%. **LENGTH:** 14.5 inches. **WINGSPAN:** 26 inches. **ADULT:** Slate-gray overall. White collar with greenish iridescence below the collar. Bill is yellow with dark tip. Tail is long and square. The lower half of the tail is a paler gray. In flight, the wings are pointed and bicolored with darker flight feathers. **JUVENILE:** (June–Nov.) Similar to adults but lacks the white collar. **VOICE:** Song is a hoarse owl-like *huh-whoo huh-whooo*. On takeoff, the wings make a loud clapping sound. **BEHAVIORS:** Walks short distances and will move as a flock with small groups leapfrogging to the front. Flight is swift and strong. **HABITAT:** Found in oak-juniper and juniper-piñon forests at high elevations. **STATUS:** Uncommon to rare summer resident and rare winter resident of the Davis and Chisos mountains. It is rare in summer in the Guadalupe Mountains.

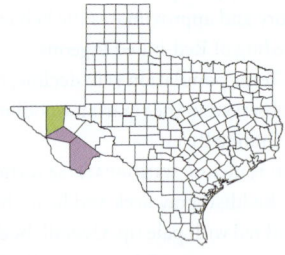

Eurasian Collared-Dove

(Streptopelia decaocto)

IUCN RED LIST STATUS (2019): Least Concern
POPULATION TREND: Increasing

The Eurasian Collared-Dove is originally from India and nearby China. Until about the 1930s, it was never found west of the Balkans. Suddenly, over the next forty years or so, it expanded across Europe. In the 1970s, a group of about fifty birds were released in the Bahamas and these birds are largely responsible for the expansion across North America. Human alteration of the landscape created the open habitat they require. The rapid expansion and its directional peculiarity (north then west) suggest that some unknown genetic mutation may have been a catalyst. **LENGTH:** 13 inches. **WINGSPAN:** 22 inches. **ADULT:** Head is pale gray with dark bill. Black band on the nape of the neck. Body is pale gray-brown. Wingtips are dark. Gray undertail coverts surrounded by dark. Tail has broad white band. Upper tail has broad white corners. In flight, shows a light gray band across upper wing coverts. **JUVENILE:** (Feb.–Nov.) Similar to adult but lacks the dark collar on the nape. **VOICE:** Song is a resonant, repeated *woohoo-woop woohoo-woop*. Call is a raspy high-pitched *haaaay haaaay*. **BEHAVIORS:** Flight is strong and direct. Walks with head bobbing. **HABITAT:** Heavily modified habitats, such as suburban and older urban habitats and farmland. Avoids heavily wooded areas and areas with predominately concrete and asphalt surfaces. **STATUS:** Common and locally abundant in urban areas and farmsteads statewide.

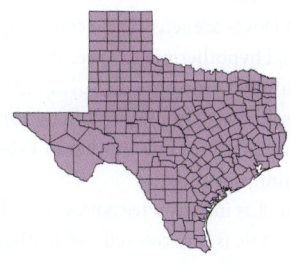

Inca Dove

(Columbina inca)

IUCN RED LIST STATUS (2016): Least Concern
POPULATION TREND: Increasing

The diminutive Inca Dove has expanded from Mexico into North America, from California to the Mississippi River. Originally known only from the Lower Rio Grande Valley, its plaintive *no hope* call can be heard in all but the Panhandle of Texas. The dun-colored feathers are dark-tipped, giving it a scaly appearance. In flight, large rufous wing patches are visible and the white outer tail feathers show.

Inca Doves are sensitive to cold, employing nocturnal hypothermia to lower their body temperature. During the day, pyramid roosting is used: five to twelve birds in two or three rows roost on top of one another in a sheltered, sunny location.

LENGTH: 8.25 inches. **WINGSPAN:** 11 inches.
ADULT: Male is clay-colored. All feathers are dark-edged, giving it a scaly pattern.

Tail is long and narrow with white outer tail feathers. In flight, primaries and the entire underwing are rufous. Female is similar to males with less-pronounced scaly pattern. **JUVENILE:** (Mar.–Nov.) Similar to adults, but feather edges are browner. **VOICE:** High resonant *whoa-WOAP* often described as *no hope*. **BEHAVIORS:** Walks often, head held high and nodding. Strong and rapid wingbeats in flight. **HABITAT:** Associates with humans in small towns, suburbs, older urban neighborhoods, and farms. **STATUS:** Common resident in the southern two-thirds of the state. Rare and local farther north to the Red River and in the Pineywoods.

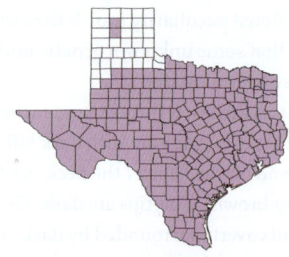

Common Ground Dove

(Columbina passerina)

IUCN RED LIST STATUS (2020): Least Concern
POPULATION TREND: Stable

The Common Ground Dove is the smallest of Texas's native doves. It is less scaly in appearance than an Inca Dove, and in flight, its short square tail is almost black. Like the Inca Dove, it has large rufous patches on the wings. Common Ground Dove is the most northerly occuring species of the tropical doves in Texas. **LENGTH:** 6.5 inches. **WINGSPAN:** 10.5 inches. **ADULT:** Eastern adults are pale sandy brown with a ruddy tint. Crown is pale gray. Bill has a pinkish base. Head and breast are scaly. Upper wings have dark brownish spots on them. Tail is short and in flight looks dark with small white corners. Above, the wings have rufous primaries, below the entire underwing is rufous. Western females are like eastern males but without the ruddy tint to the body. Scaly pattern can be faint. **JUVENILE:** (Mar.–Nov.) Like eastern female with faint all-over scaly pattern. **VOICE:** High-pitch *wo-oop woi-oop* repeating about once a second. **BEHAVIORS:** Very terrestrial, walks quickly while bobbing the head. Flight is short and low. **HABITAT:** Natural and modified habitats. Prefers open early successional habitat. Seems to prefer sandy soils. **STATUS:** Uncommon to locally common in Texas, south of Galveston Bay and west in the southern Trans-Pecos. Summer resident along the eastern and southern edge of the Edwards Plateau.

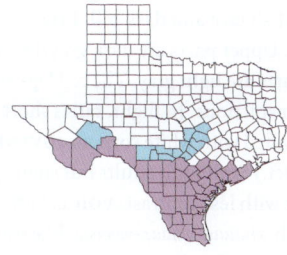

White-tipped Dove

(Leptotila verreauxi)

IUCN RED LIST STATUS (2020): Least Concern
POPULATION TREND: Stable

The White-tipped Dove was formerly known as the White-fronted Dove for its pale forehead. White-tipped refers to the white corners of its broad square tail. Pale gray, with darker gray-brown wings and back, it is strikingly plain. Perhaps its most standout feature is its strong, bright red legs. It is more often seen walking than flying, and flight is short. **LENGTH:** 11.5 inches. **WINGSPAN:** 18 inches. **ADULT:** Head is small-looking for its body. Pale whitish face and dark bill. Bright red eye ring. Upper parts are dark gray/brown, underparts are light gray. The red legs are often conspicuous. In flight, tail is short and rounded with white corners. **JUVENILE:** (May–Oct.) Similar to adults but slightly duller with less contrast. **VOICE:** Soft, low-pitch *whaaaa-waaaa-wooooo*, like someone blowing over a bottle. **BEHAVIORS:** Almost entirely terrestrial. When it does fly, it is low to the ground with long glides. **HABITAT:** Historically used the native riparian habitat along the Rio Grande and oxbow lakes known as resacas in the Lower Rio Grande Valley. Clearing of the habitat forced them into citrus groves and now urban neighborhoods. **STATUS:** Common resident in the Lower Rio Grande Valley and uncommon north to Maverick, Zavala, and Atascosa counties. Slowly expanding north and now resident to Nueces County. Expanding into the Edwards Plateau in Uvalde County.

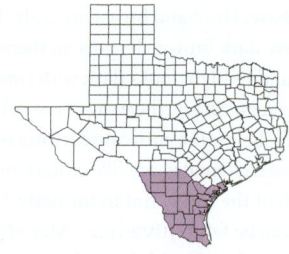

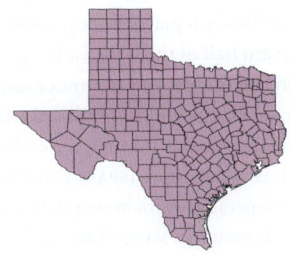

White-winged Dove

(Zenaida asiatica)

IUCN RED LIST STATUS (2016): Least Concern
POPULATION TREND: Increasing

The White-winged Dove in Texas was originally confined to the Rio Grande Valley and was heavily hunted. The season was long and often overlapped with the breeding season. The conversion of the native scrub forests to agriculture forced the population to use old citrus orchards for nesting until a series of freezes destroyed them. Younger orchards have few trees large enough for nesting, and the application of chemicals is prevalent.
LENGTH: 11.5 inches. **WINGSPAN:** 19 inches.
ADULT: Slightly larger and heavier than a Mourning Dove. Overall, gray-brown. Upper wings have large white patches and black primaries and secondaries. Tail has white terminal band. White wing patch is almost always visible as a white stripe on the folded wing. Face is plain with dark malar stripe, blue orbital ring, and red iris. **JUVENILE:** (Mar.–Oct.)

Like adults but has light cheek patch and forehead. **VOICE:** Hooting *whooo-who-ha-hoooo* or *who-cooks-for-you* with the same cadence as a Barred Owl. **BEHAVIORS:** Walks on limbs and on the ground. Flies singly, in pairs, and in small flocks. Flight is swift, silent, and direct. **HABITAT:** Historically uses dense, thorny woodlands in South Texas. Now very urbanized and uses the urban forest extensively. **STATUS:** Historically, its core range was the Texas Rio Grande Valley. It has now occupied the entire Texas coast and is rapidly expanding statewide.

Mourning Dove

(Zenaida macroura)

IUCN RED LIST STATUS (2016): Least Concern
POPULATION TREND: Increasing

Mourning Doves are one of the most abundant and widespread species of Texas birds. This habitat generalist can be found virtually statewide. A popular game bird, the Texas Cooperative Extension of the Texas A&M University System estimated in 2006 that about 400,000 hunters pursued mostly Mourning Doves in the annual Texas dove hunting season.

Large numbers of out-of-state migrant Mourning Doves replace the population in the northern half of the state. Fall migration begins in late August and continues until early November. Spring migration starts in February and extends through the end of May. **LENGTH:** 12 inches. **WINGSPAN:** 18 inches. **ADULT:** Generally, warm brown-gray overall. Black spots on upper wings. Face is plain with dark malar mark. Tail is long and pointed,

with white tips and black subterminal band. Males have a bluish gray cap and pinkish rosy tint to the face, throat, and breast. **JUVENILE:** Can occur in any month. White feather tips can give it a scaly appearance. Face is faintly patterned. **VOICE:** Resonant, low-pitched *hooAAAA hooo hooo hoo*. **BEHAVIORS:** Flight is rapid and swift. Often glides and can give the impression of a mini raptor. **HABITAT:** Very adaptable and nests in a variety of habitats, from ground nesting to ledges and trees. Avoids dense forests, preferring open woodlands and open edges. **STATUS:** Common to abundant summer and winter resident throughout the state.

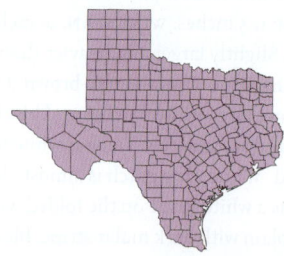

CUCULIDAE
(CUCKOOS)

The cuckoos are a large and ancient family of birds with no known close relatives. They are diverse and range from the fowl-like Greater Roadrunner, to the secretive Yellow-billed Cuckoo, to the parrot-like Groove-billed Ani. There are about 147 species of cuckoos worldwide and 6 species are found in Texas and North America. Two of those are rare visitors to Texas, the Mangrove Cuckoo and the Smooth-billed Ani.

The cuckoo family is well known for brood parasitism, or laying eggs in the nest of another species. This mostly occurs in the Old World species, but three New World species in Central and South America are also nest parasites.

Cuckoos feed primarily on insects and small vertebrates, but there are fruit-eating species. Many will eat hairy caterpillars that are ignored by other species. Yellow-billed Cuckoos have been observed consuming as many as 100 tent caterpillars in a single sitting. The hairs from the caterpillars are stored in the stomach and eventually regurgitated as a pellet, not unlike the pellets owls regurgitate.

Groove-billed Ani

(Crotophaga sulcirostris)

IUCN RED LIST STATUS (2021): Least Concern
POPULATION TREND: Increasing

Groove-billed Anis have one of the most unusual breeding systems of any bird. They form groups of one to five pairs and have a defended group territory. All females lay their eggs in a single nest. Individuals lay unequal numbers of eggs, and females have been observed removing other eggs before laying their own, resulting in a more equal distribution.
All members incubate the eggs, including males, and males guard incubating females. One male is the nocturnal incubator. This male's testes rapidly regress to 20% of their original size.
LENGTH: 13.5 inches. **WINGSPAN:** 17 inches. **ADULT:** Shiny jet-black overall. Bill is thick and bulbous, parrotlike with evenly spaced grooves. Skin around the eye is bare. **JUVENILE:** (Apr.–Sept.) Bill is smaller and not as bulbous; grooves may be absent or indistinct. Plumage is less glossy and brownish. **VOICE:** Sharp piercing *TEEP-weep* often repeated rapidly. **BEHAVIORS:** Flies in short glides and lands awkwardly, like the bird is front-heavy. Moves smoothly and squirrel-like through vegetation. Frequently suns. **HABITAT:** Open and partly open habitat such as brushy pastures, savannah, and orchards—usually dry. **STATUS:** Common summer resident in the Lower Rio Grande Valley. Uncommon to rare in the South Texas Brush Country north to the Balcones Escarpment. Uncommon to rare fall and winter visitors to the Lower Rio Grande Valley and along the coast north to Jefferson County.

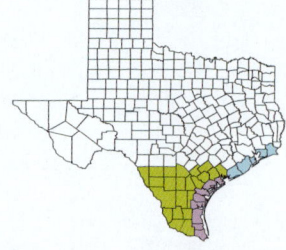

Greater Roadrunner

(Geococcyx californianus)

IUCN RED LIST STATUS (2022): Least Concern
POPULATION TREND: Increasing

Does any bird epitomize the Southwest better than the Greater Roadrunner? It is the only ground-dwelling member of the cuckoo family in North America. An omnivore and opportunistic predator, the Greater Roadrunner will eat nearly anything it can catch. As development fragments habitat, especially in the Pineywoods of East Texas, roadrunners are becoming more common.

LENGTH: 23 inches. **WINGSPAN:** 22 inches.
ADULT: All ages similar. Long-tailed and shaggy/streaked overall. Dark streaks on body feathers are dark glossy green. A short crest is often raised. Pale eye line behind the eye. When wings are raised, underwings are dark. **VOICE:** Call is a soft, dove-like *cooo-cooo-cooo-cooo-cooo-cooo*. Also makes a hollow, clattering bill rattle. **BEHAVIORS:** Does not fly often, but when it does, it mostly glides for short distances. Prefers to walk or run. Runs with the neck extended low to the ground. **HABITAT:** Generally found in arid and semiarid habitat with scattered shrubs. In East Texas, can be found in the Pineywoods. **STATUS:** Common to locally uncommon resident throughout the state. In the Pineywoods and northeastern corner of the state, it is a rare to locally uncommon resident.

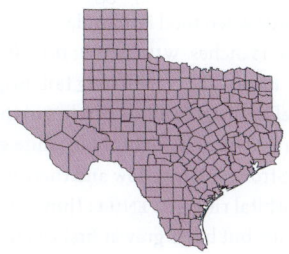

Yellow-billed Cuckoo

(Coccyzus americanus)

IUCN RED LIST STATUS (2021): Least Concern
POPULATION TREND: Decreasing

Yellow-billed Cuckoos require only seventeen days to go from egg-laying to fledging. An abundant local food supply triggers breeding, which is rapid. Bursting feather sheaths allow nestlings to be fully feathered just two hours after hatching. Yellow-billed Cuckoo and the closely related Black-billed Cuckoo are the only known brood parasites that are triggered by circumstances among the birds with dependent young. They will lay eggs in the nests of other birds when food is abundant.
LENGTH: 12 inches. **WINGSPAN:** 18 inches.
ADULT: Brown above with long tail. Bright white below. Bright rufous patches in the wings are obvious in flight. Bold white spots on tail. Strong bill is yellow and there is a yellow orbital ring. **JUVENILE:** (June–Aug.) Like adults but bill is gray at first when young, turning more yellow as the bird ages. **VOICE:**
Decelerating *coo-coo-coo-cooo-coooo* or rapid clicking *kek-kek-kek-kek-kek*. **BEHAVIORS:** Skulking behavior when perched. Avoids movement when under observation. Moves quietly through branches. **HABITAT:** Open woodlands, often near water. In the west, it is associated with riparian areas along streams. **STATUS:** Uncommon migrant and summer resident in the eastern two-thirds of the state. Locally uncommon to rare migrant and summer resident in Trans-Pecos. Spring migrants arrive in the east from early April to late May, and in the west from mid-May to mid-June. Fall migrants are found from mid-August to mid-October.

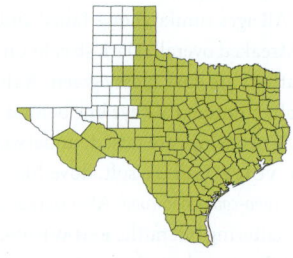

Black-billed Cuckoo

(Coccyzus erythropthalmus)

IUCN RED LIST STATUS (2021): Least Concern
POPULATION TREND: Decreasing

The Black-billed Cuckoo is a declining species in North America. They are known for consuming large numbers of tent caterpillars. There are old accounts of flocks of Black-billed Cuckoos, but these days, single birds are the norm. Tent caterpillar are now largely controlled with pesticides, and lowered prey availability may be impacting mortality and nesting success negatively.
LENGTH: 12 inches. **WINGSPAN:** 17.5 inches.
ADULT: Brown above and off-white below. Tail is long with small white tips. Little to no rufous in the wings. Undertail is gray. Bill is dark and eye has a red orbital ring.
JUVENILE: (June–Sept.) Lower contrast than adults. Throat is buffy. Wing coverts have pale fringes. Orbital ring is pale greenish. **VOICE:** Hollow rapid *ohp-ohp-ohp-ohp*. **BEHAVIORS:** Skulks and slips quietly through dense vegetation. Avoids movement when observers are present. Flight is graceful and swift, often low over ground. **HABITAT:** Found in woodlots, often along streams and ponds, and dense meadow borders. **STATUS:** Mostly encountered as a rare spring and very rare fall migrant in the eastern half of Texas. Most often found in coastal woodlots during migration. Spring migrants are seen from mid-April to late May, and fall migrants are seen from mid-August to late October.

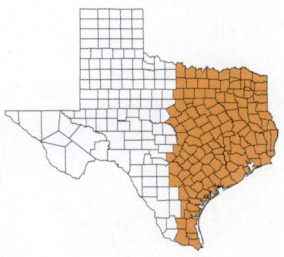

CAPRIMULGIDAE (NIGHTJARS AND ALLIES)

The nightjars are a family of crepuscular and nocturnal birds that are poorly understood. The term nightjar is a European term coined because the calls were said to be "jarring the night." The other common name for the family, goatsuckers, comes from the ancient belief that they would fly into a barn at night and suck the udder of a milking goat dry. In fact, the family name Caprimulgidea means "milker of goats" in Latin.

All nightjars are insectivores and have small bills and wide gapes, or in simpler terms, big mouths for catching insects. Nighthawks are the crepuscular hunters in the family, catching insects on the wing at dawn and dusk and in urban areas around bright lights. Others in the family, like the poorwills, hunt from a low perch or the ground, bolting up to catch a flying insect.

Roosting by day on the ground or a low branch, nightjars are some of the best-camouflaged birds. Their intricate earth-toned patterns perfectly match their roost sites. Many a birder has been amazed that they could not see a Common Pauraque roosting on the ground just inches off the trail in the Texas Rio Grande Valley.

Lesser Nighthawk

(Chordeiles acutipennis)

IUCN RED LIST STATUS (2016): Least Concern
POPULATION TREND: Increasing

Lesser Nighthawks aren't a well-studied species. They are known for their tolerance for temperature extremes. While nesting, they can gular flutter, meaning they will open their mouths and "flutter" their neck muscles to enhance evaporative cooling, the avian equivalent of panting. During cold spells, Lesser Nighthawks can remain torpid for long periods, similar to a hibernating mammal.
LENGTH: 9 inches. **WINGSPAN:** 22 inches.
ADULT: Gray-sandy pattern above and buffy speckled pattern below. Pale throat and notched tail. Wing has a pale bar about a third of the way from the wingtip to the bend in the wing, making an equilateral triangle. Females are buffier and the pale throat and wingbar are buffy. On a resting bird, the wingbar shows beyond the tertials. **JUVENILE:** (June–Sept.) Grayer and lacking the white throat and wingbars. **VOICE:** Low whistled trill up to ten seconds long, like the trill of an Eastern Screech-Owl. **BEHAVIORS:** Flies low and silent. Less urban than Common Nighthawks. **HABITAT:** Brushy lowlands and scrubby areas. **STATUS:** Common to uncommon migrant and summer resident in Trans-Pecos and Brush Country. Rare in summer on the Edwards Plateau. Rare to locally uncommon winter resident in the Brush Country. Very rare to rare migrant on the upper Texas coast. Spring migrants are found from late March to mid-May, fall migrants from August to late October.

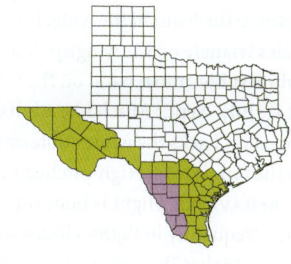

Common Nighthawk

(Chordeiles minor)

IUCN RED LIST STATUS (2021): Least Concern
POPULATION TREND: Decreasing

Called the "bullbat" for its flight style, the Common Nighthawk is familiar to almost everyone who spends time outside on a summer evening. Despite its name, it's most active at dawn and dusk. **LENGTH:** 9.5 inches. **WINGSPAN:** 24 inches. **ADULT:** Buffy-gray mottle on back, buffy barred below. Varies from grayish to rufous geographically. Notched tail. Pale throat. White wingbar is located about halfway from the wingtip to the bend in the wing, forming an isosceles triangle on the wingtip. Males have a white subterminal band on the tail. **JUVENILE:** (July–Sept.) Much like adults but lighter in color. Lacks the white subterminal band on the tail. **VOICE:** High-pitched buzzy *breeeppp*. **BEHAVIORS:** Flight is buoyant and high. Calls frequently in flight. Glides with raised wings. Makes "booming" display fights where males dive over a female and create a jet-engine-like roar with the wings at the bottom of the dive. **HABITAT:** Open areas, beaches, marshes, and agricultural areas. Much more urban than Lesser Nighthawks. Frequents large parking lots with lights at night. Makes use of flat gravel roofs for nesting and roosting. **STATUS:** Uncommon to common migrant and summer resident, except in the Pineywoods. Common migrant and rare summer resident in the Pineywoods. Regularly lingers until about mid-December. Spring migrants begin arriving in the east in early April to mid-May, in the west, late-April to late-May.

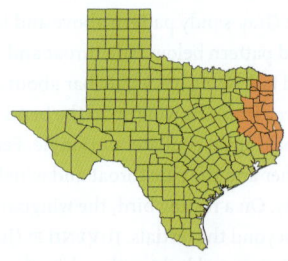

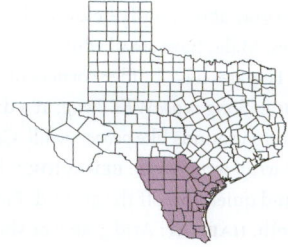

Common Pauraque

(Nyctidromus albicollis)

IUCN RED LIST STATUS (2020): Least Concern
POPULATION TREND: Stable

Common Pauraque is almost an enigma in South Texas, most often heard and rarely seen. It has some of the best camouflage in Texas, and many a birder has struggled to see a roosting bird just feet in front of them at known roost sites.
LENGTH: 11 inches. **WINGSPAN:** 24 inches.
ADULT: Mostly gray. Scapulars edged in pale lines give it a look of dead leaves when at rest. Males have a bold white wingbar and extensive white in the tail. Females have only white corners to the tail and the wingbar is more buffy.
VOICE: Emphatic *wrep-wrep-wrep-wheeercheee*.
BEHAVIORS: Leaps from the ground to catch low flying insects. Not often seen in flight but flies low from day roost to foraging sites. When flushed, flies through branches and brush, not over, until it settles. **HABITAT:** The Brush Country amid mesquite, ebony, and live oak. **STATUS:** Common resident of the South Texas Brush Country north to a line from about Val Verde County to Nueces County.

Common Poorwill

(Phalaenoptilus nuttallii)

IUCN RED LIST STATUS (2021): Least Concern
POPULATION TREND: Stable

Common Poorwills are known for their ability to enter almost daily torpor in winter and to tolerate high daytime temperatures. They can remain inactive for long periods of time, from ten days up to twenty-five days in one case. Daily torpor is often used by Common Poorwill in the cool spring and fall weather after foraging at dusk.

LENGTH: 7.75 inches. **WINGSPAN:** 17 inches.
ADULT: Short-tailed and rounded wings. Speckled gray above, buffy below with buffy primaries. Males have small white corners on the tail. Females have buffier corners on the tail. **VOICE:** Song is *pooWEEPwup*; at a distance it often sounds like *poooor-willll*. Call is a hoarse *wrup-wrup-wrup*. **BEHAVIORS:** Flight is slow and quiet, low off the ground. Flutters like a moth. **HABITAT:** Arid grassy or shrubby areas, often hillsides. **STATUS:** Common to uncommon migrant and summer resident in South Texas Brush Country, Edwards Plateau, and Trans-Pecos. Uncommon to rare summer resident on the High Plains and Rolling Plains. Rare to uncommon winter resident on the southern third of their Texas range. Spring birds arrive in mid-March to mid-May and fall migrants are found from early September to late-October.

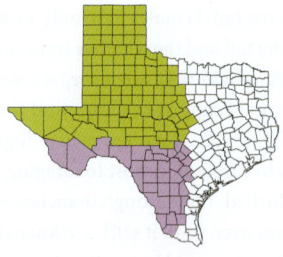

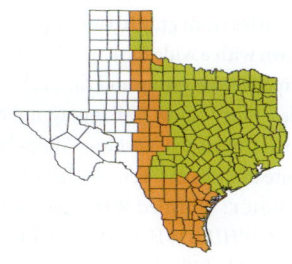

Chuck-will's-widow

(Antrostomus carolinensis)

IUCN RED LIST STATUS (2021): Near Threatened **POPULATION TREND:** Decreasing

The largest of the North American nightjars is easily detected, but much is still unknown about this species, including its nesting habits, breeding success, and habitat use. While known to nest in suburbia, there is fear that intensifying development and modern agricultural practices may be producing a population decline that has not yet been detected.

Chuck-will's-widows are crepuscular, most active in the dawn and dusk twilight. Several individuals will often begin calling almost at once when the fading or early light gets just right, and calls end almost as abruptly. **LENGTH:** 12 inches. **WINGSPAN:** 26 inches. **ADULT:** Reddish-brown or grayish-brown pattern. Rufous below. Upper tail is rufous, not gray like whip-poor-wills. Males have white inner webs on the tail giving the impression of diffuse white corners. Females have buffy corners. **VOICE:** Loud *chek-WHILLO-WHILL.* **BEHAVIORS:** Flight is typically silent with lots of flaps and glides. When migrants are flushed, they fly a long distance and land on large branches. Known to forage from the ground, making short hops for prey. **HABITAT:** Occurs in a variety of forests. Forest gaps may be important in habitat choices. **STATUS:** Uncommon to common migrant and summer resident in the eastern two-thirds of Texas. Commonly encountered in coastal migrant traps. Spring migrants are found from late March to mid-May, fall migrants from early August to late October.

Eastern Whip-poor-will

(Antrostomus vociferus)

IUCN RED LIST STATUS (2021): Near Threatened **POPULATION TREND:** Stable

Like many of the nightjars, there is still a lot to learn about this species. Easily heard in spring when present, their crepuscular and nocturnal habits and woodland habitat preferences make them a difficult species to study.

Until 2010, Eastern Whip-poor-will and Mexican Whip-poor-will were considered one species. There are known differences in calls and some minor differences in appearance. **LENGTH:** 9.75 inches. **WINGSPAN:** 19 inches. **ADULT:** Varies from gray to rufous pattern. Gray crown with a wide dark median strip. Gray stripe between wing and the back on both sides. Buffy below. Tail is gray, not rufous as in a Chuck-will's-widow. Males have large white tail corners. Females have buffy corners. **VOICE:** Whistled *WHIP-purrWHILL WHIP-purrWHILL WHIP-purrWHILL.* **BEHAVIORS:** Flight is slow and noiseless, through or over the canopy or near the ground. Flaps then glides, often making 180-degree turns. **HABITAT:** Dry forests with little or no underbrush. The composition of the forest appears to be less important than the amount of understory. **STATUS:** Rare to uncommon migrant in the eastern half of the state west to the eastern part of the Edwards Plateau. Spring migrants are present from mid-March to mid-May, and they can be vocal at dusk and dawn during spring migration. Fall migrants are silent and harder to detect and pass through from mid-August to early November.

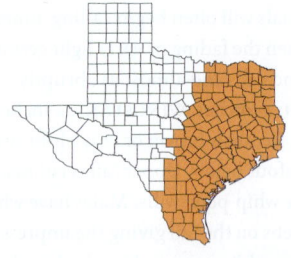

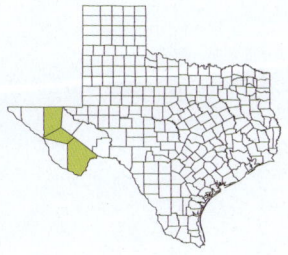

Mexican Whip-poor-will

(Antrostomus arizonae)

IUCN RED LIST STATUS (2020): Least Concern
POPULATION TREND: Decreasing

Mexican Whip-poor-will was split from Eastern Whip-poor-will in 2010. It is one of the least studied species in North America and one of the least accessible species in Texas. Its biology is essentially unstudied. Found almost exclusively at high elevations, there are very few locations where this species can be found without a long hike in the dark.
LENGTH: 9.75 inches. **WINGSPAN:** 19 inches.
ADULT: Grayish with faint buffy collar. Gray crown with wide dark median stripe. Tail is gray, not rufous like Chuck-will's-widow. Two gray stripes on the back where the wings meet the body. Males have white tail corners, females have buffy tail corners.
VOICE: *WHIP-puRRRLLLL-WHILL* repeated, lower pitched and rougher than Eastern Whip-poor-will. **BEHAVIORS:** Flight is floppy and buoyant. **HABITAT:** Thickets

and pine-oak woods on brushy canyon and mountain slopes. **STATUS:** Common summer resident at the upper elevations of the Guadalupe, Davis, and Chisos mountains. Almost unknown from lower elevations. Birds arrive in mid-April and depart between late August and mid-September.

APODIDAE (SWIFTS)

The swift family is a worldwide family consisting of 112 species, 4 of which have occurred in Texas: Black Swift and White-collared Swift are rare in the state, while Chimney Swift and White-throated Swift are regulars. A fifth species, Vaux's Swift, is considered by many to pehaps have occurred in Texas, but the challenge in separating Vaux's Swift from Chimney Swift in the field has made gathering the necessary documentation difficult.

Swifts have traditionally been considered closely related to hummingbirds based on morphological traits, and hummingbirds and swifts were originally placed in the same order. A more recent analysis shows that swifts are most closely related to the treeswifts of Southeast Asia and some western Pacific islands. Hummingbirds are a sister family to the swifts and treeswifts, but they have been moved to a new and separate order.

Swifts generally have long, sickle-shaped wings and compact, elongated bodies that have led to a common description of "cigars with wings." Their feet are small and far back on their bodies and almost useless on land. The family name Apodidea comes from a greek word meaning "footless." They are masters of the air, though, and live almost their entire lives on the wing. Swift nests are built on vertical surfaces, glued on using their salivary cement. One species, the Edible-nest Swiftlet, builds a nest entirely from salivary cement, and the nests are collected to make the legendary bird's nest soup in China.

Chimney Swift

(Chaetura pelagica)

IUCN RED LIST STATUS (2018): Vulnerable
POPULATION TREND: Decreasing

Chimney Swifts live almost entirely on their wings: feeding, drinking, even bathing. They only stop flying to roost and nest. Populations likely increased due to human-caused changes to North America, but changing building practices that have resulted in fewer chimneys have led to a population decline.

Fall-migrating swifts often form roosts in large chimneys. In the evening, foraging swifts gather in increasing numbers around a large roost site. As the light fades, they swirl tighter and tighter around the entrance until they form a "Swift Tornado" and swirl into the roost in a matter of a few seconds.
LENGTH: 5.25 inches. **WINGSPAN:** 14 inches.
ADULT: Dark gray overall with no contrasting markings. Like a cigar with wings. Short tail and stiff wings in flight. **VOICE:** High-pitched chips run together in a rapid twittering.

BEHAVIORS: Flies almost constantly except at roost or when nesting. Flight is swift with lots of rapid flapping. **HABITAT:** Access to nest sites is the biggest factor in its range. Once they used hollow trees in mature forests; they are now almost completely dependent on artificial hollow trees such as chimneys. **STATUS:** Common migrant and summer resident in all but the Trans-Pecos. Spring birds appear in late March, and fall migrants are seen from early September to mid-October, sometimes present into mid-November.

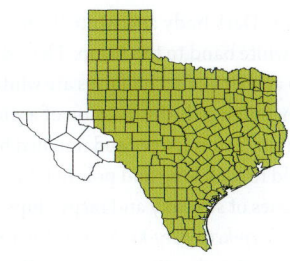

White-throated Swift

(Aeronautes saxatalis)

IUCN RED LIST STATUS (2022): Least Concern
POPULATION TREND: Decreasing

These highly social birds forage in groups, chasing at high speeds with abrupt changes in direction. In the breeding season, they make spectacular "courtship falls"—two birds cling to each other, pinwheeling and falling hundreds of feet before releasing each other just above the ground.
LENGTH: 6.5 inches. **WINGSPAN:** 15 inches. **ADULT:** Dark body and white throat. Narrow white band to the rump. The sides of rump and tips of secondaries are white, sometimes giving the impression of a wholly white rump. Tail is longer and notched but often held to look long and pointed. **VOICE:** Rapid series of squeaky and raspy chips: *kerp-kerp-kerp-kerp-kerp-kerp.* **BEHAVIORS:** Flight is rapid and swift. Flies in small groups along cliffs, up to 300 feet in the air. **HABITAT:** Nests in crevices on cliffs, canyon walls, and mountainsides. Will use man-made environments, overpasses, bridges, and highway cuts. **STATUS:** Common summer resident and locally uncommon winter resident in the Trans-Pecos, east to the edge of Val Verde County.

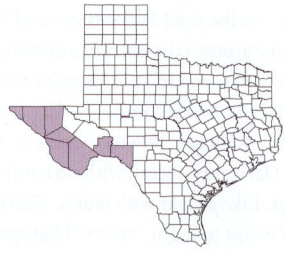

TROCHILIDAE (HUMMINGBIRDS)

The family of hummingbirds is restricted to the Americas and consists of 363 species. Eighteen species have been recorded in Texas. Eleven are considered to occur regularly in Texas while seven are considered rare vistors. Hummingbirds are the smallest birds in the world, and in their supercharged, sugar-buzzed world, males compete ferociously for females. In some species males compete for mates in leks, groups of courting males. Some males defend prime food sources to attract a mate.

Hummingbird plumage is a kaleidoscope of variables. All manner of fancy tail plumes exist and many have feathers with angle-sensitive iridescence, particularly iridescence on the throat of males as a jewel-like gorget.

Hummingbirds are the champions of aerial maneuverability. This extreme agility gives them an air of invincibility, and one can often approach closer to hummingbirds than other species. Hummingbirds need a high-energy diet to do this. They are effective at gathering tiny insects, but 90% of their diet consists of nectar. Even this huge amount of sugar is not always enough, and some species go into a state of torpor to preserve energy and avoid starving overnight.

Rivoli's Hummingbird

(Eugenes fulgens)

IUCN RED LIST STATUS (2022): Least Concern
POPULATION TREND: Stable

Rivoli's Hummingbird was named in honor of François Victor Masséna, third duke of Rivoli, in 1829. Masséna was a dedicated amateur ornithologist who amassed a collection of 12,500 specimens, now part of Philadelphia's Academy of Natural Sciences collection.

Rivoli's Hummingbird's name was changed to Magnificent Hummingbird in the 1980s and then restored in 2017, when the species was split into Rivoli's and Talamanca Hummingbird.
LENGTH: 5.25 inches. **WINGSPAN:** 7.5 inches.
ADULT BREEDING: Male has violet crown that is often raised. Emerald green gorget. Small postocular spot. Belly is blackish. All dark green back and tail. Females have dark green back and tail, and tail has small white corners. Postocular spot is elongated into a thin line. Pale green throat and breast feathers have pale edges, giving a scalloped appearance. **JUVENILE:** Like females but slightly darker. **VOICE:** Song is a sharp *stwipt stwipt stwipt*. Chase call is a rapid *stit-stit-stit-stit*. **HABITAT:** Mixed oak, pine, and juniper forest at high elevations. **STATUS:** Uncommon and local summer resident in the Davis Mountains and rare in the Guadalupe Mountains. Spring birds arrive in mid-March, fall migrants depart in mid-September and most have departed by mid-October.

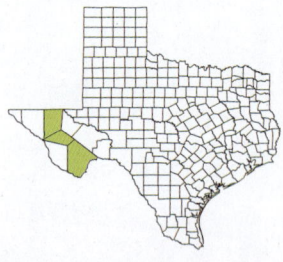

Blue-throated Mountain-gem

(Lampornis clemenciae)

IUCN RED LIST STATUS (2021): Least Concern
POPULATION TREND: Stable

The Blue-throated Mountain-gem is North America's largest hummingbird, coming in at three times the weight of the common Ruby-throated Hummingbird. **LENGTH:** 5 inches. **WINGSPAN:** 8 inches. **ADULT:** Males have green crown and blue gorget. White eye line and white malar stripe. Underside is dirty gray. Back is light green, and rump is somewhat bronzy green. Tail is large and dark with bold white spots that flash in flight. Females are like males but lack a blue gorget. **VOICE:** Call is a loud *SEEK SEEK*. **HABITAT:** High-elevation moist woodlands, usually pine-oak. **STATUS:** Locally uncommon summer resident in the Chisos Mountains of Big Bend National Park from mid-April to early October.

Male

Female

Lucifer Hummingbird

(Calothorax lucifer)

IUCN RED LIST STATUS (2021): Least Concern
POPULATION TREND: Stable

While the Lucifer Hummingbird's biology isn't well known, there are some things unique about it. It is the only North American hummingbird with a strongly decurved bill. It is also the only hummingbird known to display at the nest. Other hummingbirds do courtship displays, but these displays are not at the nest. **LENGTH:** 3.5 inches. **WINGSPAN:** 4 inches. **ADULT:** Male has distinctive long magenta gorget. Postocular spot. Long decurved bill. Tail is forked. Green back and dirty green undersides. Female lacks gorget, and the breast is cleaner with a buffy band. Tail is much shorter and less forked than males. Buffy supercilium. **JUVENILE:** (May–Oct.) Like females. Subadult males begin to develop longer tail feathers and a gorget between August and November. **VOICE:** Call is a dry, sharp, irregular twitter. **BEHAVIORS:** Male courtship display to females on the nest is unique among hummingbirds. It consists of several rapid side-to-side flights facing the female, then a vertical climb and powerful dive. **HABITAT:** Upper-elevation Chihuahuan desert scrub. **STATUS:** Locally common summer resident in the foothills of the Chisos and Christmas mountains in Brewster County. Post-breeding, rare and local in the Davis and Chinati mountains. Spring birds begin arriving in mid-April and are present through early October.

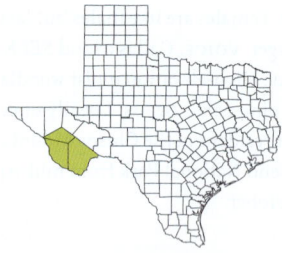

Male

Female

Immature male

Ruby-throated Hummingbird

(Archilochus colubris)

IUCN RED LIST STATUS (2021): Least Concern
POPULATION TREND: Increasing

Ruby-throated Hummingbirds are the most familiar hummingbirds in most of Texas, expected statewide except for the most western part of the Trans-Pecos.

This most widely distributed of North America's hummingbird species pours through Texas in fall, concentrating on the coast. Seeing as many as a hundred hummingbirds swarm feeders at the annual Rockport-Fulton HummerBird Celebration is one of the great avian spectacles in Texas.

LENGTH: 3.75 inches. **WINGSPAN:** 4.5 inches.
ADULT: Male has green back and crown with bright ruby-red gorget. Tail has wide black subterminal band and white tips. Small postocular spot and narrow black chin strap. Breast is dusky green. Wings are more pointed at the tips than the similar Black-chinned Hummingbird. Female lacks the red gorget

and black chin strap. **JUVENILE:** (June–Dec.) Like females but with a more contrasting face pattern. **VOICE:** Chase call is a rapid *tsi-tsi-tsi-tsi-tsi*. **BEHAVIORS:** Courtship display begins when a female enters a male's territory, with a U-shaped looping dive display. Once the female perches, the male shifts to a shuttle display back and forth in front of the female. **HABITAT:** Mixed woodlands and eastern deciduous forest. **STATUS:** Common summer resident in the eastern third of Texas, breeds south to Victoria. Uncommon migrant on the Edwards Plateau and in South Texas Brush Country. Spring migrants arrive in early March and leave between early August and late October.

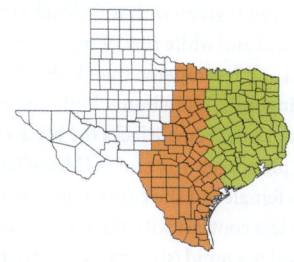

Male

Female

Black-chinned Hummingbird

(Archilochus alexandri)

IUCN RED LIST STATUS (2021): Least Concern
POPULATION TREND: Increasing

Black-chinned Hummingbirds are generalists not easily described with superlatives. They are not the largest nor the smallest. They're not in the largest or smallest range. They're not particularly showy. They occur along with other species of hummingbirds.
LENGTH: 3.75 inches. **WINGSPAN:** 4.75 inches.
ADULT: Male has green crown and back, dusky green on the chest. Wide black chin and iridescent purple gorget. Small postocular spot. Tail is green with wide black subterminal band and white tips. Wings are more rounded and less tapered than Ruby-throated Hummingbirds, like a butter knife compared to a steak knife. Female lacks the black chin and purple gorget. **JUVENILE:** (May–Oct.) Like the female but the crown is grayer and the face has less contrast with the throat. **VOICE:** Chase call is a rapid *tsi-tsi-tsi-tsi-tsi*, similar to

a Ruby-throated Hummingbird. Adult male's wings produce a soft low whistle in flight.
BEHAVIORS: When at flowers or feeders, it flips its tail almost continuously, unlike the similar Ruby-throated Hummingbird which holds its tail mostly still. The male diving display is done over a female and consists of a high, narrow U pattern about 3 feet above a female, repeated six to ten times. **HABITAT:** Xeric (dry) woodlands, often near a stream. **STATUS:** Common to locally abundant summer resident in the western two-thirds of the state. In winter, rare along the coast.

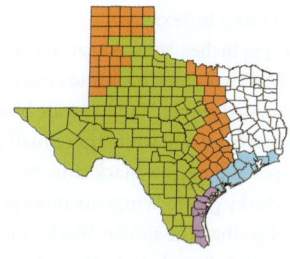

Male

Female

Immature male

Anna's Hummingbird

(Calypte anna)

IUCN RED LIST STATUS (2021): Least Concern
POPULATION TREND: Increasing

Anna's Hummingbird was named after
Anna Debelle, duchess of Rivoli. The duke,
her husband, is the namesake of Rivoli's
Hummingbird.

Anna's Hummingbird has been expanding its range since the 1930s. Once restricted
to coastal Baja and the Southern California
coast, it now breeds into Canada and west to
the New Mexico border with Arizona. Anna's
is a winter-breeding hummingbird, a strategy
likely to coincide with increased nectar sources from winter rains and decreased competition from other species.
LENGTH: 4 inches. **WINGSPAN:** 5.25 inches. **ADULT:** Male has red crown and throat.
Throat or gorget is elongated, but not as much
as in a Costa's Hummingbird. Thin white
line over the eye. Back of the crown and nape
are dark. Green tail and back. Tail has black

subterminal band and white tips. Undertail
feathers are gray-edged. Belly is dusky green.
Female has green crown. Throat has a red central patch. **JUVENILE:** (Feb.–Sept.) Like female
but without a central throat patch. **VOICE:**
Chase call is a rapid, dry chatter. **BEHAVIORS:**
Flies and hovers well, able to back away while
hovering. **HABITAT:** In Texas, habitat varies
wherever suitable flowers and feeders are
available. **STATUS:** Uncommon to rare fall
migrant and winter resident in the western
and central Trans-Pecos. Rare to very rare and
irregular visitor to the rest of the state, mostly
in late fall.

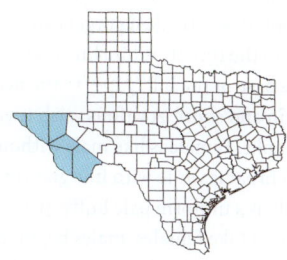

Male

Female

Calliope Hummingbird

(Selasphorus calliope)

IUCN RED LIST STATUS (2021): Least Concern
POPULATION TREND: Increasing

Calliope Hummingbird is the smallest of North America's hummingbirds. Known for being a fierce defender of territories in its breeding range, Calliope Hummingbirds in Texas are more retiring and probably often overlooked in migration in Texas. There is good evidence that their wintering range is expanding eastward.

LENGTH: 3.25 inches. **WINGSPAN:** 4.25 inches.
ADULT: Male has small short wings and short square tail. Rosy-streaked gorget extending down onto the throat. White line is visible from the base of the bill (gape) to the neck. Bill is thin and short. A rufous wash is visible on the flanks. Female is like male without the gorget. Throat is white with fine grayish spotting. Belly is a uniform pale buffy. **JUVENILE:** (June–Oct.) Like females, males begin to get gorget feathers in October. **VOICE:** Call is a quiet, high, musical chip. **BEHAVIORS:** Usually shy around larger hummingbirds and waits for a lull in activity to come into a feeder. This contrasts with breeding birds out of state that are usually quite aggressive in defending a territory. **HABITAT:** Mostly higher elevations. **STATUS:** Casual spring and uncommon fall migrant in western half of Trans-Pecos. Becoming rare and casual further east. Most records away from Trans-Pecos are fall records. Males begin to pass through in mid-July, followed by females and younger birds through September. A small number have overwintered at feeders.

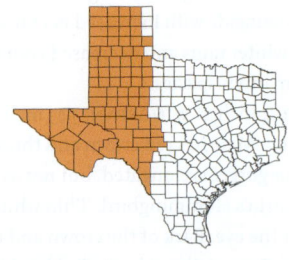

Male

Female

Rufous Hummingbird

(Selasphorus rufus)

IUCN RED LIST STATUS (2020): Near Threatened

POPULATION TREND: Decreasing

Rufous Hummingbirds are known to be one of the most aggressive hummingbirds. Their anatomy makes them better at flying at lower altitudes than at higher altitudes, where the closely related Broad-tailed Hummingbird takes over. They have one of the largest breeding ranges in North America, breeding well into Alaska. Rufous Hummingbirds are known for overwintering nearly anywhere a hummingbird feeder is available.

LENGTH: 3.75 inches. **WINGSPAN:** 4.5 inches. **ADULT:** Male has orange-rufous gorget. Extensive rufous below and on the back in typical adult males. White chest band. Up to 5% of adult males are reported to have some green in the back. Tail feathers have notched tips. Female has green crown and back and white breast. Slightly notched next to the central tail feather. Non-rufous-backed birds are not safely distinguished from Allen's Hummingbird unless the tail feather shape can be assessed. **JUVENILE:** (June–Nov.) Like females. **VOICE:** Chase call is a sharp buzz followed by a series of sharp *tzup tzup tzup*. **BEHAVIORS:** Territorial migration stopovers. **HABITAT:** Several types of habitats at high and medium elevations. **STATUS:** Very rare in spring and common to abundant fall migrant in Trans-Pecos. Uncommon to common migrant on High Plains and western Rolling Plains south to Edwards Plateau. Rare to locally uncommon winter resident in western and southern Trans-Pecos, coastal plains, and Lower Rio Grande Valley.

Allen's Hummingbird

(Selasphorus sasin)

IUCN RED LIST STATUS (2021): Least Concern
POPULATION TREND: Increasing

Allen's Hummingbird is named for Charles Allen, a California collector who noted differences between Allen's and the cospecific Rufous Hummingbird in 1877, naming it *Selasphorus alleni*. It turns out that the species was described earlier, in 1829 by René Lesson, who believed he was renaming the Rufous Hummingbird, then called the Nootka Hummingbird for Nootka Island near Vancouver. He named it *Ornismya sasin*. Sasin means hummingbird in the language of the Indigenous people of Nootka. Because the name sasin was described first, it takes precedence and the modern name is *Selasphorus sasin*, but the common name is still the eponymous Allen's Hummingbird.
LENGTH: 3.75 inches. **WINGSPAN:** 4.25 inches.
ADULT: Male is essentially identical to Rufous Hummingbird except for a green back. A small number of Rufous Hummingbirds have partial green backs, and green-backed birds are not safely identified except by assessing details of the tail feathers. Female is identical to female Rufous except for small differences in tail feather shape. **VOICE:** Like Rufous Hummingbird. **BEHAVIORS:** Territorial with other species during migration. **HABITAT:** Higher elevations in montane chaparral, open coniferous forest, and mixed woodland habitats. **STATUS:** Rare to uncommon fall migrant through the Trans-Pecos. Allen's Hummingbird is a rare winter resident on the coastal prairies and casual inland resident elsewhere in the state.

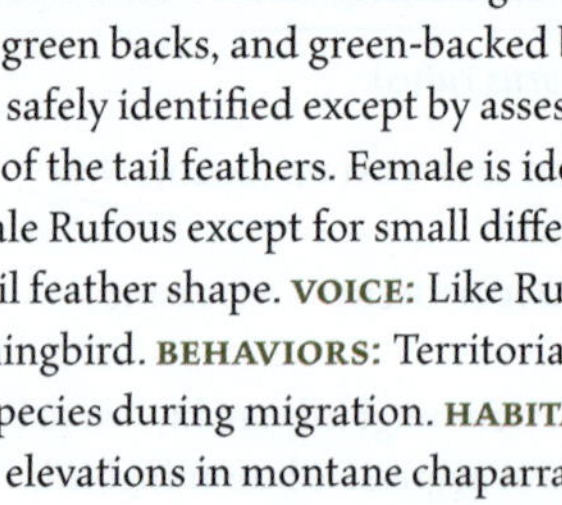

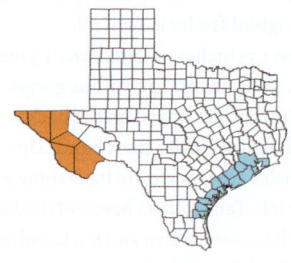

Male

Female

Broad-tailed Hummingbird

(Selasphorus platycercus)

IUCN RED LIST STATUS (2022): Least Concern
POPULATION TREND: Decreasing

Broad-tailed Hummingbird is the major breeding hummingbird of higher elevations, and it is a summer resident on the Trans-Pecos sky islands.
LENGTH: 4 inches. **WINGSPAN:** 5.25 inches.
ADULT: Males have rosy red gorget. Green and buffy flanks, green back and crown. White line from chin to eye ring, down to the neck. Rufous only on the outer few tail feathers. Tail extends well beyond the wingtips when perched. Female lacks gorget but has spotted throat and cheeks. Flanks are more buffy than green. **JUVENILE:** (June–Nov.) Like females.
VOICE: Call is a rapid, sharp *chript chipt chipt*. The loud wing trill is probably a substitute for a song. **BEHAVIORS:** Display flight is a series of high climbs, then hovering before a rapid powered dive accompanied by a ringing, musical wing trill audible up to 120 feet

away. The male pulls out of the dive in front of a lower-perched female and performs a hovering display in front of her. **HABITAT:** Cypress-pine-oak and piñon-juniper-oak montane woodlands. **STATUS:** Uncommon to locally common summer resident in the Chisos, Davis, and Guadalupe mountains. Common to uncommon migrant in rest of Trans-Pecos. Rare spring and uncommon fall migrant on High Plains and western part of Edwards Plateau. Spring birds arrive as early as late February and migration ends in early May. Fall migration begins mid-August and ends in mid-October.

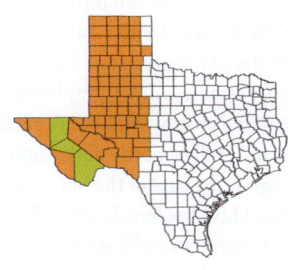

Broad-billed Hummingbird

(Cynanthus latirostris)

IUCN RED LIST STATUS (2021): Least Concern
POPULATION TREND: Stable

The male Broad-billed Hummingbird is one of the most spectacular of Texas's hummingbirds. Irridecent blues and greens and a bright red bill make the male unmistakable. Summer birds in Texas and North America are restricted to remote areas, so their biology is poorly studied.

LENGTH: 4 inches. **WINGSPAN:** 5.75 inches.
ADULT: Male has glittery blue throat and dark blue-green underparts. Bill is thick bright red and slightly decurved with a dark tip. Green back with dark blue-green tail, light undertail coverts. Female is dingy gray below with green back and crown. Dark cheek patch and white eye line. Bill is mostly dark; some birds have orange at the base of the bill. **JUVENILE:** (Mar.–Aug.) Like the female with all-dark bill. **VOICE:** Song is a rapid sputtering followed by a squeaky twitter. Call is a sharp double *chet-chet* like a Ruby-crowned Kinglet. **BEHAVIORS:** Males may form leks during the breeding season, taking turns singing with no aggressive behavior. Males perform a "pendulum" display. **HABITAT:** Montane canyons. **STATUS:** Rare to very rare summer visitor to the Davis Mountains. Casual to very rare winter visitor to El Paso County. Casual visitor to virtually all areas of the state, with records for all seasons. Only one case of evidence of nesting in Texas. Summer residents arrive in the Davis Mountains in late March and are present through mid-October.

Buff-bellied Hummingbird

(Amazilia yucatanensis)

IUCN RED LIST STATUS (2022): Least Concern
POPULATION TREND: Unknown

Buff-bellied Hummingbirds are one of the specialty birds for South Texas, eagerly sought out. Despite birders' keen interest, it is one of the least-studied hummingbirds in North America. Most information is from older studies in Northern Mexico. It is perhaps unique among North American hummingbirds due to its strong movement north along the Gulf Coast during fall and winter.
LENGTH: 4.25 inches. **WINGSPAN:** 5.75 inches. **ADULT:** Male has green back and crown with red bill. Green iridescent throat. Buffy belly is separated by a clean line from the throat, but varies from buffy to buffy gray. Tail is rufous. Female is essentially the same as males but bill is less bright red. **JUVENILE:** (May–Sept.) Buffy belly is less pronounced, with demarcation between throat and belly very diffuse. **VOICE:** Song is a high-pitched descending *seelp seelp seelp.* Call is a rapid sharp *prt-prt-prt.* **BEHAVIORS:** Typical hummingbird habits, no breeding display. **HABITAT:** Typically in thorn forests with understory of hummingbird-friendly plants like sage, Turk's-cap lily, and coral bean. Common in residential and park areas in cities. **STATUS:** Uncommon to locally common summer resident in the Lower Rio Grande Valley and north along the coast to Victoria County. Rare spring and summer visitor to Bastrop and Washington counties. There are scattered winter and fall records along the coast to the Pineywoods.

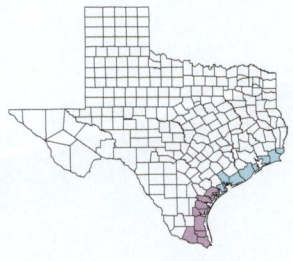

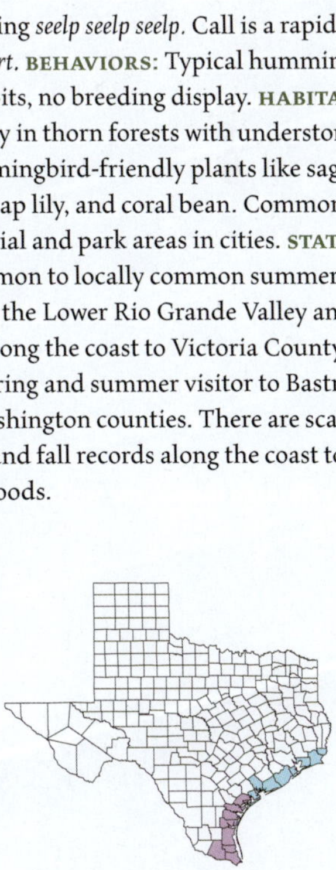

RALLIDAE (RAILS, GALLINULES, AND COOTS)

Rails, gallinules, and coots have long been placed in the order Gruiformes with cranes and limp-kins. The rail family is composed of 155 species worldwide. Ten species occur in Texas, though two, the Spotted Rail and Paint-billed Crake, are considered very rare.

They are wetland specialists but occupy almost every type of wetlands, from isolated Pacific islands to mountain lakes. Many species seem to be barely able to get off the ground in flight yet are long-distance migrants. Some species have evolved to flightlessness on remote islands, a trait that in many cases has led to their extinction from inability to deal with introduced predators. Twenty-four species of rails have become extinct, and many more species are threatened and endangered.

In Texas, the Black Rail is listed as threatened, yet several species of rails are abundant enough to be hunted as game birds. There are legal hunting seasons for six species of the rail family in the state. Even with generous daily and possession limits, they are not a popular game bird due to their mostly secretive nature and the challenging habitat for a gangly human to stalk prey in.

Within the family of rails, coots and gallinules live on open wetlands much like ducks and have short bills that extend up the forehead in a shield. Crakes include the Sora, Yellow Rail, and Black Rail and are secretive wetlands birds with short bills. Species like the King, Clapper, and Virginia Rail have long decurved bills with laterally compressed bodies that allow them to slip quietly and invisibly through their dense wetland habitat. The phrase "skinny as a rail" refers to the slim bodies of rails.

Adult

Chick

Clapper Rail

(Rallus crepitans)

IUCN RED LIST STATUS (2018): Least Concern
POPULATION TREND: Decreasing

Clapper Rails have historically been a species of commercial value. They were once hunted for both meat and eggs, which were collected by commercial eggers. Clapper Rails lay large clutches of up to nine eggs on the Gulf Coast and have been recorded laying up to five clutches a year after nest failures. Eggers have collected as many as 100 dozen in a day, it's been reported.

While not especially popular, there is still a Texas sport hunting season for rails. The season is during fall, and the difficulty of movement in Clapper Rail habitat makes the impact of hunting light in Texas. Loss of habitat to development drives population decline. **LENGTH:** 14.5 inches. **WINGSPAN:** 19 inches. **ADULT:** Heavy long decurved bill, orange to gray. Cheeks are gray. Breast and foreneck are drab gray. Back feathers are brown with white buffy edges, giving the birds a shaggy look. Fine white streaking on the lower body. **JUVENILE:** (July–Sept.). Gray-brown overall with little distinct pattern. Chicks small, all downy black. **VOICE:** Call is a series of *kek kek kek kek*, often given in chorus with others. Also a slower-paced *kerrrk kerrk kerrk*. **BEHAVIORS:** Runs and walks when it can. Flies when it must, weakly and seldom very far. **HABITAT:** Salt and brackish marsh, seldom far from salt water. **STATUS** Common resident on the entire Texas coast.

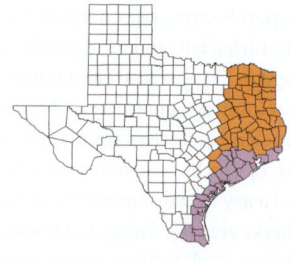

King Rail

(Rallus elegans)

IUCN RED LIST STATUS (2021): Near Threatened
POPULATION TREND: Decreasing

As befits its species name, this secretive bird is best described as elegant. The largest rail in Texas, its status is declining in spite of a large distribution in North America. Though a game bird, there is little pressure on the population from hunting. Instead, loss of habitat as wetlands are modified for development and changes in rice production have the greatest negative impacts.

LENGTH: 15 inches. **WINGSPAN:** 20 inches.
ADULT: Warm brown overall; face is always brown, not gray. Flanks and lower body have bold white bars. Back feathers are dark, almost black with brown edges. Less shaggy looking than Clapper Rail, always appears neat and dapper. **JUVENILE:** (July–Sept.) Mottled speckling below, less contrast on the back. Lacks white barring on flanks. **VOICE:** Similar to Clapper Rail but calls longer, deeper, more resonant, and slower-paced. **BEHAVIORS:** Mostly walks and runs, flying when flushed or provoked. May swim to cross water. Flies during long-distance movement, usually close to the ground but at greater heights during migration. **HABITAT:** Freshwater marshes, wetlands, retention ponds, and irrigation canals. **STATUS:** Uncommon to locally common resident in freshwater marshes, irrigation ditches, and weedy lakes from the Louisiana border to Baffin Bay on the coastal prairie. Uncommon southward to the Lower Rio Grande Valley. A seldom-observed migrant throughout the state. Migrants move from September to October and March through April.

Virginia Rail

(Rallus limicola)

IUCN RED LIST STATUS (2019): Least Concern
POPULATION TREND: Increasing

Virginia Rails are secretive but common, and often abundant in the proper habitat. **LENGTH:** 9.5 inches. **WINGSPAN:** 13 inches. **ADULT:** Sexes are identical, chicken-like in posture with long, decurved, reddish-brown bill. In spring the lower mandible is more reddish and can appear bright red or scarlet. Crown is dark, face is gray. Iris is reddish-brown in adults. Breast is a rich reddish-brown and flanks are dark, with white barring. Feathers on the back are dark with rich reddish-brown edges. Upper wing coverts are reddish-brown with dark flight feathers. Underwings are slate gray with bright white patagial bars on the shoulder. The white patagial bar is often quite striking in flight at close range. Legs and feet are reddish-brown. **IMMATURE:** Much like adults but lacking the rich reddish colors. Breast is pale with dark charcoal blotches. Throat is white. **VOICE:** Makes a nasal pig-like grunting, usually descending in pitch and accelerating in timing. Also makes *kiddick kiddick kiddick* and *ki ki keer* calls. **BEHAVIORS:** Prefers to run or swim when disturbed, but when flushed the flight is short and low, usually lower than 20 feet. **HABITAT:** Shallow water and mud with cattails, reeds, or tall grasses. In Texas, wintering birds will use fresh, brackish, and saltwater wetlands. **STATUS:** Winter resident of Texas on the coastal plain. Uncommon to locally common resident in the northern Panhandle.

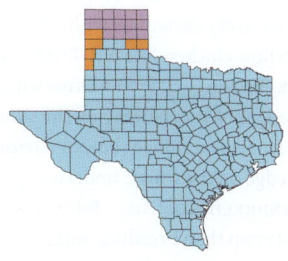

Sora

(Porzana carolina)

IUCN RED LIST STATUS (2016): Least Concern
POPULATION TREND: Increasing

Sora are the most widely distributed of rails. They are more often heard than seen, often giving their distinctive whinny call almost randomly, particularly during migration. Sora breed in freshwater marshes but winter in a variety of wetland marshes and can be especially abundant in rice fields during migration.
LENGTH: 8.75 inches. **WINGSPAN:** 14 inches.
ADULT BREEDING: (Mar.–Aug.) Black face and throat. Gray auriculars with a brown patch. Yellow chicken-like bill. Brown crown and neck. Brown back with narrow white feather edges. White barring on the sides. In flight, a small percentage shows a narrow pale trailing edge on the wings; underwings are gray. **NONBREEDING:** (Aug.–Feb.) Less distinct and crisp than breeding. Bill is more of a dull yellow-green. **JUVENILE:** (July–Feb.) Dull buffy-colored white, speckled not streaked on the back. **VOICE:** Whinny call is a long descending *whee-hee-hee-hee-hee-hee*. *Soree* call is a piercing *per-weeee*. **BEHAVIORS:** Prefers to run over flying. More likely to be seen on wetlands edges than other crakes. Difficult to flush, but when it does fly, it flies longer and stronger than other crakes. **HABITAT:** Fresh, brackish, and salt marshes. Most important to them is shallow water and emergent vegetation. Often found in cattails and rushes. **STATUS:** Uncommon to locally common migrant statewide. Locally abundant winter resident on the coastal prairies. Rare to uncommon inland during winter.

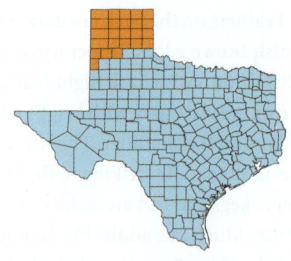

Adult

Immature

Common Gallinule

(Gallinula galeata)

IUCN RED LIST STATUS (2016): Least Concern
POPULATION TREND: Stable

Once widely known as Florida Gallinule, in 1957 the name Common Gallinule was adopted. The name changed to Common Moorhen in 1983, when the species was lumped in with its European relative. In 2011, it was split from what is now known as Eurasian Moorhead and became Common Gallinule again.
LENGTH: 14 inches. **WINGSPAN:** 21 inches.
ADULT BREEDING: (Feb.–Sept.) Red face shield and bill with yellow bill tip. Head is charcoal-colored with a matching eye that at times seems to disappear. Body is a lighter shade of gray, with extensive white flanks and a bright white undertail. **NONBREEDING:** (Sept.–Feb.) Similar to breeding but face shield and bill are darker. Paler gray on the throat. **JUVENILE:** (July–Feb.) Drab light gray, brownish back. Bright white undertail coverts. **VOICE:** Call is a whining series

slowing down in cadence: *eh-eh-eh-eh-eeww-eew w-ewwww-ewwwwww.* **BEHAVIORS:** Walks on floating vegetation and mud. Swims well. **HABITAT:** Variety of freshwater wetlands with floating-leaved and submerged plants. **STATUS:** Uncommon to locally abundant resident from the Lower Rio Grande Valley and coastal prairies inland to Bexar and Brazos counties. Rare to locally uncommon summer resident throughout the eastern three-quarters of the state. Increasingly common winter resident in the same range. It is not yet known if breeding birds vacate the breeding grounds and are replaced or if these wintering birds are resident birds.

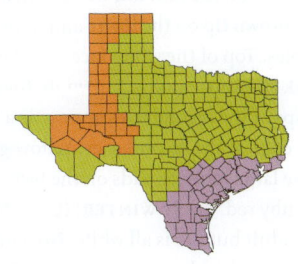

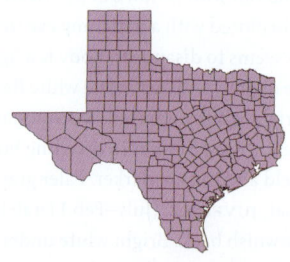

American Coot

(Fulica americana)

IUCN RED LIST STATUS (2016): Least Concern
POPULATION TREND Decreasing

American Coot is the most visible species in the rail family. In winter, large numbers can gather on open water. Their plaintive calls seem to tell a story of misery, complaining of their lot in life. Perhaps this is where the term "old coot," to describe a crotchety old man, comes from.
LENGTH: 15.5 inches. **WINGSPAN:** 24 inches.
ADULT: Overall, dark charcoal gray. Small white marks on the tail. Bill white with a gray to dark-brown tip on the lower and upper mandibles. Top of the white face shield is usually dark chestnut or olive brown. In flight, the white tips of the secondaries appear as a white trailing edge of the secondaries. Yellow-green legs have large lobes or pads on the feet. Eye is a dark ruby red. **FIRST WINTER:** (Oct.–Mar.) Like an adult but bill is all white. No patch at top of face shield. **JUVENILE:** (July–Oct.)
Dull gray below, dark brown above. **VOICE:** Makes a variety of plaintive notes: *krrp* or *prik.* **BEHAVIORS:** Flies with difficulty, must run on the water some distance to take off. Flight is weak and crashes to the water to land. Walks readily and often will leave the water in a group to graze on grass. **HABITAT:** Variety of open-water habitats. Artificially warm water attracts large numbers in winter. **STATUS:** Uncommon to common resident statewide.

Purple Gallinule

(Porphyrio martinicus)

IUCN RED LIST STATUS (2016): Least Concern
POPULATION TREND: Decreasing

The exotic-looking Purple Gallinule is the most vibrant of the rails in Texas. It is highly prone to vagrancy and can feed on exotic plants like water hyacinth and hydrilla. These traits probably make it more effective as a pioneer colonizing new wetlands.
LENGTH: 13 inches. **WINGSPAN:** 22 inches.
ADULT: Purple head and underparts. Bill is bright red with yellow tip. Bright sky-blue facial shield. Back is blue blending into greenish. Bright white undertail. **FIRST WINTER:** (Jan.–Apr.) Similar to adults but head is brown and bill is dusky red. **JUVENILE:** (July–Dec.) Pale brown with green back and wings.
VOICE: Call a series of *pek pek pek pek pek pek* or *eeck eeck eeck eeck eeck*. **BEHAVIORS:** Walks with ease on floating vegetation with very long toes. **IMMATURE:** Flies weakly with legs hanging down. **HABITAT:** Freshwater marshes with dense stands of floating vegetation. **STATUS:** Rare to uncommon migrant in the eastern half of the state. Rare to locally common summer resident in the eastern third of the state. Very rare winter visitor on the coast. Spring birds arrive in early April and fall migration begins in August, with most birds departing by mid-October.

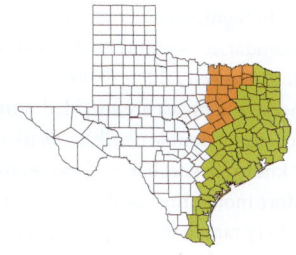

Yellow Rail

(Coturnicops noveboracensis)

IUCN RED LIST STATUS (2020): Least Concern
POPULATION TREND: Decreasing

Yellow Rail is perhaps the most secretive of the birds of Texas. While it has a distinctive call, it is not known to vocalize in Texas, ever. Confusingly, there are insects and frog species present in their habitat that make calls nearly identical to Yellow Rails.
LENGTH: 7.25 inches. **WINGSPAN:** 11 inches. **ADULT BREEDING:** (Mar.–Aug.) Yellow-buff-colored overall. Bright buffy stripes on back. Dark lores and ear patch. Bill is yellow. In flight, no other rail shows bright white secondaries. **NONBREEDING AND JUVENILE:** (Aug.–Feb.) Bill is darker, duskier. Head, neck, and breast are speckled. **VOICE:** Call is a distinct double *tik-tik tik-tik tik-tik*. It is not known to vocalize in Texas. **BEHAVIORS:** More mouse than bird. Walks and flees on foot. Very rare to see one in the open. Flies only when forced to. Flight is low and often short, 10 feet or lower at times, rarely more than 50 feet. **HABITAT:** Primarily in drier portions of cordgrass prairies on the coast. Occasionally found in wet stands of other grasses like bluestem. **STATUS:** Rare migrant through the eastern state. Rare to locally common winter resident on the upper and central coast. Fall migrants arrive in mid-October and most birds have left by the first week of May.

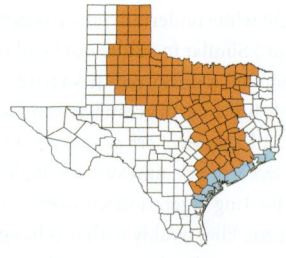

Black Rail

(Laterallus jamaicensis)

IUCN RED LIST STATUS (2021): Endangered
POPULATION TREND: Decreasing

The secretive Black Rail is one of the most sought-after species by birders in Texas. Limited in its distribution, it's one of the species many have to count as "heard only." Fortunately for birders, when present in late April through June, it is often quite vocal, giving its distinctive call for extended periods.

Black Rails have an uncanny ability to move around undetected. Many birders have had a vocalizing rail in a small isolated stand of grass under observation. While intently watching for the birds, they suddenly hear the vocalizing from an adjacent patch of grass—the bird has moved from one to the other with what seems like magic.

LENGTH: 6 inches. **WINGSPAN:** 9 inches.
ADULT: Charcoal gray with red eye. Rufous nape and speckling on the back. **JUVENILE:** (Aug.–Feb.) Like adults but eyes are brown.

VOICE: Loud *kee-kee-dew*. Less commonly a low growling *krr-krr-krr-krr*. **BEHAVIORS:** Secretive and rarely seen. Walks and runs to avoid detection. When flushed, flight is short and low, dropping quickly to the ground and running. **HABITAT:** Cordgrass-dominated salty prairie. Higher portions where the ground is damp but not flooded. **STATUS:** Rare migrant in the eastern third of the state. Resident on the upper and central coast in salty cordgrass prairies. Since 2020, has been federally listed as threatened.

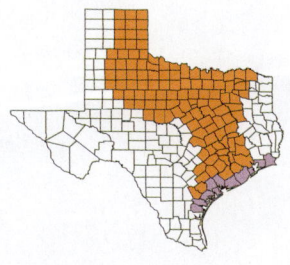

ARAMIDAE (LIMPKINS)

The limpkin family is a monotypic family, meaning it has only one species. About 70% of its diet consists of apple snails, a large freshwater species of snail. Its bill has even evolved to make scooping the snail from the shell easier. Limpkins are also known to consume mussels, crustaceans, frogs, and even grasshoppers, but the presence of apple snail is the best indicator for the likely presence of Limpkins. In Florida, the Florida apple snail is their primary food, but the snail population is in decline due to changes in water management. In Texas there is an explosion of exotic apple snails, and most sites where Limpkins have been found have an abundance.

The Limpkin gets its name from its slow, strolling gait that appears to some as if the bird is limping when it walks.

Limpkin

(Aramus guarauna)

IUCN RED LIST STATUS (2016): Least Concern
POPULATION TREND: Stable

When the Limpkin arrived in Texas, it was a bit of a surprise for birders. In a recent poll of knowledgeable birders on what species they thought might occur next, Limpkins did not receive a single vote.
LENGTH: 26 inches. **WINGSPAN:** 40 inches. **ADULT:** Brownish overall. Head and neck are heavily speckled with white, making them look pale. Large brown eye. Bill is long and stout, straight compared to an ibis. Back is brown, with variable white tips on the mantle feathers. Breast is brown with white speckles. Flight feathers are unmarked brown. Legs protrude behind the bird in flight. **JUVENILE:** (First year) Like adults but the white marks on the wing coverts are elongated and not triangular. **VOICE:** Loud screaming *SKREEAAWW* often repeated many times. **BEHAVIORS:** Flies with flat, shallow, stiff wingbeats. Walks with a slow limping gait. **HABITAT:** Freshwater marshes, riparian swamp forests, and lake and pond shorelines. The presence of apple snails is critical to habitat selection. **STATUS:** The first record in Texas was in Fort Bend County in 2021. Since that time, there have been hundreds of records in the eastern half of Texas west to the eastern edge of the Edwards Plateau. The status and range of Limpkin in Texas is not yet well known, but there appears to be a resident population in southeast Texas.

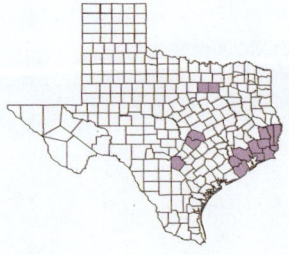

GRUIDAE
(CRANES)

Cranes are a small family of 15 species with a worldwide distribution. Three species occur in Texas. The Common Crane is a very rare visitor. Most of the endangered Whooping Cranes in the wild spend winters on the central Texas coast in a region anchored by the Aransas National Wildlife Refuge. Sandhill Cranes are the most widely distributed of the crane species in Texas.

In general, cranes are tall wading birds with long necks, long legs, and stout bills. Most have bare skin on the head. Elongated tertial feathers give cranes the look of having a bustle when they are walking. Most species have elaborate mating rituals and most mate for life. Their mating dances are often accompanied by resonant trumpeting made possible by a very long windpipe coiled inside the breastbone.

Cranes use a wide variety of open habitats worldwide, but most prefer a wet or wetland habitat. Loss of wetland habitat is a contributing factor to the threatened status of more than 70% of the crane species worldwide.

Sandhill Crane

(Antigone canadensis)

IUCN RED LIST STATUS (2021): Least Concern
POPULATION TREND: Increasing

The huge flocks of Sandhill Cranes leaving a night roost at Muleshoe National Wildlife Refuge at dawn is one of the great avian spectacles of Texas. Flocks of hundreds disperse to feed in the agricultural fields around the refuge.

Three subspecies occur in Texas: two of the larger Greater Sandhill Crane group occur statewide and the Lesser Sandhill Crane occurs on the coastal prairie. Formerly, the subspecies *A. canadensis pratensis*, known as the Florida Sandhill Crane, was a rare resident on the upper Texas coast.

LENGTH: 41–46 inches. **WINGSPAN:** 73–77 inches. **ADULT:** Tall gray bird. Prominent bare red crown patch. Summer birds have a rusty color. **JUVENILE:** (First year) Brown color without the prominent crown patch. **VOICE:** Loud resonant hollow rattle or rolling bugle. **BEHAVIORS:** Flies with head and legs outstretched, never with the neck coiled like a Great Blue Heron. Walks with an upright posture. Gregarious, often forming large flocks. **HABITAT:** Shallow open wetlands. Makes heavy use of agricultural fields for foraging. **STATUS:** Common to uncommon migrant statewide, rare in the Pineywoods. There are two main wintering populations: the High Plains and the Rolling Plains in the north and the coastal prairies in the south. Cranes begin to arrive in Texas in early October and most have departed by mid-March.

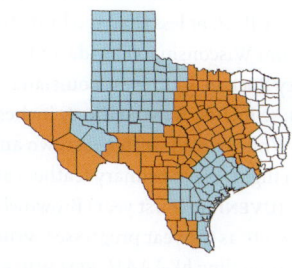

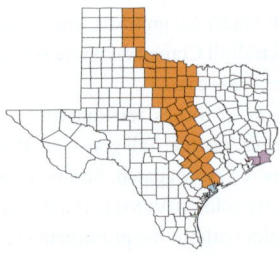

Whooping Crane

(Grus americana)

IUCN RED LIST STATUS (2020): Endangered
POPULATION TREND: Increasing

All Whooping Cranes alive today are descendants of the last remaining 14 that wintered in Texas in the late 1930s. This population travels from Wood Buffalo National Park in Canada to Aransas National Wildlife Refuge near Tivoli, Texas. As of January 2023, the US Fish and Wildlife Service counted 536 Whooping Cranes in that population.

Attempts to establish additional flocks are ongoing with mixed results. There is currently a migratory flock of less than 100 birds that travels from Wisconsin to Florida and a non-migratory flock in Southwest Louisiana. **LENGTH:** 52 inches. **WINGSPAN:** 87 inches. **ADULT:** White overall, with red crown and red malar. In flight, black primary feathers are obvious. **JUVENILE:** (First year) Brownish fading into white as the year progresses. **VOICE:** Loud clear bugling *bKAAAH.* **BEHAVIORS:** Walks upright, showing its full height. Back stays relatively flat, making the silhouette unique among white birds in Texas. **HABITAT:** Primarily winters in coastal marshes on the central coast. Will use crawfish farms. **STATUS:** Uncommon winter resident in Aransas and Calhoun counties. In recent years, birds from a flock being established in Louisiana have taken up residence in Jefferson and Chambers counties.

RECURVIROSTRIDAE (STILTS AND AVOCETS)

The family of stilts and avocets is made up of nine species, two of which occur in Texas. All have slender bills, long necks, and legs so long and thin it's hard to imagine how they function. The family name comes from the Latin word *recurvas*, meaning "bent back," and the Latin word *rostrum*, meaning "bill," referring to the bent-back bill of avocets. Avocets generally have upcurved bills and stilts have long, needlelike bills.

Stilts and avocets use a wide range of shallow wetlands, from freshwater to hypersaline lakes. Their very long legs allow them to use water much deeper than other shorebirds. Avocets have webbed feet and are one of the few shorebirds regularly seen swimming ducklike in water too deep to stand in. Stilts generally feed by picking individual prey items from the water, while avocets feed by scything, or sweeping their bills back and forth, often in unison with others in the flock.

Stilts and avocets are opportunistic breeders, waiting until water levels are right to create the small isolated islands they prefer to nest on as protection from predators. Pairs are monogamous and both sexes incubate the eggs. Stilts and avocets often breed and winter together.

Black-necked Stilt

(Himantopus mexicanus)

IUCN RED LIST STATUS (2019): Least Concern
POPULATION TREND: Increasing

The Black-necked Stilt is tall, bold, and elegant, making it an irresistible subject for bird photographers. They form loose colonies to breed in freshwater wetlands and aggressively defend those territories. Their yipping small dog–like call has earned them the nickname "Marsh Poodle."

The taxonomy of Black-necked Stilt isn't entirely settled. While North American ornithologists place it as a full species, outside of North America most regard it as a subspecies of Black-winged Stilt (*Himantopus himantopus*).

LENGTH: 14 inches. **WINGSPAN:** 29 inches. **ADULT:** Black above, clean white below. Needlelike black bill. Bold white eyebrows and red eyes. Very long red legs. White back is visible in flight. Females have a brown tinge to the back. **JUVENILE:** (July–Nov.) Like adults but dark grayish above. **VOICE:** Strident yapping *keef keef keef*. **BEHAVIORS:** Walks delicately while calling. Tips forward often to pick food carefully. **HABITAT:** Freshwater marshes with emergent vegetation. Uncommon to rare summer resident north to the High Plains. **STATUS:** Common summer resident on the coastal prairies. Locally common in the South Texas Brush Country and Blackland Prairies. Spring birds begin arriving in Texas in late March. Fall migration begins in late August, and some birds may linger as late as early December.

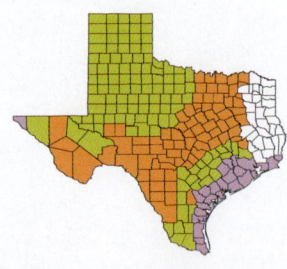

Breeding

Winter plumage

American Avocet

(Recurvirostra americana)

IUCN RED LIST STATUS (2021): Least Concern
POPULATION TREND: Stable

The name avocet comes from the Italian word *avosetta*, meaning "graceful," which perfectly describes this bird and its actions. A large group of avocets feeding together, sweeping their bills in search of food, is mesmerizing to watch.

American Avocets breed in ephemeral wetlands. In Texas, they sometimes breed in Panhandle playas, the very definition of ephemeral wetlands. Modifications to the landscape resulting in the loss of Panhandle playas are certainly impacting their breeding in Texas. LENGTH: 18 inches. WINGSPAN: 31 inches. ADULT BREEDING: (Mar.–Aug.) Peach-colored head and neck. Long, fine, upward-curved bill. Female bills average much straighter than males. Two wide black stripes on the back. Wings are black with white secondaries. Legs are light blue-gray. Feet partially webbed.

NONBREEDING: (Sept.–Feb.) Head and neck are white. VOICE: Call is a sharp *kweep kweep kweep*. BEHAVIORS: Flies with head and legs extended. When foraging, sweeps bill back and forth in the water. Often large wintering flocks feed in an undulating mass in shallow water. Swims well, upright. HABITAT: Shallow water bodies with fine-sediment or mud bottoms. STATUS: Common to abundant winter resident on the coast. Common summer resident on the High Plains. Uncommon to rare summer resident on the Texas coast south of Galveston Bay. Migrant statewide in spring between mid-March and early June, in fall from late July through early November.

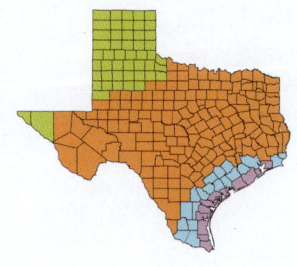

HAEMATOPODIDAE (OYSTERCATCHERS)

The family of oystercatchers currently contains 12 species. Almost all species are coastal. Species that live on rocky coasts are generally all black and species that live on sandy or muddy coasts are generally pied, or black-and-white. Oystercatchers are closely related to stilts and avocets and probably evolved from a plover-like ancestor.

Oystercatchers feed mostly on bivalve shellfish like oysters, clams, and mussels. Some species, and even individuals within a species, may adopt a diet mostly of worms. Oystercatchers that feed by a quick stab into an open bivalve to sever the abductor muscle have longer, thinner bills. Those that feed by hammering into a closed bivalve have shorter, stouter bills. Those that feed on worms have thinner, more pointed bills. Experiments with captive birds have shown that individual birds bills can change structure in a matter of weeks to better hunt the type of prey available.

American Oystercatcher

(Haematopus palliatus)

IUCN RED LIST STATUS (2016): Least Concern
POPULATION TREND: Stable

An unmistakable bird, American Oystercatcher can usually be found in its characteristic range wherever there are saltwater mollusks. The orange bill that makes it look like it's smoking a carrot is actually a key tool it uses like an oyster knife. A keen observer can observe an oystercatcher driving it into an oyster and twisting to open its prey, not unlike the way one would shuck an oyster.

LENGTH: 17.5 inches. **WINGSPAN:** 32 inches.
ADULT: Black head and neck. Bright orange bill and yellow eye, with an orange orbital ring that gives it a permanently startled look. White below with dark brown back and upper wings. Wings have a bold white wing stripe. Wide white rump and wide black band on the end of the tail. Legs are a pale fleshy pink.
JUVENILE: (Aug.–Feb.) Like adults, but bill is dusky orange with a darker tip. Legs are less pink. **VOICE:** The call is a loud whistling *queep queep*. **BEHAVIORS:** Flight is direct, with deep rapid wing beats. Walks and runs rather than flying. Forms flocks in winter. **HABITAT:** Nests primarily on shell and sand beaches. Makes use of dredge spoil islands. **STATUS:** Locally common along the upper and central coasts, rare to uncommon on the lower coast. Rarely ever found far from salt water.

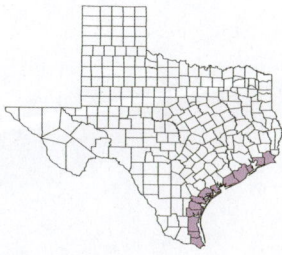

CHARADRIIDAE (PLOVERS AND LAPWINGS)

Plovers and lapwings are a large worldwide family of 69 species that live on open shores. They run and stop suddenly to pick small prey from the sand or mud. They feed primarily on insects, worms, and small crustaceans.

Plovers are small to medium-size shorebirds. They have upright posture, medium-length legs, and feet with only a vestigial hind toe. Their necks are short and thick and their heads are adorned with short, thick bills. Plovers' eyes are usually large.

Plovers and lapwings had long been thought to be closely related to sandpipers, but recent studies place them nearer to gulls and terns and to stilts and avocets. Lapwings have wider, more rounded wings than plovers. None of the 23 species of lapwings regularly occur in North America. The first record of Southern Lapwing in Texas occurred in April 2024.

There are two groups of plovers: The smaller *Charadrius* plovers like the Snowy and Piping Plovers, likely evolved in the tropics and southern hemisphere. The larger *Pluvialis* plovers like the Black-bellied and American Golden-Plover evolved in the northern hemisphere.

Eleven species of plovers have been recorded in Texas. Three species—the Pacific Golden-Plover, Collared Plover, and Southern Lapwing—are rare vagrants to Texas.

Breeding

Nonbreeding

Black-bellied Plover

(Pluvialis squatarola)

IUCN RED LIST STATUS (2019): Least Concern
POPULATION TREND: Decreasing

Also known as the Gray Plover, likely because of its all-gray nonbreeding plumage, the Black-bellied Plover has near-worldwide distribution.
LENGTH: 11.5 inches. **WINGSPAN:** 29 inches.
ADULT BREEDING: (Apr.–Sept.) Black bill, face, neck, and belly. Undertail is clean white. Gray on the crown, nape, and back. Bold white supercilium extends down to the sides of the neck. In flight, shows a white tail and bold wing stripe. **NONBREEDING:** (Aug.–Apr.) Light gray overall with white belly and undertail. In flight, white tail, wing stripe, and black axillaries, or wing pits, obvious. **JUVENILE:** (July–Nov.) Much like nonbreeding, but with streaked breast and checkered back. Supercilium is more pronounced than adults. **VOICE:** Sad-sounding *pleeooee*. **BEHAVIORS:** In migration, flies mostly nocturnally in small groups.

Forages by running and pecking food. Nocturnal feeding is common and may be almost 50% of foraging. They are rarely with other large plover species. **HABITAT:** In Texas, uses coastal mudflats hard enough to support running. **STATUS:** Uncommon to common as a migrant through the eastern half of the state, becoming less common westward. In winter, common on the coast, rare to uncommon in the interior. A small number of birds usually in nonbreeding plumage may summer along the coast. Fall migrants are present mid-August through October. In spring, migrants can be found from late March to late May.

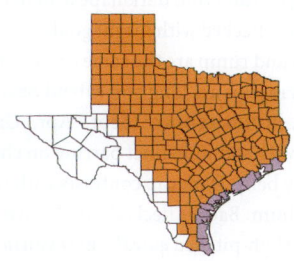

Breeding

Nonbreeding

American Golden-Plover

(Pluvialis dominica)

IUCN RED LIST STATUS (2016): Least Concern
POPULATION TREND: Decreasing

The striking American Golden-Plover is the only one of the three species of golden-plovers regularly occuring in Texas.
LENGTH: 10.5 inches. **WINGSPAN:** 26 inches.
ADULT BREEDING: (Apr.–Sept.) Boldly marked with black face and bill, the black extends down the throat, chest, and belly all the way to the vent. A bold white supercilium extends down the sides of the neck. A contrasting cap and thin dark nape lead to the gray back, flecked with bright golden speckles. Tail and rump are the same color as the back. Wingtips are long and extend beyond the tail. **NONBREEDING:** (Sept.–Apr.) Grayish overall with soft indistinct barring on chest and gray belly. Dark cap contrasts with white supercilium. Back is flecked with brown tones.
VOICE: High-pitched *queedle*. **BEHAVIORS:** In flight, very fast. Flocks have been measured at 45 mph and faster. A radio-tracked bird averaged 34 mph over several hundred miles. Forages by sight, running, scanning, and quickly picking up prey. **HABITAT:** During migration, prefers upland areas with low or sparse vegetation. Frequently found on sod farms, large mowed lawns, golf courses, and close-grazed pastures. **STATUS:** Common migrant in the eastern half of Texas, expected to arrive mid-March; some birds present through mid-May. Very rare west of the Balcones Escarpment. In fall, very rare migrant in the eastern half of the state, barely more common in the west.

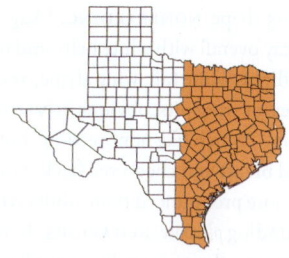

Adult

Chick

Killdeer

(Anarhynchus vociferus)

IUCN RED LIST STATUS (2016): Least Concern
POPULATION TREND: Decreasing

The most widespread of shorebirds, Killdeer are present year-round in Texas.
LENGTH: 10.5 inches. **WINGSPAN:** 24 inches.
ADULT BREEDING: The only ringed plover in North America with two breast bands. Black bill and white forehead front of the crown, which can be almost black. Crown and auriculars are brown. Eye is dark with bright red orbital ring. Bold white supercilium. White throat and two neck rings. Back and upper wings are brown. Birds from the south tend to have more rufous on the upper wing coverts. Clean white underneath the tip of the tail. Legs are pinkish. Tail extends beyond the wingtips. In flight, shows a rufous rump with dark subterminal band and white terminal band. Bold white wing stripe. Trailing edge of the wings are dark, with primaries being dark at the base. **JUVENILE:** (June–Sept.) Much

like an adult but brownish between breast bands. Supercilium is creamy-colored. **VOICE:** Drawn-out *deee, deeyee tyeeeee deew deew* or repeated *tewddew.* **BEHAVIORS:** Runs and picks prey in typical plover fashion. Frequently bobs head. Breeds close to human activity. Feigns injury to draw predators away from the nest. **HABITAT:** Breeds in open areas like sandbars, mudflats, heavily grazed pastures, golf courses, athletic fields, and graveled parking lots and rooftops. **STATUS:** Common to abundant statewide. They become more common in winter in central and southern parts of the state.

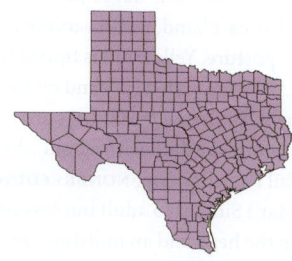

Semipalmated Plover

(Anarhynchus semipalmatus)

IUCN RED LIST STATUS (2016): Least Concern
POPULATION TREND: Stable

Semipalmated Plover gets its name from its partially webbed feet. They breed in North America's open near-arctic lakes, marshes, and rivers, and inland on dry, gravel, or sparsely vegetated sites. Their open habitat and ease of capture make them an ideal species for study on their breeding ground.
LENGTH: 7.25 inches. **WINGSPAN:** 19 inches.
ADULT BREEDING: (Mar.–Sept.). Dark brown on back and crown, like wet sand. Single complete dark breast band, width varying with the bird's posture. Yellow bill is tipped black. Forehead is white with dark band on the forward crown. Dark face mask. Legs are yellow. In flight, wings are dark with a bold white stripe. Tail has a dark tip. **NONBREEDING:** (Sept.–Mar.) Similar to adult but less crisp; whiter on the head and an indistinct supercilium. **JUVENILE:** (Aug.–Oct.) Like nonbreeding,

with some fine barring on the back. **VOICE:** Short, soft *chuWEE.* **BEHAVIORS:** Walks quickly and runs with head up. Powerful flight and flies with ease in strong winds. **HABITAT:** Beaches, mudflats, agricultural fields. Common in rice fields in spring migration. **STATUS:** Common winter resident along the Texas coast. Uncommon to common migrant in the eastern third of the state, becoming more uncommon westward. Large flocks sometimes form on the coast in migration. Spring migration is early March–late May, fall migration early July–late October.

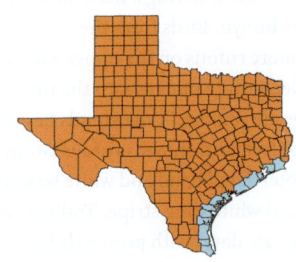

Breeding

Winter plumage

Piping Plover

(Anarhynchus melodus)

IUCN RED LIST STATUS (2020): Near Threatened
POPULATION TREND: Increasing

The threatened and endangered Piping Plover breeds along the northern Atlantic coast and inland along rivers and wetlands of the northern Great Plains. It winters along the Gulf Coast and southern Atlantic coast, forced to share its habitats with humans.
LENGTH: 7.25 inches. **WINGSPAN:** 19 inches.
ADULT BREEDING: (Feb.–Aug.) Light gray above, crown and back the color of dry sand. Orange bill with dark tip. Forward crown is dark. Single dark breast band, usually broken. Yellowish legs. White upper tail with dark tip. Flight feathers dark with white wing stripes. **NONBREEDING:** (Sept–Feb.) Like breeding, bill dark. The dark breast band and forward crown is absent, breast band is pale. **JUVENILE:** (July–Sept.) Nearly identical to nonbreeding, underwing paler. **VOICE:** Clear gentle *peep* or *peep-lo.* **BEHAVIORS:** Spends much of its time walking or running to remain inconspicuous. **HABITAT:** Winters on beaches, mudflats, and sand flats on the coast. In migration, uses beaches and flats in the interior away from the coast. **STATUS:** Listed as endangered species. While they breed in the northern Great Plains, birds are not often recorded inland in Texas. Uncommon to common on the coast in winter. Some birds linger through summer but do not breed. Southbound birds may arrive in late June, but the main population arrives late July to early September. Spring migrants are found between late March and early May.

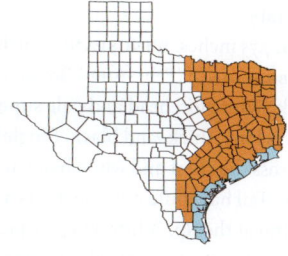

Breeding

Nonbreeding

Wilson's Plover

(Anarhynchus wilsonia)

IUCN RED LIST STATUS (2016): Least Concern
POPULATION TREND: Decreasing

Wilson's Plover is named for the Scottish-American poet and ornithologist Alexander Wilson, who collected the first specimen in the year he died, 1813, in New Jersey. Wilson's Plovers breed on the Gulf Coast. Breeding pairs are territorial and may engage in group defense of the nest. During nonbreeding season, many congregate in groups, from a few individuals to several hundred. Their primary food source is small crustaceans, particularly fiddler crabs.

LENGTH: 7.75 inches. **WINGSPAN:** 19 inches.
ADULT BREEDING: (Feb.–Aug.) Brown head and back. White forehead and pale supercilium. Bill is long, thick, and black. Single black band across the chest and white band around the nape. Tail has white outer feathers and black band at the tip. White wing stripe. **NON-BREEDING:** (Sept.–Feb.) Similar to breeding but paler with less contrast. Breast band is brown. **JUVENILE:** (June–Oct.) Like nonbreeding; breast band is less distinct. **VOICE:** Call is a high pure *queet*. **BEHAVIORS:** Feeds in typical plover fashion: runs a few steps, stops, and looks for prey. **HABITAT:** Open beaches on the coast, usually with large areas of dry sand or mud with little vegetation. Runs horizontally low to the ground. **STATUS:** Common migrant and summer resident on the coast. Casual inland. Rare and local on the coast in winter. Summer birds return in mid-February and most have departed by late September.

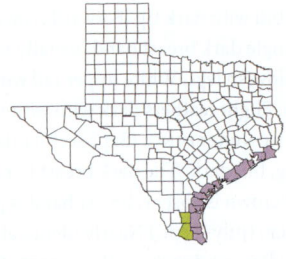

Mountain Plover

(Anarhynchus montanus)

IUCN RED LIST STATUS (2020): Near Threatened
POPULATION TREND: Decreasing

Mountain Plovers are declining rapidly and getting harder and harder to find. The North American Breeding Bird Survey suggests a rate of 2.5% over the course of 2010–2015. That would mean that the species has declined by about 26% over only three generations of about twelve years! Ironically, the planting of sunflowers and millet, in many cases for birdseed, has made some areas unsuitable for breeding. **LENGTH:** 9 inches. **WINGSPAN:** 23 inches. **ADULT BREEDING:** (Mar.–Aug.) Black lores and white supercilium. Brown crown, nape, and back. White below. Tail is brown with dark patch at the end. Primaries dark with a white patch. Underwing coverts are white. **NONBREEDING:** (Aug.–Mar.) Dusky brown on the chest. Black lores are absent and supercilium is less distinct. **JUVENILE:** (Aug.–Oct.)

Like nonbreeding adult. **VOICE:** Flight call is a course *jeert jeert jeert.* **BEHAVIORS:** Flight is low, strong, and swift. On foot, moves in an unhurried run of about 3 feet. **HABITAT:** Winters on tilled fields, sod farms, burned-over fields. Summer residents are found in shortgrass prairies and open grasslands. **STATUS:** Very rare summer resident in the northwestern Panhandle. Uncommon to rare winter resident in suitable habitat from Williamson County south to Nueces County, and Willacy and Hidalgo counties west to Kinney County. Fall migrants arrive August through October. Spring migration is from early March to early May.

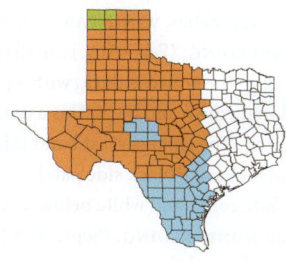

Breeding

Nonbreeding

Snowy Plover

(Anarhynchus nivosus)

IUCN RED LIST STATUS (2020): Near Threatened
POPULATION TREND: Decreasing

Like many ground-nesting birds, the declining Snowy Plover population is threatened by the degradation of their habitat from human use and invasive species. Snowy Plovers use an uncommon breeding strategy. In some populations, females will desert first broods soon after hatching (Snowy Plover chicks are precocial) and renest with new mates, sometimes hundreds of miles away.

LENGTH: 6.25 inches. **WINGSPAN:** 17 inches.
ADULT BREEDING: (Feb.–Aug.) Small dark bill. Front of the crown is dark, with a pale dun-colored crown and back. Has a dark cheek patch. The single breast band is incomplete and looks like black side patches on the neck. White collar and white below. Legs are dark gray. **NONBREEDING:** (Sept.–Feb.) Similar to breeding adult, but dark patches the same dun color as the rest of the bird. **JUVENILE:** (July–Oct.) Like nonbreeding. **VOICE:** Call is a hoarse *treep*. **BEHAVIORS:** Typically runs when disturbed, flies only when pushed. Often roosts in vehicle tracks and behind driftwood or trash on beaches. **HABITAT:** Breeds on inland saline lake flats and river sandbars. Winters primarily on coastal beaches. **STATUS:** Uncommon summer resident on the immediate coast, rare to locally uncommon summer resident of the saline lakes. Rare to locally uncommon winter resident along the coast. Spring migrants return in early March and are present through May. Fall migrants arrive in late July to early October.

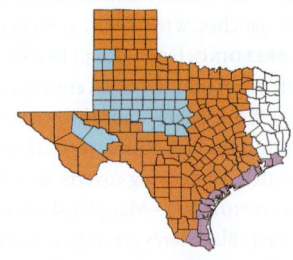

SCOLOPACIDAE (SANDPIPERS AND ALLIES)

The sandpipers are a family of small to medium-size shorebirds with long toes, long tapered wings, and necks and legs that vary considerably in length. All have a short elevated hind toe (with the exception of the Sanderling) that distinguishes them from plovers. Bills can be straight, curved, decurved, and even spatulate like the extremely rare Spoon-billed Sandpiper.

Like the plovers, sandpipers prefer wet, open, and sometimes grassy habitats. Many breed in the high arctic and subarctic tundra and winter in intertidal habitats. It was often thought that the plovers and the sandpipers were each other's closest relatives. Recent skeletal and DNA evidence suggests that painted-snipes and jacanas are the closest relatives to sandpipers.

There are ninety-seven species of sandpipers worldwide. Forty-one have been recorded in Texas. Twelve of those are very rare in Texas, and one species, the Eskimo Curlew, is now widely believed to be extinct.

Upland Sandpiper

(Bartramia longicauda)

IUCN RED LIST STATUS (2021): Least Concern
POPULATION TREND: Increasing

Few things say spring to a Texas birder like the mellow *wip-wip wip-wip* call of an Upland Sandpiper flying high overhead. Although primarily a night migrant, it seems these birds can be heard and spotted flying overhead in spring when in open country. Upland Sandpiper is very closely related to the curlews.

LENGTH: 12 inches. **WINGSPAN:** 26 inches.
ADULT: Small yellow bill with dark tip on a small head. Large eye and pale face give them a dovelike face. Thin neck with fine streaking giving way to dark barring or large chevrons on the flanks. White underneath. Back feathers are dark-centered with buffy edges. In flight, inner portion of the wing is pale, and the primaries and end of the wing are dark.
JUVENILE: (July–Nov.) Similar to adults. A pale supercilium is present. Back scapular feathers are brown with thin buffy edges.
VOICE: Flight call is a loud, musical *wip-wip wip-wip* or *wip-wip-wip*. **BEHAVIORS:** Often strides gracefully, but also moves plover-like by dash-and-pause. Flight in migration is swift and strong, often quite high. **HABITAT:** Mainly makes use of mowed fields and low grasslands, some in plowed fields, during migration.
STATUS: Common spring migrant and uncommon fall migrant in eastern three-quarters of the state. Rare spring migrant in Trans-Pecos, and uncommon fall migrant. Spring migration is mid-March–mid-May, fall mid-July–September.

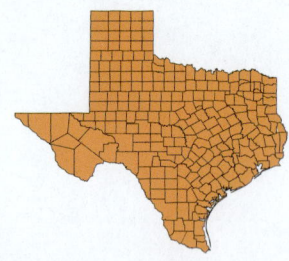

Whimbrel

(Numenius phaeopus)

IUCN RED LIST STATUS (2016): Least Concern
POPULATION TREND: Decreasing

In Texas, Whimbrels are far more common in spring than in fall. They use different routes: northbound in spring and southbound in fall, when they often make 4000-mile flights from coastal eastern Canada and New England directly over the Atlantic to the Carribean and South America. In spring, one can see birds on migration stopover streaming into freshly flooded rice fields.
LENGTH: 17.5 inches. **WINGSPAN:** 32 inches.
ADULT: Long decurved bill; most of the downward curve is in the end of the bill. Dark crown stripes. Speckled gray-brown overall with little contrast. **JUVENILE:** (Aug.–Feb.) Nearly identical to adult but head and neck slightly darker and less contrast in the crown stripes. **VOICE:** Call is a rapid *QuipQuipQuipQuip* of a single pitch. **BEHAVIORS:** Walks and runs rapidly. Migration flight and flights to and from night roosts form a loose V or U, sometimes in long lines. **HABITAT:** In migration, uses a variety of habitats including meadows, fields, sandy and rocky beaches. On the upper and central Texas coast during spring migration, makes heavy use of rice fields. Roosts communally at night in shallow water. **STATUS:** Uncommon to rare migrant along the coast, rare inland in eastern half of the state except for the Pineywoods. Very rare migrant in western half of the state. Spring migration is mid-March–late May, fall mid-July–late October.

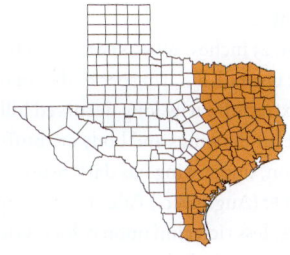

Long-billed Curlew

(Numenius americanus)

IUCN RED LIST STATUS (2016): Least Concern
POPULATION TREND: Decreasing

Long-billed Curlew was once commercially hunted. Like the now-extinct Eskimo Curlew, it was a popular menu item in many fancy restaurants circa 1900. Antique Long-billed Curlew decoys are a sought-after prize for decoy collectors. The eastern North American population was negatively impacted by hunting and conversion of habitat to agriculture and has never fully recovered. Now, rising sea levels are a concern for wintering coastal habitat.
LENGTH: 23 inches. **WINGSPAN:** 35 inches.
ADULT: Overall, a warm speckled cinnamon color. Extraordinarily long decurved bill. In flight, underwings and body are buffy cinnamon, upper wing has dark primaries.
JUVENILE: (Aug.–Dec.) Pale crown compared to adults; less rich cinnamon colors. **VOICE:** Call is a musical *whe-eeep*. **BEHAVIORS:** Usually

flies by flapping a few times, then gliding. When landing, glides in and runs a few feet.
HABITAT: In winter in Texas, almost exclusively uses shallowly inundated mudflats. Habitat use for interior birds in Texas is little studied.
STATUS: Common to abundant winter resident on the coast, common to uncommon in winter on the coastal prairies. They are winter visitors to most of the state except for the Pineywoods. Locally common summer resident in the northwestern Panhandle. Spring migration mid-March– mid-May, fall early July–mid-November.

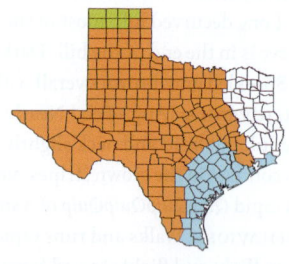

Hudsonian Godwit

(Limosa haemastica)

IUCN RED LIST STATUS (2020): Least Concern
POPULATION TREND: Decreasing

Hudsonian Godwit is one of the least-known birds. Most of the population is concentrated in remote sites and little-observed. Its migration routes are complicated. Hudsonian Godwits make migration flights of up to 5000 miles, and are thus rarely observed between the breeding grounds in the Canadian subarctic and wintering grounds in southern South America. In spring, most of the population passes through the eastern part of Texas. **LENGTH:** 15.5 inches. **WINGSPAN:** 29 inches. **ADULT BREEDING:** (Apr.–Sept.) Male has dark crown and light face and neck with fine streaks. Long, slightly upturned bill with the base two-thirds light pinkish. Chest and belly are dark rufous with dark barring. Back is dark and speckled. Distinctive white rump and dark tail. Flashes of white in the wings. Female is like male but chest is light with dark rufous barring. **NONBREEDING:** (Oct.–Mar.) Plain gray overall, neck and upper chest gray, belly and undertail light. **JUVENILE:** (Aug.–Nov.) Neck and chest washed with dirty brown. Distinct pale supercilium. Back is gray with dark spotting. **VOICE:** Call is a single *weet* or high-pitched *we-weat*. **BEHAVIORS:** Flies with legs stretched out and bill horizontal. **HABITAT:** In migration, found in rice fields and shallow, flooded soil. **STATUS:** Uncommon to rare spring migrant in the eastern half of Texas, mostly on the coastal prairies. Spring migrants pass through between mid-April and mid-May.

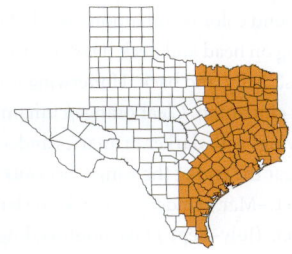

Marbled Godwit

(Limosa fedoa)

IUCN RED LIST STATUS (2016): Least Concern
POPULATION TREND: Decreasing

Marbled Godwit is one of the largest shore-birds. Like Hudsonian Godwits, they sometimes forage almost entirely on plant tubers during migration. Both parents share in incubating eggs, but the incubating parent does not flush easily and can sometimes even be picked from the nest.

LENGTH: 18 inches. **WINGSPAN:** 30 inches.
ADULT BREEDING: (Apr.–Sept.) Very long, slightly upturned bill, pale pink at the base. Background color is cinnamon overall. Fine streaking on head and neck. Dark barring on breast, belly, and back. Underwing is unmarked cinnamon. Clear orange/cinnamon upper wings with darker primaries and a dark patch near the end of the wings. **NONBREEDING:** (Oct.–Mar.) Underparts lack any barring. **JUVENILE:** (July–Sept.) Like nonbreeding but pattern on back is less distinct. **VOICE:** Call a hoarse *ekk ekk ekk* or hoarse two-note *eu-euk euk-euk*. **BEHAVIORS:** Strong direct flight, bill pointed forward and feet stretched out behind. Walks and runs rapidly. **HABITAT:** Migrants found in shallow wetlands, pastures, and native grasslands. Wintering birds are primarily coastal at mudflats and adjacent uplands, estuaries, beaches, and sand flats. **STATUS:** Common to uncommon winter resident along the coastal prairies and immediate coast. Rare to locally uncommon migrant statewide except for the Pineywoods and Edwards Plateau, because of a lack of habitat. Fall birds begin passing through Texas late June through November. Spring migrants are found late February to mid-May.

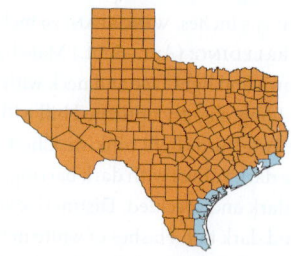

Breeding

Nonbreeding

Ruddy Turnstone

(Arenaria interpres)

IUCN RED LIST STATUS (2019): Least Concern
POPULATION TREND: Decreasing

Named for its habit of turning over small stones and shells looking for prey, Ruddy Turnstone is one of the showiest and most distinctive of Texas shorebirds. Short and stocky, it drives away intruders much larger than itself in its arctic breeding grounds.
LENGTH: 9.5 inches. **WINGSPAN:** 21 inches.
ADULT BREEDING: (Apr.–Sept.) Male has small pointed bill, pale crown, and black-and-white harlequin face pattern. Breast is black, belly is white. Upper wings and sides of the back are rufous brown. Upper back is black, lower back white. Black rump and single wide black subterminal band on the white tail. Female is like male but duller overall. **NONBREEDING:** (Sept.–Apr.) Rufous colors of breeding become dull brown and the black becomes dark gray. Head is brown-gray. **JUVENILE:** (Aug.–Nov.) Dark back and upper wing feathers edged in rusty brown. Strongly patterned head in gray-brown. Rounded dark patch on breast. White undersides. **VOICE:** Call is a rapid, mellow *pik-pik-pik-pik*. **BEHAVIORS:** Active, rapid runner on rocks and shore in pursuit of small prey like fiddler crabs. **HABITAT:** Mostly coastal rocky and stony shorelines. Also common on mudflats, sand flats, and hard mudbanks. **STATUS:** Common migrant and winter resident along the coast. Rare in summer on the coast. Migrants are rare inland in the eastern half of the state. Fall migration late July–mid-October, spring late March–late May.

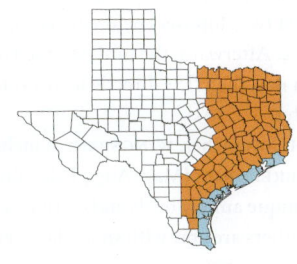

Breeding

Nonbreeding

Red Knot

(Calidris canutus)

IUCN RED LIST STATUS (2018): Near Threatened

POPULATION TREND: Decreasing

Red Knots have declined dramatically in the last thirty years. Much of this decline is attributed to the greatly increased harvest of female horsecrabs in Delaware Bay during the 1990s, as horseshoe crab eggs are a major food source. Red Knots that winter in Texas have also declined precipitously in the last few decades. A recent study using radio locators found that the wintering birds left Texas and made just two stopovers before breeding in the arctic. Afterward, the birds stopped on Hudson Bay, then flew directly to Texas for the winter.

LENGTH: 10.5 inches. **WINGSPAN:** 23 inches.

ADULT BREEDING: (May–Aug.) Pale salmon color, unique among birds and distinctive. Back feathers are gray with small dark centers. **NONBREEDING:** (Sept.–Apr.) Gray with indistinct spotting on breast. Has indistinct supercilium. White flanks with dark barring. **JUVENILE:** (Aug.–Oct.) Feathers have narrow pale edges, giving it a fine scaly pattern. Barring in the flanks is broken almost into chevrons. Dull yellow legs. **VOICE:** Generally quiet away from breeding areas. **BEHAVIORS:** Usually walks rather than runs. Movements when foraging are steady and methodical like a grazing animal. **HABITAT:** Sandy beaches. **STATUS:** Uncommon migrant along the coast, primarily the upper Texas coast. Very rare migrant in the eastern third of the state, and very rarely detected. Rare and local winter resident on the coast.

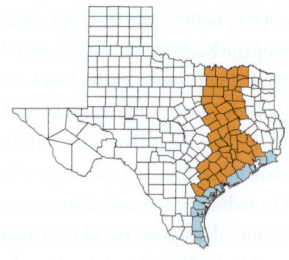

Breeding

Stilt Sandpiper

(Calidris himantopus)

IUCN RED LIST STATUS (2016): Least Concern
POPULATION TREND: Increasing

Although described to science in 1826 by Charles Bonaparte (nephew of Napoléon Bonaparte), it was not well known in the field to either ornithologists or market gunners. It was called Bastard Yellowlegs because it was suspected to be a hybrid. Other names used by the gunners were Mongrel and Frost Snipe.

An oddity about Stilt Sandpiper is that a long-term study in Manitoba found that longer-billed females would pair with shorter-billed males before birds with medium bills would pair up.
LENGTH: 8.5 inches. **WINGSPAN:** 18 inches.
ADULT BREEDING: (Apr.–Aug.) Distinct red crown and ear patch, white supercilium. Bill is thick and droops slightly. Heavily barred below. Dark blotchy back is speckled white. White rump is speckled with fine dark spots. Tail is pale gray. **NONBREEDING:** (Sept.–Apr.)

Pale gray, light below without barring below. White supercilium. **JUVENILE:** (July–Sept.) Finely streaked head, neck, and breast. Pale supercilium. Scaly pattern on back. **VOICE:** Call a soft *kute* or buzzy *krrrp*. **BEHAVIORS:** Flight is fast and direct. Forms tight flocks in migration. Walks, often wading slowly when feeding. **HABITAT:** A pond-foraging species, mostly found in freshwater pools, marshes, and rice fields. **STATUS:** Uncommon to locally common migrant statewide. Rare to uncommon winter resident on the coast but becomes more common southward along the coast.

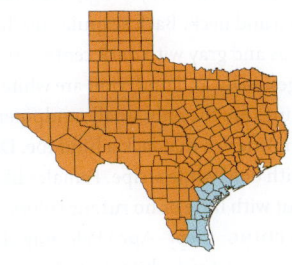

Nonbreeding

Breeding

Juvenile

Sanderling

(Calidris alba)

IUCN RED LIST STATUS (2016): Least Concern
POPULATION TREND: Unknown

Sanderlings are one of the most cosmopolitan species. Outside of breeding season, they can be found on sandy beaches anywhere in the world. Despite this, regional populations are in decline and much needs to be learned about their breeding biology. Increased recreational use of sandy beaches is thought to be one cause of decline.

LENGTH: 8 inches. **WINGSPAN:** 17 inches.

ADULT BREEDING: (May–Aug.) Male has rufous head and neck. Back scapular feathers are rufous and gray with dark centers and pale edges. Belly and undertail are white. Sanderlings are the only small sandpipers in North America that lack a hind toe. Dark wings with bold white stripe. Females like males but with little or no rufous colors.

NONBREEDING: (Sept–Apr.) Pale gray above, white below. Distinct white shoulder mark.

JUVENILE: (July–Nov.) Like nonbreeding but spangled on the back. **VOICE:** Call is a short high *kwit* or *kwit kwit*. **BEHAVIORS:** Runs well and fast. Spends long periods standing on one leg. Known for chasing waves down for exposed food, then running back in front of the next advancing wave. **HABITAT:** Primarily sandy beaches. **STATUS:** Common to abundant migrant and winter resident along the immediate coast. Rare migrant inland in the eastern two-thirds of the state, very rare in the western third. Spring migration late March–late May, Fall mid-July–mid-October.

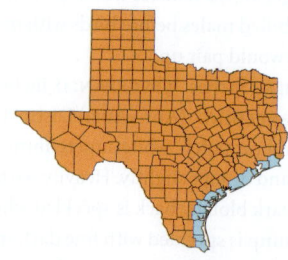

Breeding

Nonbreeding

Dunlin

(Calidris alpina)

IUCN RED LIST STATUS (2016): Least Concern
POPULATION TREND: Decreasing

Dunlin is striking in its bold breeding plumage, with its inky black belly and bright red rufous back. In winter, they are much more incognito, molting to an inconspicuous dun color. There are nine subspecies of Dunlin worldwide, though much is unknown about the Texas subspecies, *C. alpina hudsonia*. **LENGTH:** 8.5 inches. **WINGSPAN:** 17 inches. **ADULT BREEDING:** (Apr.–Aug.) The most distinctive feature of breeding plumage is the large square black patch on the belly. Long, slightly drooping bill, averaging slightly longer in females. Head, neck, and breast are finely streaked with rufous. Back is rufous with dark spots. Tail is light with gray center. Bold wing stripe on dark wings. **NONBREEDING:** (Aug.–Mar.) Dull brownish gray above and white below. Head and breast are a pale brownish. **JUVENILE:** (July–Sept.) Similar to breeding

adult; the black patch is speckled. Back feathers have larger black centers. **VOICE:** Call is a raspy buzzing *pjeev*. **BEHAVIORS:** Generally walks and only occasionally runs. Roosts on one leg. Flies swiftly in tight flocks. **HABITAT:** Coastal bays, estuaries, mudflats. Makes heavy use of rice fields. **STATUS:** Common to abundant winter resident on the coast. Rare inland. Rare to uncommon migrant in the eastern half of the state, rare in the west. More common in fall migration. Fall migrants pass through late August–mid-November, spring mid-March–late May.

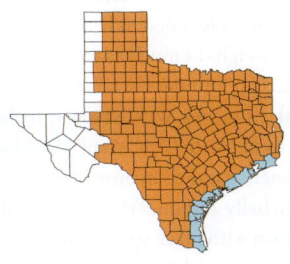

Baird's Sandpiper

(Calidris bairdii)

IUCN RED LIST STATUS (2018): Least Concern
POPULATION TREND: Stable

Baird's Sandpiper was one of the last sandpipers described in North America. It was named in 1861 for Spencer Fullerton Baird, then assistant secretary and later second secretary of the Smithsonian Institution. Most Baird's Sandpipers make a fall migration of 3700 miles, directly from southern Canada and the northern United States to northern South America.

The female performs an amazing feat not yet well understood: it lays a clutch of eggs that weighs as much as 120% of the bird's weight in just four days.

LENGTH: 7.5 inches. **WINGSPAN:** 17 inches.

ADULT BREEDING: (Apr.–Sept.) Head and breast finely streaked in brown. Clear white below on belly and undertail. Back is silvery gray-brown with black spots. Wings are long and extend beyond the tail. In flight, wing stripe is weak. Tail is gray. **NONBREEDING:** (Oct.–Apr.) Brown, finely streaked head. Back is brown with darker brown centers to the back scapulars. **JUVENILE:** (Aug.–Nov.) Brown-headed with buffy breast, scaly brown on the back, light belly and undertail. **VOICE:** Rough *treeep*. **BEHAVIORS:** Walks deliberately, feeding with rapid motion of the bill. **HABITAT:** Mostly freshwater wetlands and rain-soaked fields. On beaches, keeps high above the waves. **STATUS:** Uncommon to common migrant through the state in spring and fall. Rare to uncommon in the Pineywoods. Fall migration mid-July–mid-October, spring mid-March–mid-May.

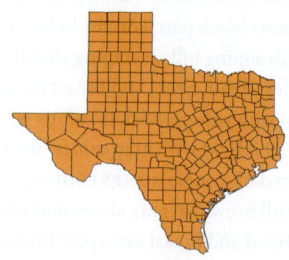

Breeding

Nonbreeding

Juvenile

Least Sandpiper

(Calidris minutilla)

IUCN RED LIST STATUS (2016): Least Concern
POPULATION TREND: Decreasing

The Least Sandpiper is the smallest shorebird. Away from the coast, it is the most common "peep" encountered, but its small size and habit of foraging near or in vegetation make it disappear from view. On the coast itself, it is usually outnumbered by other small sandpipers and may be completely absent from its flock out in the open on mudflats.

LENGTH: 6 inches. **WINGSPAN:** 13 inches.
ADULT BREEDING: (Apr.–Sept.) Head and breast are finely streaked brown. Bill is fine and slightly drooped. Brown-backed with dark spots. Belly and undertail are clear white. In flight, the sides of the rump are white, tail has a dark center, and wings show little or no wing stripe. The only small sandpiper with yellow legs. **NONBREEDING:** (Oct.–Mar.) Head, breast, and back are mostly dark and brownish. White below. **VOICE:** Call is a high trilling *prreep.* **BEHAVIORS:** Walks slowly and hunched when feeding. Motions are best described as dainty. **HABITAT:** Uses inland habitat more than other small sandpipers. On the coast, sticks to channels on mudflats and the edges of flats. **STATUS:** Common to abundant migrant statewide and winter resident on the coast. Uncommon to locally common winter resident inland. Spring migration is mid-March–mid-May, fall early July–mid-November.

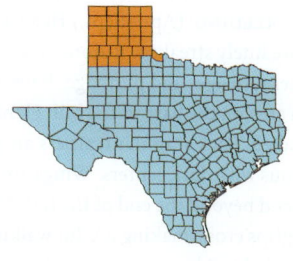

White-rumped Sandpiper

(Calidris fuscicollis)

IUCN RED LIST STATUS (2016): Least Concern
POPULATION TREND: Decreasing

White-rumped Sandpipers make one of the longest migrations in the western hemisphere. They breed in the high arctic and winter at the southern tip of South America. They migrate in several long-distance flights, some up to 60 hours in length. To do this, they rely on body fat for fuel acquired at key stopover sites, making them vulnerable to disruptions at these sites.

LENGTH: 7.5 inches. **WINGSPAN:** 17 inches.
ADULT BREEDING: (Apr.–Sept.) Head and breast are finely streaked. Rufous on crown and cheek. All ages and plumages have a reddish area on the base of the bill. Flanks are streaked. Back scapular feathers are gray and rufous with dark centers. Wings are long and extend beyond the end of the tail. Often the wingtips cross, making a V for walking birds. In flight, it has an obvious white rump.

NONBREEDING: (Aug.–Apr.) Grayish with faint gray streaking on the flanks. There is an obvious white supercilium. **VOICE:** Call is a high mouse-like squeak. **BEHAVIORS:** Walks steadily, more deliberate than other small sandpipers. Flight is swift and undulating. Swerves right and left in flocks. **HABITAT:** Tidal marshes, flooded fields, shallow wetlands, and rice fields. **STATUS:** Uncommon to common spring migrant in the eastern half of the state. Very rare in the Trans-Pecos. Spring migration late April–early June. Essentially absent in Texas during fall migration.

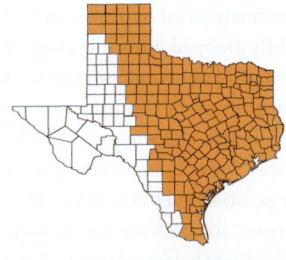

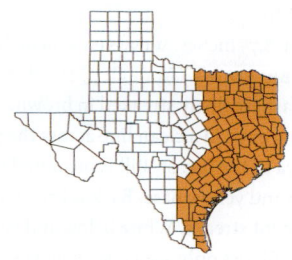

Buff-breasted Sandpiper

(Calidris subruficollis)

IUCN RED LIST STATUS (2016): Near Threatened

POPULATION TREND: Decreasing

Buff-breasted Sandpipers are another of the "grasspipers" that shun the normal haunts of shorebirds and prefer shortgrass habitats. They are frequently found on sod farms and overgrazed fields. This suggests some compatibility with human activities, although this also exposes them to potentially harmful chemicals.

Formerly abundant, this species has declined significantly due to habitat loss on its migratory route and wintering ground in South America and due to commerical market hunting. The Buff-breasted Sandpiper will return to a wounded flock member, making them vulnerable to market hunters in the late 1800s; the species has never recovered. **LENGTH:** 8.25 inches. **WINGSPAN:** 18 inches. **ADULT BREEDING:** (Apr.–Oct.) Face is plain and buff colored, darker on the crown.

Back scapular feathers are dark-centered. Breast and flanks are clear buffy-colored. Legs yellow. No wing stripe. Dark rump. **NONBREEDING:** (Oct.–Mar.) Essentially identical to breeding plumage. **JUVENILE:** (Aug.–Nov.) Pale edges to back scapular feathers give a scaly appearance. **VOICE:** Flight call is a quiet *greep greep*. **BEHAVIORS:** Flies low and fast in a tight flock with many turns and zigzags. Walks (high-stepping) or runs, then pecks during feeding. **HABITAT:** Short grass uplands sites. Commonly seen at sod farms, golf courses, bare or newly planted fields. **STATUS:** Rare to uncommon migrant in the eastern half of the state. Spring migration is early April–mid-May, fall late July–early October.

Pectoral Sandpiper

(Calidris melanotos)

IUCN RED LIST STATUS (2016): Least Concern
POPULATION TREND: Stable

Pectoral Sandpipers are champion long-distance migrants. They breed in the high arctic and winter in southern South America. Some birds make an annual round trip of 30,000 miles.

Males are promiscuous breeders. A male will keep a harem of females in a guarded territory and mate with as many females as it can. The females then leave to perform all the nesting alone. Males migrate earlier than females do.

LENGTH: 8.75 inches. **WINGSPAN:** 18 inches.
ADULT BREEDING: (Apr.–Oct.) Head, neck, and breast are finely streaked in brown. Streaking ends in a clean line across the pectoral area of the breast. Bill is bicolored with dark tip and yellow base. Back is brown with several light streaks. White below and yellow legs. Shows only a very weak wing stripe.

NONBREEDING: (Nov.–Mar.) Like breeding but less pattern in the back; streaks on the back are mostly absent. **JUVENILE:** (July–Nov.) Like breeding with a more obvious supercilium. **VOICE:** Call is a harsh *chuurt chuurt*. **BEHAVIORS:** Walks slowly, stands upright when alert. Flight is rapid and turning, like a snipe. **HABITAT:** Wet grasslands and marshy grassy areas. **STATUS:** Locally common migrant in the eastern two-thirds of Texas. Rare to uncommon in the western third of the state. Spring migration is early March–early June, fall mid-July–mid-October.

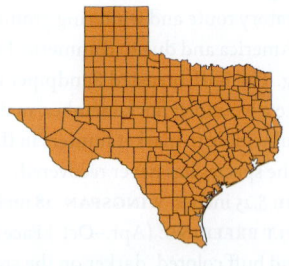

Semipalmated Sandpiper

(Calidris pusilla)

IUCN RED LIST STATUS (2016): Near Threatened

POPULATION TREND: Decreasing

Semipalmated Sandpiper is perhaps the most abundant of the small "peep" shorebirds. Flocks of thousands gather in migrations. These large flocks mask the fact that the species is declining rapidly. While not as popular as larger shorebirds with market hunters in the 1800s, they were considered good eating and a dozen could be taken with single blast of a shotgun into a flock. Wintering habitat is being converted to agriculture, but a dependance on coastal habitats and a lack of agricultural chemicals in studies of their carcasses suggest that continued legal and illegal hunting is a major factor in their decline.

LENGTH: 6.25 inches. **WINGSPAN:** 14 inches.

ADULT BREEDING: (Mar.–Sept.) Gray-brown color, lacks warm colors. Streaking on the breast is fine. Flanks mostly clear. Legs are dark. Bill is short and straight. A slight bulge in the tip of the bill is often noticeable. **NONBREEDING:** (Oct.– Mar.) Dingy gray, including on the breast. **JUVENILE:** (July–Nov.) Upper parts are scaly and lack rufous. Breast is dingy. Dark cap. **VOICE:** Call is a husky *churf*. In feeding flocks, a scolding *he-he-he-he-he*. **BEHAVIORS:** Outside of breeding season, forms large flocks. **HABITAT:** Areas of shallow water, muddy or soft silt/clay mudflats. **STATUS:** Uncommon to locally common migrant east of Trans-Pecos. Does not winter in Texas or North America. Spring migration March–early June, fall early July–mid-October.

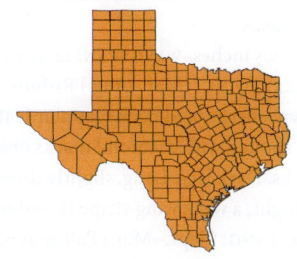

Breeding

Nonbreeding

Western Sandpiper

(Calidris mauri)

IUCN RED LIST STATUS (2016): Least Concern
POPULATION TREND: Decreasing

Western Sandpiper is one of the most abundant shorebirds in the western hemisphere. In 2000, the population was estimated to be 3.5 million birds. Remarkably, they have one of the most restricted breeding ranges of all shorebirds, breeding on the western edge of Alaska on the tundra. Even though only about 10% of the population winters and migrates to the Gulf and southern Atlantic coasts, it is also one of the most abundant shorebirds in Texas.
LENGTH: 6.5 inches. **WINGSPAN:** 14 inches.
ADULT BREEDING: (Mar.–Aug.) Rufous on the crown, cheeks, and back scapular feathers. Heavily spotted with diamond marks on chest and flanks. Dark legs. Long, slightly drooping bill. In flight, a weak wing stripe is visible.
NONBREEDING: (Sept.–Mar.) Pale grayish overall; breast is mostly white. **JUVENILE:**

(July–Sept.) Pale face and breast. Lacks the spotting of breeding plumage. Upper scapulars are bright rufous, lower ones are gray with dark anchor-shaped centers. **VOICE:** Call is a high-pitched squeaky *cheet*. **BEHAVIORS:** Walks or runs while foraging. Stands on one leg for long periods; hops on a single leg. **HABITAT:** Primarily coastal sand flats and mudflats. In migrations, inland margins of ponds and lakes; favors salt lakes. **STATUS:** Common to abundant migrant along the coast, common winter resident. Fall migration is early July–November, spring March–late May.

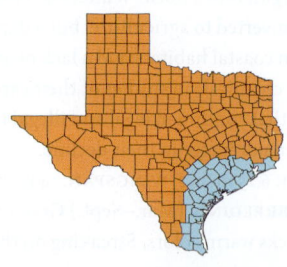

Breeding

Nonbreeding

Short-billed Dowitcher

(Limnodromus griseus)

IUCN RED LIST STATUS (2016): Least Concern
POPULATION TREND: Decreasing

The Short-billed Dowitcher is a bit of an enigma. They are so similar to Long-billed Dowitchers that they were not accepted as a separate species until the 1950s. In Texas, much can still be learned about its distribution away from the coast. Difficulties in distinguishing it from Long-billed Dowitchers make establishing distribution of migrants difficult. In fall, juveniles are likely overreported in relative abundance because they are more easily identified. **LENGTH:** 11 inches. **WINGSPAN:** 19 inches. **ADULT BREEDING:** (Apr.–Aug.) Long straight bill. Neck is always clean orange, with some spotting on breast. Underside is mostly orange. Back scapular feathers are barred golden. In flight, back is white and tail is narrowly barred with nearly even amounts of black and white. Back appears mostly flat when feeding. **NONBREEDING:** (Sept.–Apr.) Gray overall. Face is streaked and flanks are paler than Long-billed Dowitcher. **JUVENILE:** (July–Nov.) Buffy breast. Back feathers have broad buffy fringes. **VOICE:** Call is a liquid *kewtututu* or *tututututu*. **BEHAVIORS:** Feeds in water up to the belly with a sewing machine motion. **HABITAT:** Prefers saltwater habitat, tidal flats, beaches, and salt marshes. **STATUS:** Uncommon to rare migrant in the eastern half of Texas, except for the Pineywoods. Strong preference for salt water. Locally common winter resident on the coast. Spring migration is late March–mid-May, fall early July–mid-October.

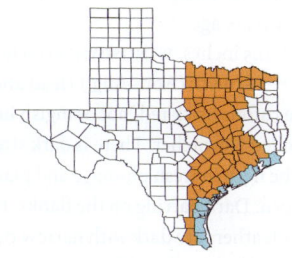

Breeding

Nonbreeding

Long-billed Dowitcher

(Limnodromus scolopaceus)

IUCN RED LIST STATUS (2016): Least Concern
POPULATION TREND: Unknown

Differentiating Long-billed Dowitcher from Short-billed Dowitcher is one of the great challenges of Texas field identification. Differences are subtle and take some experience to appreciate. Long-billed are dark and more reddish. Short-billed have flatter backs. The voices are different. Mitochondrial DNA between the two species are some of the most divergent between congeners in the bird world; the two species likely separated as far back as 4 million years ago.
LENGTH: 11.5 inches. **WINGSPAN:** 19 inches.
ADULT BREEDING: (Apr.–Aug.) Head and neck are rufous brick red, not orangish like a Short-billed Dowitcher. There is dark streaking on the neck in fresh plumage and plain when worn. Dark barring on the flanks. Back scapular feathers are dark with narrow rufous bars and white tips. White rump and lower back visible in flight. **NONBREEDING:** (Sept.–Apr.) Plain grayish overall. The back of Long-billed Dowitcher is rounded, like it swallowed a grapefruit. **JUVENILE:** (July–Nov.) Gray breast, dull overall. Back scapular feathers are dark-centered with narrow rufous edges. **VOICE:** Call is a sharp *chit chit*. **BEHAVIORS:** Feeds in water up to the belly with a sewing machine motion. **HABITAT:** Most common in freshwater habitats, shallow ponds, marshes, and rice fields, for example. **STATUS:** Uncommon to common migrant statewide. Common winter resident along the coast, locally common inland. Spring migration is early March–late May, fall early July–mid-November.

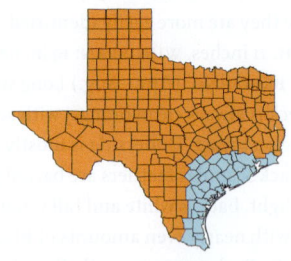

American Woodcock

(Scolopax minor)

IUCN RED LIST STATUS (2020): Least Concern
POPULATION TREND: Decreasing

American Woodcock, sometimes known as the Timberdoddle, is technically a shorebird but about as different from a shorebird as it can be. It lives in forests. Its shape is more like a small dove with long bill. It is one of the two shorebirds still hunted in Texas, although its popularity as a game bird is waning.
LENGTH: 11 inches. **WINGSPAN:** 18 inches.
ADULT: All birds are similar year-round. Belly and lower breast are pale orangish buff. Neck is grayish, extending to a collar and into two gray stripes on the back. Crown is dark with narrow pale stripes across it. Tail is very short. **VOICE:** Call is a buzzy *preemp*.
BEHAVIORS: Walks on the ground often with a peculiar rocking motion. Males call from the ground to attract mates. Nocturnal.
HABITAT: Wide variety of forests. **STATUS:**

Rare to locally common winter resident in the eastern half of the state. Rare but regular breeder in the eastern third of the state.

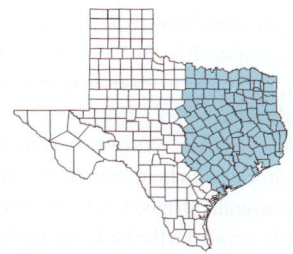

Wilson's Snipe

(Gallinago delicata)

IUCN RED LIST STATUS (2020): Least Concern
POPULATION TREND: Increasing

Many new to birding and birdwatching who are familiar with the classic prank of taking someone on a "snipe hunt" are surprised to find that the snipe is a real bird and there is a real hunting season for it. As a game bird, it's considered one of the most challenging. A flushed Wilson's Snipe can fly more than 60 mph in a zigzag pattern. British soldiers in India during the 1770s would hunt Common Snipe (*Gallinago gallinago*). A good sniper would have to be a good marksman, but also stealthy. By World War I, the term sniper evolved to mean a stealthy military marksman.

LENGTH: 10.5 inches. **WINGSPAN:** 18 inches.
ADULT: Stocky with long straight bill. Dark brownish overall. Striped head with bold light-buffy stripes on the back. Strong bars on white flanks. **JUVENILE:** (July–Oct.)

Essentially the same as adults. **VOICE:** Call is a raspy *keeesh*. **BEHAVIORS:** When flushed, will fly in a zigzag pattern; at other times flies very fast. **HABITAT:** Requires wet organic soils with clumps of vegetation for cover. Found in marshes, swamps, wet meadows and pastures, marshy edges of streams and ponds, and recently burned areas. **STATUS:** Common migrant statewide. Common winter resident in the eastern two-thirds of Texas. In the western third, an uncommon to rare winter resident. Spring migration is mid-March–mid-May, fall early September–early November.

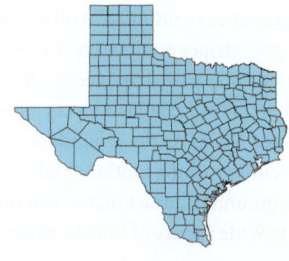

Breeding

Nonbreeding

Spotted Sandpiper

(Actitis macularius)

IUCN RED LIST STATUS (2016): Least Concern
POPULATION TREND: Decreasing

Spotted Sandpiper is the most widespread breeding sandpiper in North America. In Texas, it has been recorded in all 254 counties. A generalist, Spotted Sandpiper can feed on a great variety of animal matter and can use a great variety of shoreline habitats. They are among a small group of birds that have reversed many sex roles, from establishing territory to incubating and brooding their young. **LENGTH:** 7.5 inches. **WINGSPAN:** 15 inches. **ADULT BREEDING:** (Apr.–Aug.) Boldly spotted breast extending down onto the flanks. Brown-backed with dark spots. Bill is orangish-yellow with dark tip. Thin supercilium over a dark eye line. In flight, primaries and secondaries are dark, almost black. Shows a bold white wing stripe. **NONBREEDING:** (Aug.–Mar.) Lacks spotting in the breast. Sides of the breast are brown. Bill is gray.

JUVENILE: (July–Oct.) Essentially the same as nonbreeding. **VOICE:** Call is a two-note *pee-pit*. **BEHAVIORS:** Walks with a constant teetering and lowering of the body. In short flights, alternates burst of rapid fluttering wingbeats and short glides. **HABITAT:** Water's edging, often open mud, concrete, or rocks. **STATUS:** Common migrant throughout the state. Common winter resident in southern half of the state, becoming uncommon to south of the High Plains. Can be found irregularly in summer in the northern half of the state. Spring migration late March–late May, fall early July–October.

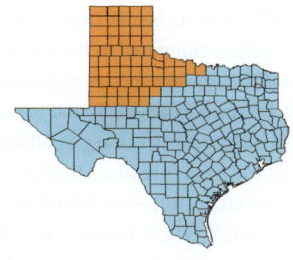

Solitary Sandpiper

(Tringa solitaria)

IUCN RED LIST STATUS (2018): Least Concern
POPULATION TREND: Decreasing

The Solitary Sandpiper lives up to its name; it is not often found around others of the same species. This might be because of its unusual preference for forested and wooded bodies of water.
LENGTH: 8.5 inches. **WINGSPAN:** 22 inches.
ADULT BREEDING: (Apr.–Sept.) Head and neck are finely streaked. White eye ring stands out. Back and upper wings are dark gray with fine white spots. In flight, outside of the tail is white. Clean white below. Legs are pale greenish-yellow compared to the bright yellow of both Yellowlegs species. **NONBREEDING:** (Oct.–Mar.) Head, neck, and breast are gray with little to no streaking. Spotting on the wings and back is reduced to a few white flecks. **JUVENILE:** (July–Nov.) Spotting is like a breeding adult. White lores and eye ring give the impression of spectacles. Head and breast are more like nonbreeding adults. **VOICE:** Call is a high-pitched *pee-WEET* or a *weet-weet-weet*.
BEHAVIORS: Often wades up to its belly. Moves more deliberately than Yellowlegs. When flushed, flies almost straight up. On landing, extends wings above the body before folding them. **HABITAT:** Prefers unclosed wet or muddy habitats, forested ponds, and lakes. Can be attracted to drainage ditches. **STATUS:** Uncommon to common migrant statewide. Rare winter resident on the coastal prairie. Spring migration is early March–early May, fall mid-July–late October.

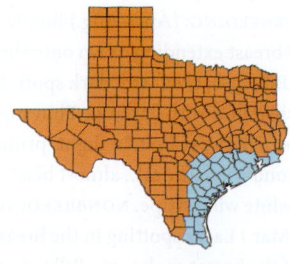

Breeding

Nonbreeding

Lesser Yellowlegs

(Tringa flavipes)

IUCN RED LIST STATUS (2016): Least Concern
POPULATION TREND: Decreasing

Lesser Yellowlegs was a popular species with market hunters in the early 20th century, a probable cause of population decline. As market hunting ended with the passage of the Migratory Bird Treaty Act in 1927, the species recovered. Several thousand are still taken by hunting in the Carribean, particularly in Barbados, where as recently as 2012 the annual harvest was estimated to be 7000–15,000 Lesser Yellowlegs. **LENGTH:** 10.5 inches. **WINGSPAN:** 24 inches. **ADULT BREEDING:** (Apr.–Sept.) Boldly patterned gray down to the belly. Belly white with a small amount of spotting in the flanks. Bill is fine and about the same length as the head is deep. Bright yellow legs. **NONBREEDING:** (Oct.–Mar.) Less boldly patterned in gray. **JUVENILE:** (July–Nov.) Head and neck are patterned like a nonbreeding adult. White patches on the edges of the scapulars give it a spotted appearance, lighter gray than a Solitary Sandpiper. **VOICE:** High-pitched *tu-tu* or *tu-tu-tu-tu*. **BEHAVIORS:** Walks quickly when foraging. Movements are delicate as it picks food. Flight is more buoyant than Greater Yellowlegs and less powerful. **HABITAT:** Salt, brackish, and freshwater wetland; shallow pools with vegetation. **STATUS:** Common migrant throughout the state. Locally common to uncommon winter resident on the coast. During peak migration, concentrations of hundreds are regular. Spring migration is mid-March–mid-May, fall late June–late October.

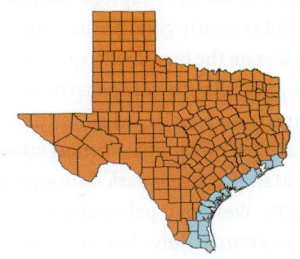

Western, nonbreeding

Eastern, breeding

Willet

(Tringa semipalmata)

IUCN RED LIST STATUS: Least Concern
POPULATION TREND: Stable

There are two subtly distinct subspecies of Willets that have been proposed as separate species—the western subspecies *T. semipalmata inornata* and the eastern *T. semipalmata semipalmara*.

LENGTH: 15 inches. **WINGSPAN:** 26 inches.
ADULT BREEDING: (Apr.–Aug.) Eastern subspecies is boldly barred over the whole body in browns on a gray background. Stout gray bill, darker at the tip. Gray legs. Narrow white rump. Tail is mostly gray. Western subspecies is paler on the breast with sparse barring. Pale gray overall with little pattern. In flight, rump and tail are mostly white, with pale gray terminal band on the tail. **NONBREEDING:** (Sept.–Mar.) Eastern is dark slate gray with white belly. Western is pale gray with a whitish throat. **JUVENILE:** (July–Nov.) Eastern is plain dark gray. Western is paler gray than eastern,

with whitish center to the breast. Forehead is also whitish. More crisply marked on the coverts than eastern birds. **VOICE:** Eastern song is a loud repeated *oh-WEE-WEEP*. Eastern call is *kept-kept-kept-kept-kept*. Western call is *kit-kit-kit-kit-kit*. **BEHAVIORS:** Wades while feeding, often up to the belly. **HABITAT:** Breeding eastern birds exclusively in saltwater habitats, salt marshes. Wintering western birds are mostly coastal. **STATUS:** Common to abundant year-round along the coast. Western birds occasionally inland migrants. Western birds are present mid-July–late March but may linger into May. Eastern breeding birds present late March–August.

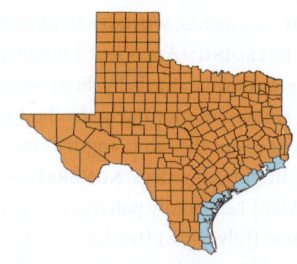

Breeding

Greater Yellowlegs

(Tringa melanoleuca)

IUCN RED LIST STATUS (2016): Least Concern
POPULATION TREND: Stable

Greater Yellowlegs is one of the largest of Texas shorebirds, and one of the most energetic when dashing about chasing prey. It is very similar to Lesser Yellowlegs but distinguished by a larger, longer bill that is longer than the head is deep. It can also be identified by its different call.
LENGTH: 14 inches. **WINGSPAN:** 28 inches.
ADULT BREEDING: (Mar.–Aug.) Streaked head and neck. Bill is slightly upturned and longer than the head is deep. Barring on the flanks. Fine white speckles on back. Bright yellow legs. **NONBREEDING:** (Aug.–Mar.) Marking is less detailed, less crisp. White speckles are absent from back. **JUVENILE:** (July–Nov.) Like nonbreeding adult. **VOICE:** Call is a high-pitched, strident *DU-DU-DU-DU*.
BEHAVIORS: Walking is rapid and jerky.
HABITAT: Wide variety of wetland habitats.

STATUS: Uncommon to common migrant statewide. Uncommon to locally common winter resident along the coastal prairies. Can be found as migrants in Texas in June and July, but they are not breeders in the state.

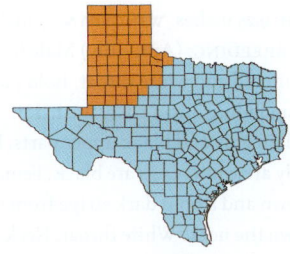

Female

Male

Wilson's Phalarope

(Phalaropus tricolor)

IUCN RED LIST STATUS (2016): Least Concern
POPULATION TREND: Increasing

The name for the genus *Phalaropus* comes from ancient Greek *phalaris*, meaning "coot," and *pous*, meaning "foot." Like coots, phalaropes have lobed toes. Females compete for males and will mate with more than one male in a season.

Most shorebirds don't swim much, but phalaropes engage in spinning: swimming in tight circles, drawing prey to the surface in the vortex.

LENGTH: 9.25 inches. **WINGSPAN:** 17 inches.
ADULT BREEDING: (Apr.–July) Male has a small thin bill and white throat. Bold pale supercilium fading into a rusty wash down the nape. Plain gray and black upper parts. Neck and belly are white. Legs are black. Female has gray crown and a bold dark stripe from the face down the neck. White throat. Neck and breast are rusty. Back is gray and rusty. Belly is white, legs are black. **NONBREEDING:** (Aug.–Mar.) Sexes are the same. Pale gray and white below. White face. **JUVENILE:** (July–Aug.) Dark crown and white face. Back is gray and white below. Legs are yellow. **VOICE:** Generally silent in migration. **BEHAVIORS:** Looks awkward when walking. Phalarope spinning feeding is distinctive. **HABITAT:** Mainly shallow open-water habitats. **STATUS:** Common to abundant migrant across the state. Rare to casual in winter on the lower coast and in the Lower Rio Grande Valley. Spring migration is late March–late May, fall late June–late October.

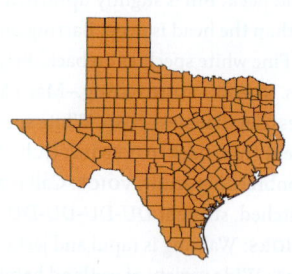

Nonbreeding

Red-necked Phalarope

(Phalaropus lobatus)

IUCN RED LIST STATUS (2019): Least Concern
POPULATION TREND: Decreasing

Like the closely related Wilson's Phalarope, Red-necked Phalarope is polyamorous. The more showy female will abandon the male to incubate and care for the young while she seeks additional mates. In winter, this species is mostly pelagic, meaning it spends most of winter in the open ocean, sometimes in large flocks.

LENGTH: 7.75 inches. **WINGSPAN:** 15 inches. **ADULT BREEDING:** (Apr.–July) Male is gray with a red wash on the neck and a white throat. Pale face with dark ear patch. Legs are dark. Female's sides and front of the neck are brick-red with a distinctive white throat. Rusty streaking on back. **NONBREEDING:** (Aug.–Apr.) Sexes are the same. Streaked gray on back, white below. White face and neck with dark ear patch. Forehead is pale. **JUVENILE:** (July–Oct.) Dark crown and white face

with dark ear patch. Boldly streaked back with rust and gray. Belly is pale. **VOICE:** Call is a *kip* or *pip* note. **BEHAVIORS:** Virtually always on water away from nesting. Often feeds by swimming in tight circles to create a vortex to suck prey to the surface. **HABITAT:** Almost any wetland in migration. **STATUS:** Uncommon to rare fall migrant throughout the western third of the state, rare to very rare migrant in the remainder. More numerous in fall. Spring migration is mid-April–late May, fall mid-August–mid-November.

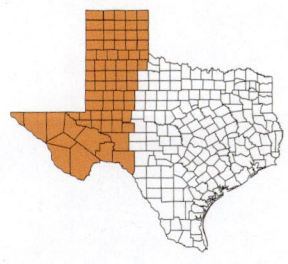

STERCORARIIDAE (SKUAS AND JAEGERS)

Skuas and jaegers are very similar in appearance to gulls and have long been placed in the same family. But recent DNA and skull characteristics studies suggest skuas and jeagers are a sister family to the auks, murres, and puffins.

The name jaeger is German for "hunter," and as the name suggests skuas and jaegers are like raptors of the seas. They are relentless predators and have no qualms stealing their next meal from other birds. They nest on the tundra or bare open areas of islands. Outside of the breeding season they spend their time far offshore, mostly out of sight of land, and are rarely observed from the shore.

Worldwide there are seven species of skuas and jaegers. None of these species are a conservation concern. Four have been recorded in Texas. Two, the Long-tailed Jaeger and the South Polar Skua, are very rare visitors to Texas.

Pomarine Jaeger

(Stercorarius pomarinus)

IUCN RED LIST STATUS (2019): Least Concern
POPULATION TREND: Stable

Pomarine Jaegers may be unique among birds in that they are dependent on a single species to breed: They will only breed in the arctic when there are sufficient brown lemming. **LENGTH:** 18.5 inches. **WINGSPAN:** 52 inches. **ADULT BREEDING:** (Apr.–Oct.) Light morph is dark brown on back and undertail. Belly pale with dark breast band. Creamy-colored collar extends up the nape. Dark face and crown. The two spoon-shaped central tail feathers extend outward and twist when flying. In flight, base of primaries is light and makes a distinctive light patch on the underside of the wings. Pale shafts of the primaries may be visible on upper wing. Dark morph is very dark brown except for wing patches. Loses the distinctive spoon-shaped tail feathers soon after leaving the arctic breeding grounds. **NONBREEDING:** (Oct.–Mar.) Similar to breeding; light belly and flanks are barred. Creamy collar is dingier. **JUVENILE:** (Aug.–Feb.) All morphs are like adults, but underside is barred and underwings have a checkered pattern. **VOICE:** Call is a raspy barking *ow-ow-ow*. **BEHAVIORS:** Larger and therefore less agile in flight than other species of jaegers. **HABITAT:** Outside of breeding, spends most of its life in open ocean. **STATUS:** Uncommon migrant and winter resident at sea in the Gulf of Mexico. Rare migrant and winter resident on the immediate coast, mostly late September–late April.

Light adult

Light immature

Parasitic Jaeger

(Stercorarius parasiticus)

IUCN RED LIST STATUS (2018): Least Concern
POPULATION TREND: Stable

Parasitic Jaegers get most of their food by stealing from other birds in migration and in winter.
LENGTH: 16.5 inches. **WINGSPAN:** 46 inches.
ADULT BREEDING: (Mar.–Aug.) Light morph has dark-brown cap, pale neck, and collar back to the nape. Dusky breast and a white belly. Undertail is gray-brown. The two central tail feathers extend as narrow points beyond the main tail. In flight, light primary patches on underside. White primary shafts visible above. Rump is brownish. Wide dark breast band and dark tail and undertail. Dark morph is all dark except for reduced wing patches like light morph. **NONBREEDING:** (Nov.–Mar.) Light morph is like breeding plumage but belly patch is reduced. Barring on flanks. Dark morph is like breeding. **JUVENILE:** (Aug.–Mar.) Light morph is like nonbreeding but

barred underneath. Underwings have checked brown-and-white pattern. Dark morph has feathers edged in brown on dark. Only four or five primary shafts are white. **VOICE:** Nasal *EY-YA* similar to a Gull-billed Tern. **BEHAVIORS:** Usually flies near the surface, up to 30 feet above. Agile when chasing another species to steal food. **HABITAT:** Offshore and near-shore waters of the Gulf of Mexico. **STATUS:** Uncommon migrant and rare winter resident offshore. Rare migrant and very rare winter resident on the coast. Spring migrants have been found early March–mid-May, fall early October–late November.

LARIDAE (GULLS, TERNS, AND SKIMMERS)

The gulls, terns, and skimmers are a large and diverse family. They range in size from the tiny Least Tern to the hulking Great Black-backed Gull with a wingspan greater than a Red-tailed Hawk. All are web-footed and there are three groups: Gulls have stout rounded bills. Terns average smaller than gulls and have pointed bills. Skimmers are unique, with a lower mandible longer than the upper mandible. All are shades of white, gray, and black.

Gulls, terns, and skimmers are usually found near water. Most species are found near the coast, but they often use freshwater marshes, open tundra, and even the open ocean. They have adapted to humans and can be found using agricultural fields, urban ponds, and garbage dumps.

There are 100 species of gulls, terns, and skimmers worldwide. Thirty-seven have been recorded in Texas. Fifteen of these are very rare visitors to Texas.

Black-legged Kittiwake

(Rissa tridactyla)

IUCN RED LIST STATUS (2019): Vulnerable
POPULATION TREND: Decreasing

The status of Black-legged Kittiwake has been in flux in Texas. Until 1999, the Texas Bird Records Committee had the species on its review list. From 1999 to 2005, it was considered a regularly occuring species, then it was added back to the review list and removed again in 2018. Its pelagic habits make it a hard species to detect.

LENGTH: 17 inches. **WINGSPAN:** 36 inches.
ADULT BREEDING: (Apr.–Sept.) White head and small yellow bill. Mantle and upper wings are uniform pale gray with dark wingtips. White below and all-white tail. Legs are black.
NONBREEDING: (Aug.–Mar.) Like breeding with dusky or dark nape; often shows a dark ear patch. **JUVENILE:** (Aug.–Apr.) Dark outer primaries and dark bar on secondary coverts, making a bold M pattern on upper wings. White secondaries and gray back. Dark collar on neck and ear patch. Tail is white with black terminal band. Legs are black. **VOICE:** Generally silent in Texas. **BEHAVIORS:** Highly maneuverable. Grabs small prey from the surface or shallow-dives for them. **HABITAT:** Mostly pelagic away from breeding and does not normally come to land. Juvenile or first-year birds are more likely to come to near-shore waters. **STATUS:** Very rare winter visitor and migrant along the coast and at scattered inland locations. Most records are December–February and are of juvenile or first-winter birds.

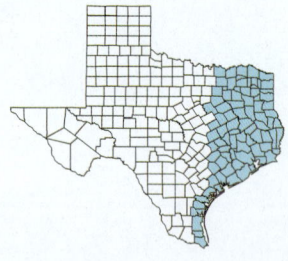

Sabine's Gull

(Xema sabini)

IUCN RED LIST STATUS (2020): Least Concern
POPULATION TREND: Stable

Sabine's Gull is one of the least-common regularly occurring gulls in Texas. Only a small percentage of the population migrates southbound from the arctic directly across North America to winter in the South Atlantic. An even smaller number migrates north in spring along this same route. There are some indications that Sabine's Gulls are high-altitude migrants and pass very stealthily through Texas. **LENGTH:** 13.5 inches. **WINGSPAN:** 33 inches. **ADULT BREEDING:** (Apr.–Sept.) Dark gray hood with black lower border. Small dark bill with yellow tip. In flight, primaries form a long dark triangle, white secondaries form a second triangle, and gray mantle makes a third triangle of color. **NONBREEDING:** (Oct.– Mar.) Like breeding but the hood is reduced to just a partial hood on the rear of the head. **JU-VENILE:** (Aug.–Dec.) Brownish back and neck.

Tail shows a black terminal band. **VOICE:** Mostly silent in Texas. **BEHAVIORS:** Graceful, buoyant, and tern-like in flight. Often hovers over objects of interest with its head pointed to one side or bill pointed down. Rarely if ever soars. **HABITAT:** Winters in offshore waters, but often found on large lakes inland during migration. **STATUS:** Rare to casual fall migrant in all regions of the state. Most records are of juvenile birds. Fall migrants appear in Texas beginning in late August to early December.

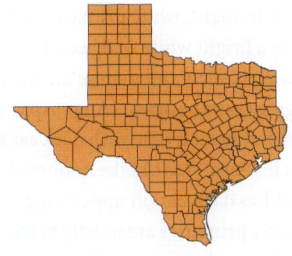

First winter

Nonbreeding

Bonaparte's Gull

(Chroicocephalus philadelphia)

IUCN RED LIST STATUS (2018): Least Concern
POPULATION TREND: Increasing

The delicate Bonaparte's Gull, almost tern-like when in flight, is named for French ornithologist Charles Bonaparte, nephew of Napoléon. It's one of the few gull species that nests in trees. This and its habit of nesting highly dispersed have made it one of the least-studied gulls on the breeding grounds. **LENGTH:** 13.5 inches. **WINGSPAN:** 33 inches. **ADULT BREEDING:** (Apr.–Aug.) Small black bill and black head. Mostly white below with white tail. In flight, white primaries with dark tips make a bright white wedge on the wing. Upper wing coverts and mantle are light gray. Pink legs. **NONBREEDING:** (Aug.–Apr.) Like breeding but head is pale with dark ear spot. **JUVENILE:** (Sept.–Mar.) Like nonbreeding adult but has dark bar on upper wing coverts. Outer primaries are mostly black and, with the bar, make a narrow M pattern on the upper wing. White tail has a black terminal band. **VOICE:** Hoarse *keek keek* or *kew kew kew*. **BEHAVIORS:** Light and buoyant in flight. Swims gracefully and buoyantly. **HABITAT:** In migration, stops on lakes, rivers, marshes, and sewage lagoons. Usually found overwintering on lakes, rivers, coastal bays, and harbors. **STATUS:** Uncommon to common migrant and winter resident along the coast and in large reservoirs. More common in the eastern two-thirds of Texas. Fall migrants arrive in late October and most have left by late March.

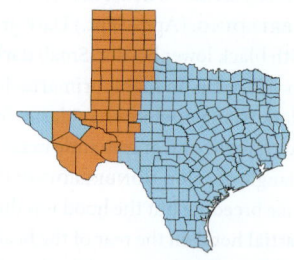

First winter

Little Gull

(Hydrocoloeus minutus)

IUCN RED LIST STATUS (2018): Least Concern
POPULATION TREND: Increasing

The smallest gulls in the world are a bit of a mystery. Have they always been in North America in small numbers, or did they colonize it recently via the Bering Sea or in waves from Europe? There is much to learn. **LENGTH:** 11 inches. **WINGSPAN:** 24 inches. **ADULT BREEDING:** (Apr.–Aug.) Small black bill and hood; the hood is more extensive than on other small gulls. Upper wings are mostly uniform gray with white trailing edge. Tail is white. Underwings are dark gray. **NONBREEDING:** (Aug.–Apr.) Black hood fades and is reduced to a dark ear patch and a darkened rear crown. **JUVENILE:** First-winter birds have extensive dark on the upper wings, making a bold black M pattern. Underwing is not as dark as adults. Tail is white with dark terminal band. Second winter birds are like nonbreeding adults but have a dark bar on the primaries with a white tip. **VOICE:** Call is a soft clack like a grackle or a raspy *tew-tew-tew-tew*. **BEHAVIORS:** Resembles large terns and Bonaparte's Gulls in flight. **HABITAT:** Favors freshwater reservoirs and sewage lagoons. Is often present on beaches and bays in salt water. Often follows barge trains in the Gulf Intracoastal Waterway. **STATUS:** Rare winter resident in north-central Texas, very rare in the rest of the state. Wintering birds have been present mid-November–late March.

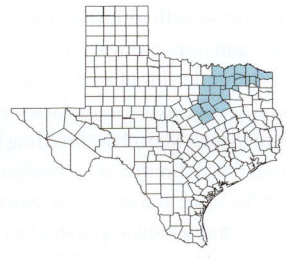

Nonbreeding

Breeding

First winter

Laughing Gull

(Leucophaeus atricilla)

IUCN RED LIST STATUS (2018): Least Concern
POPULATION TREND: Increasing

Laughing Gull is the most abundant gull on the Texas coast, where it's present year-round. It breeds along the Atlantic and Gulf coasts. At least a portion of the Texas breeding birds winter on the Pacific coast of Central America. They are replaced by migratory birds from the Atlantic coast.

LENGTH: 16.5 inches. **WINGSPAN:** 40 inches. **ADULT BREEDING:** (Mar.–Sept.) Dark gray or black hood and reddish bill. Narrow white eye arcs. White below with white tail. Dark gray on the back and upper wings. Dark wingtips including the underside of the primaries. White trailing edge to the secondaries. **NONBREEDING:** (Sept.–Mar.) Like breeding but white-headed with limited gray streaking on back of the head. Bill is dark. **JUVENILE:** (Aug.–Nov.) Dusky brown, gray-backed by winter. Forehead is pale. Base of the tail is white with a wide black band. Underwings are mottled. Second winter is like nonbreeding with more extensive gray on the back of the neck and sides of the breast. **VOICE:** Two-note *kihaw*. Also, a strident *khaw*. **BEHAVIORS:** Graceful and buoyant in flight. Hovers and soars on thermals to feed. Very agile when walking. **HABITAT:** Mostly coastal nests in a variety of habitats, from vegetated islands to sandy beaches. **STATUS:** Abundant on the coast. Uncommon to casual inland mostly in late summer and fall. Has bred on Amistad and Falcon reservoirs.

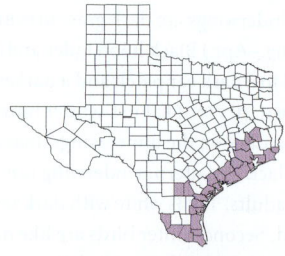

Nonbreeding

Breeding

Franklin's Gull

(Leucophaeus pipixcan)

IUCN RED LIST STATUS (2018): Least Concern
POPULATION TREND: Increasing

Franklin's Gull is named for arctic explorer John Franklin. Initially it was called Franklin's Rosy Gull for the pink wash it gets in breeding season.
LENGTH: 14.5 inches. **WINGSPAN:** 36 inches.
ADULT BREEDING: (Apr.–Aug.) Dark-hooded with broad white eye arcs. Red bill. In fresh plumage, usually has a distinctly pinkish breast and belly. Dark gray upper parts with white trailing edge on the wings. White band below a black wingtip. White tail. **NONBREEDING:** (Aug.–Mar.) Like breeding with a dark bill. The hood is reduced to only the back of the head. **JUVENILE:** (Aug.–Sept.) Brownish above with a dark hood and light forehead. By first winter, the brown has molted to gray. Lacks the white band on the wings of an adult. Tail is white with dark terminal band.
VOICE: High-pitch *ah-ha* like a Laughing Gull.

BEHAVIORS: Flight is light and buoyant. Hovers and soars on thermals, especially when feeding on insects. When feeding in a flock, may leapfrog, where the rear birds fly to the head of the group. **HABITAT:** In migration, roosts on inland lakes and bays and feeds in flooded fields, pastures, and croplands. In winter, they are found primarily along the coast and at landfills. **STATUS:** Uncommon to common migrant statewide. Rare in winter at inland reservoirs and landfills. Spring migration is late February–mid-May, fall mid-September–early December.

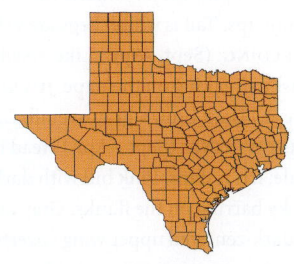

Nonbreeding

First winter

Ring-billed Gull

(Larus delawarensis)

IUCN RED LIST STATUS (2018): Least Concern
POPULATION TREND: Increasing

Ring-billed Gull could be called the French Fry Gull because it's known for opportunistically scavenging food, at times including leftover fries. This is the familiar gull of parking lots and cities away from the coast. **LENGTH:** 17.5 inches. **WINGSPAN:** 48 inches. **ADULT BREEDING:** (Apr.–Sept.) White head, pale iris with red orbital ring. Yellow bill with broad black ring on the end of the bill. Mantle is pale gray. Primaries are black with white tips. Tail is white. Legs are yellow. **NONBREEDING:** (Sept.–Apr.) Like breeding, with dusky streaking on the nape. **JUVENILE:** (July–Sept.) Dark brown-gray overall. Dark bill. **FIRST WINTER:** (Sept.–Apr.) Head is dirty pale, iris is dark. Pink bill with dark tip. Dusky barring on the flanks. Gray mantle and dark-centered upper wing coverts. Legs are pink. **SECOND WINTER:** (Aug.–Apr.)

Like nonbreeding but tip of bill is dark and not banded. **VOICE:** Call is a high-pitched, squealing *keoeee*. **BEHAVIORS:** Flight is light, graceful, and strong. Can hover and soar on thermals. **HABITAT:** Coastal and inland reservoirs. **STATUS:** Common migrant statewide and common winter resident on the coast and at inland reservoirs and landfills. Uncommon to rare summer visitor but not known to breed in Texas. Fall migrants arrive late July–early November. Spring migration is late March–early May.

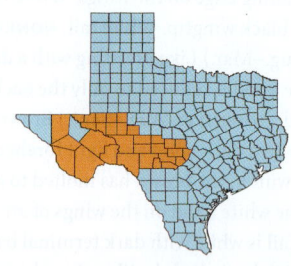

California Gull

(Larus californicus)

IUCN RED LIST STATUS (2018): Least Concern
POPULATION TREND: Decreasing

The California Gull is uncommon in Texas but a familiar inland breeding bird in western North America.
LENGTH: 21 inches. **WINGSPAN:** 54 inches.
ADULT BREEDING: (Apr.–Sept.) White head. Yellow bill with black and red marks at the end. Iris is dark with red orbital ring. Mantle and upper wings are gray. Primaries are black with white tips. Tail is white. Legs are yellow-green. **NONBREEDING:** (Oct.–Apr.) Like breeding but nape has extensive gray streaking. **JUVENILE:** (Aug.–Sept.) Variable from dark to pale, mostly brownish over-all. Primaries are all dark and tail is brown. **FIRST WINTER:** (Sept.–Apr.) Like juveniles, but greater coverts; behind, the secondaries are dark. Face and throat are pale. Bill is long and pink with dark tip. **SECOND WINTER:** (Aug.–Apr.) Mantle is gray, and rump and base of the tail turn white. Legs often bluish. More extensively pale on the face and throat, extending onto the breast. Bill is pale with a dark band at the end. **THIRD WINTER:** (Aug.–Apr.) Like nonbreeding adult but less extensive gray streaking on nape. **VOICE:** Call is high-pitched, raspy, and wheezy *reeeeek*. **HABITAT:** Reservoirs, beaches, and land-fills. **STATUS:** Rare winter resident in El Paso and Hudspeth counties. Very rare to rare migrant and winter visitor to the upper and central coast. Fall migrants begin arriving in late September, wintering birds depart by mid-March.

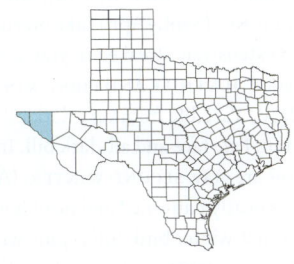

Breeding

Second winter

First winter

Third winter

Herring Gull

(Larus argentatus)

IUCN RED LIST STATUS (2019): Least Concern
POPULATION TREND: Decreasing

The American Ornithological Society regards Herring Gull as a subspecies of European Herring Gull. Many taxonomists regard the group as three species: European Herring Gull, Vega Gull, and Arctic Herring Gull. **LENGTH:** 25 inches. **WINGSPAN:** 58 inches. **ADULT BREEDING:** (Feb.–Sept.) White head and yellow bill with red spot on the end. Iris is pale. Mantle and upper wings are gray. Dark primaries with white tips. Legs are pink. **NONBREEDING:** (Sept.–Apr.) Like breeding but head extensively streaked in gray, almost gray-headed. **JUVENILE AND FIRST WINTER:** (Aug.–Nov.) Variable from very dark brown to light tan. Pale face and all-dark bill. Iris is dark. Legs are gray. **SECOND WINTER:** (Aug.–Apr.) Less neatly patterned and head is whiter than a first-winter bird. Bill is pink with dark tip. **THIRD WINTER:** (Aug.–Apr.) Like

nonbreeding but head streaking more extensive. Bill is yellow with dark tip. **VOICE:** Call is a rapid, high-pitched *que-que-que-que*. Also gives a long trumpeting *craaaaaa*. **BEHAVIORS:** Walks, runs, and glides often in flight. **HABITAT:** Large reservoirs inland, coastal and near-shore water out to the continental shelf. **STATUS:** Common migrant and winter resident on the coast. Rare migrant and winter resident inland, mostly in the northeast half of Texas. Uncommon to rare in summer on the coast and inland reservoirs. Fall migrants arrive early September–November; most leave by early April.

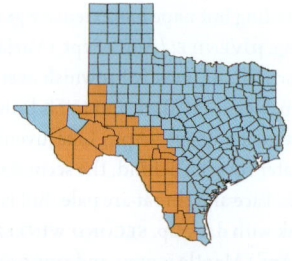

Kumlien's, second winter

Thayer's, juvenile

Iceland Gull

(Larus glaucoides)

IUCN RED LIST STATUS (2019): Least Concern
POPULATION TREND: Stable

The two subspecies of Iceland Gull that occur in Texas—Kumlien's Gull and Thayer's Gull—were considered separate species until 2017.
LENGTH: 22 inches. **WINGSPAN:** 54 inches.
ADULT BREEDING: (Apr.–Sept.) Kumlien's has white head and pale yellow to brownish iris. Small yellow bill with red spot. Mantle and upper wings are pale gray with no darkness on the primaries. White tail. Thayer's has white head and dark iris. Small yellow bill with red spot. Outer webs of primaries are dark. Gray mantle and upper wings. White tail. **NONBREEDING:** (Sept.–Apr.) Kumlien's is like breeding but bill is yellow-green. Fine limited streaking on head and neck. Thayer's is like breeding but bill is yellow-green. Head and neck often have extensive streaking.
JUVENILE: (Aug.–Apr.) Kumlien's has small

dark bill. Whitish to light brown and patterned. Primaries are never darker than the body. Thayer's is smooth gray-brown on the head. Small dark bill. Brownish gray body and wings neatly patterned. Primaries are darker than the body. **SECOND WINTER:** Kumlien's has paler plumage. Bill yellow-based. Thayer's is less neatly marked than juvenile. Bill pale-based. Head is lighter than the body. Primaries usually have pale fringes. **VOICE:** Usually silent in Texas. **HABITAT:** Almost exclusively coastal. **STATUS:** Most records in Texas are of Thayer's type birds. Rare to very rare migrant and winter resident on the coast.

Second winter

Juvenile

Adult nonbreeding

Lesser Black-backed Gull

(Larus fuscus)

IUCN RED LIST STATUS (2019): Least Concern
POPULATION TREND: Increasing

The status of Lesser Black-backed Gull is not yet well known in Texas.
LENGTH: 21 inches. **WINGSPAN:** 54 inches. **ADULT BREEDING:** (Mar.–Aug.) White head with yellow iris and red orbital ring. Bill is yellow with red spot at end. Primaries are dark with white spots at tips. Secondaries and upper wing coverts are dark gray to charcoal-colored. Secondaries have white trailing edge. Tail is white, legs bright yellow. **NONBREEDING:** (Sept.–Mar.) Like breeding but head and nape have dense dark streaking around eye. **JUVENILE AND FIRST WINTER:** (Aug.–Apr.) Pale head with duskiness concentrated around the eye that is reduced by first winter. Bill is black. Dark back contrasts with lighter head. Upper wings and mantle patterned in pale spots. Breast is streaky. Rump and upper tail are pale, and tail has wide dark terminal band. **SECOND WINTER:** (Aug.–Apr.) Like first winter but less head streaking, more contrast with the dark gray to brownish mantle. **VOICE:** Call is a hoarse *cAAW*. Long call is a screeching *AAAIAAA-auk-auk-auk-auk-auk.* **HABITAT:** Mostly coastal. Also, large inland reservoirs and landfills. **STATUS:** Uncommon to locally common winter resident on the upper coast. Rare to uncommon on the central coast and south to Cameron County. Rare but increasing winter migrant and resident inland. Rare summer resident on upper coast. Fall migration is mid-September–mid-October. Spring migration starts March–April.

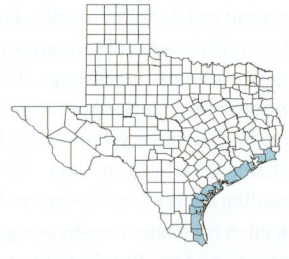

Second winter

Glaucous Gull

(Larus hyperboreus)

IUCN RED LIST STATUS (2018): Least Concern
POPULATION TREND: Stable

Glaucous Gull is the most distinct of the regularly occuring gulls in Texas, standing out for its large size and ghostly pale appearance. Most recorded in Texas have been first- or second-winter birds, which have the most distinct plumage.
LENGTH: 27 inches. **WINGSPAN:** 60 inches.
ADULT BREEDING: (Mar.–Sept.) White head with pale-yellow iris and bright yellow orbital ring. Bill is yellow with a small red spot. Very pale gray on mantle and upper wings, including primaries. Wings have a white trailing edge. Tail is white. Legs are pink. **NONBREEDING:** (Sept.–Apr.) Like breeding with limited streaking on head and nape. **FIRST WINTER:** (Aug.–Apr.) Pale creamy white to light tan with limited markings. No contrast with primaries. Legs are pink. Bill is long and pink-based with dark tip. Iris is dark. **SECOND WINTER:** (Aug.–Apr.) Palest plumage. Markings less distinct or absent. Iris is dark. **THIRD WINTER:** (Aug.–Apr.) Like nonbreeding. Bill is yellow with dark tip. Iris is yellow. **VOICE:** Generally silent in Texas. Call is a strident *clau*. **BEHAVIORS:** Walks and runs with confidence. Flight is powerful. Very good soaring and gliding ability, even into wind. **HABITAT:** Mostly coastal. **STATUS:** Rare winter resident along the coast. Most birds reported in Texas are first- or second-winter birds. Glaucous Gull arrives in Texas in late November; most have departed by early April.

Juvenile

Adult

Sooty Tern

(Onychoprion fuscatus)

IUCN RED LIST STATUS (2020): Least Concern
POPULATION TREND: Unknown

While a rare nesting species in Texas, Sooty Terns nest in some of the largest and densest colonies in the world, at times numbering 1 million birds.
LENGTH: 16 inches. **WINGSPAN:** 32 inches.
ADULT BREEDING: (Jan.–Sept.) Blackish upper parts, including crown. White forehead and black line from the bill through the eye to the black ear patch and crown. Bill is dark. Clean white on neck and breasts. Light gray under tail. Upper wings are all dark; only the trailing edge of the wing is dark under the wings. The long, deeply forked tail has white outer feathers. White bar on shoulder.
NONBREEDING: (Sept.–Jan.) Like breeding but edges of the dark areas are less crisp. **JUVENILE:** (June–Dec.) All dark, almost black with light undertail. Small white speckles on back and upper wings. **VOICE:** Deep *kreet*

and high-pitched, drawn-out *kreeerrrrr* like a Least Tern. **BEHAVIORS:** Walks easily and can run in colonies and on beaches. Powerful flapping flight in strong winds and when displaying. Glides and soars at times. **HABITAT:** Tropical and subtropical waters. Breeds on islands of sand, coral, or rock, including artificial islands. **STATUS:** Rare and local summer resident on the central and lower coast. Small numbers nest in the Laguna Madre. Present offshore in the late summer and early fall. Arrives in Texas in mid-April and leaves by early September.

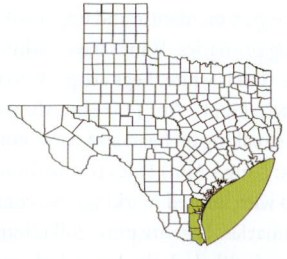

Least Tern

(Sternula antillarum)

IUCN RED LIST STATUS (2018): Least Concern
POPULATION TREND: Decreasing

The Least Tern is listed as a species of least concern by the International Union for Conservation of Nature, but the interior breeding subspecies is listed as endangered by the US Fish and Wildlife Service. **LENGTH:** 9 inches. **WINGSPAN:** 20 inches. **ADULT BREEDING:** (Mar.–Aug.) The only tern with a yellow bill and a dark tip. Dark crown and white forehead, and a dark line from the bill to the eye. Pale gray above with the outer two primaries dark. Tail is pale with a deep notch. Legs are short and yellow. **NONBREEDING:** (Sept.–Mar.). Like breeding adults but bill is dark and light forehead is more diffuse. **JUVENILE:** (July–Sept.) Faintly barred on back and upper wings, with dark primaries making an M pattern with a dark carpal bar. Crown is finely streaked in gray. Dark ear patch. Bill is dark. **FIRST YEAR:** (Sept.–July) Rear of the crown is dark with dark ear patch. White lores and dark bill. Dark carpal bar is visible. Primaries are dark. **VOICE:** Raspy *pree-dity pree-dity*. **BEHAVIORS:** Flight is light and buoyant. **HABITAT:** Nesting colonies are on sand or dried mud that is bare or sparsely vegetated. **STATUS:** Common summer resident along the coast and rare to locally uncommon summer resident at scattered inland locations. Uncommon to rare migrant in the eastern two-thirds of Texas.

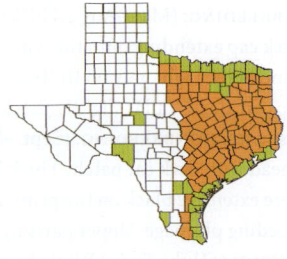

Nonbreeding

Gull-billed Tern

(Gelochelidon nilotica)

IUCN RED LIST STATUS (2019): Least Concern
POPULATION TREND: Decreasing

Gull-billed Tern is never very abundant. It nests with other terns in coastal colonies. It is less dependant on fish than other terns, feeding on insects, small crabs, and other prey it can snatch from the ground, air, and at times even bushes. They are apparently less tolerant of human disturbance than other tern species and will readily use dredge spoil areas to breed.

LENGTH: 14 inches. **WINGSPAN:** 34 inches.
ADULT BREEDING: (Mar.–Aug.) Thick black bill. Black cap extending onto the nape. White breast and belly. Light gray on the back. Primary tips are dark. Tail is slightly notched and mostly square. **NONBREEDING:** (Sept.–Mar.) White head with dark ear patch. Thick black bill. More extensive black on the primaries than breeding plumage. Upper parts light gray. **JUVENILE:** (July–Sept.) White head

and faint ear patch. Dusky on the rear of the crown. Faintly patterned grayish above. **VOICE:** Raspy *AH-rack*. **BEHAVIORS:** Swooping flight when chasing prey is distinctive. Flight less buoyant and powerful than other medium-size terns. **HABITAT:** Nests in sparsely vegetated areas on the coast, like beaches, dried mud, and dredge spoil. Often feeds over marshes. **STATUS:** Uncommon to common resident along the coast, less numerous in winter. Casual away from the coast as far inland as Williamson County.

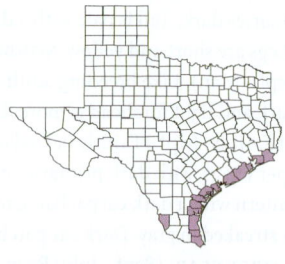

Nonbreeding

Breeding, with juvenile

Juvenile

Caspian Tern

(Hydroprogne caspia)

IUCN RED LIST STATUS (2019): Least Concern
POPULATION TREND: Increasing

Caspian Tern is the largest of the terns and occurs on every continent except Antarctica. In North America this species is increasing, benefitting from species protection, protection of traditional nesting sites, and human alterations that it can take advantage of.
LENGTH: 21 inches. **WINGSPAN:** 50 inches.
ADULT BREEDING: (Feb.–Oct.) The largest tern. Large, heavy, dark red bill with dark tip. Dark cap is smoother, less ragged than Royal Tern. Light gray on mantle and upper wings. Variable dark on primary tips. Tail is white and slightly notched. **NONBREEDING:** (Oct.–Feb.) Like breeding but crown has light streaking. Primary tips darker. **FIRST YEAR:** Like nonbreeding but whiter on the forehead. Primaries are all dark. **JUVENILE:** (July–Oct.) Dark gray crown and red-orange bill. Light gray scapulars have dark edges, making a bold chevron pattern. Dark primaries. **VOICE:** Call is a harsh *ayy-yaaa*. Also, a hoarse *CREP*. **BEHAVIORS:** Flight is graceful and strong, more gull-like than other terns. **HABITAT:** Breeds commonly on low, flat islands with sparse vegetation. **STATUS:** Common resident along the coast. Rare to uncommon spring migrant, and rare fall migrant and winter visitor, to the eastern half of the state. Spring migration is late April–mid-June, fall mid-July–late October.

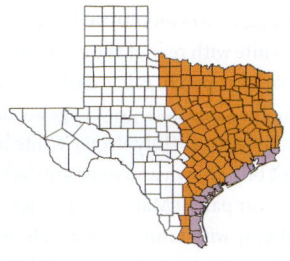

Breeding

Nonbreeding

Black Tern

(Chlidonias niger)

IUCN RED LIST STATUS (2019): Least Concern
POPULATION TREND: Decreasing

In summer breeding plumage, a Black Tern has a black head and body, contrasting with any other terns around. In winter it changes to a mostly white head and white body, looking more like a slighty darker Forster's Tern. **LENGTH:** 9.75 inches. **WINGSPAN:** 24 inches. **ADULT BREEDING:** (Mar.–Aug.) Black head and body. Vent or undertail is white. Wings are dark gray with pale leading edge, visible on a resting bird as a white bar on the shoulder. Legs are black. **NONBREEDING:** (Aug.–Feb.) Head is white with rear of the crown and nape black. Dark ear patch. Bill is dark. Mantle and wings are the same dark gray. Legs are dark reddish. **JUVENILE:** (July–Nov.) White head with dark bill. Crown and nape are dark gray with dark ear patch. Mantle and wings are charcoal-gray with faint barring. Pale below with red legs. **VOICE:** Call is a harsh *keffa*.

BEHAVIORS: Agile, buoyant, and energetic in flight. Roosts in large flocks on sand or mud. Often perches on posts or structures. **HABITAT:** Largely coastal and near shore. In migration can be found on freshwater lakes, rivers, and wetlands. **STATUS:** Uncommon to common migrant statewide. Sometimes abundant along the coast. Uncommon on the coast in summer and rare inland in summer. Does not breed in Texas. Spring migration is mid-March–late May, fall early July–late October.

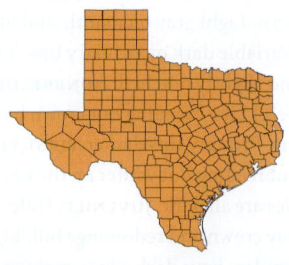

Breeding

Note uniform gray across the back and wings in all plumage.

Nonbreeding

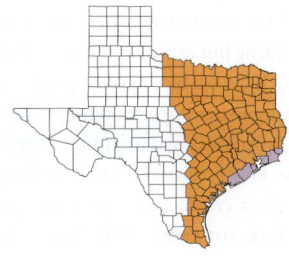

Common Tern

(Sterna hirundo)

IUCN RED LIST STATUS (2019): Least Concern
POPULATION TREND: Unknown

Widespread and familiar in North America, the Common Tern is less common in Texas than the very similar Forster's Tern. **LENGTH:** 12 inches. **WINGSPAN:** 30 inches. **ADULT BREEDING:** (Apr.–Nov.) Black cap and nape. Bill is orange-red with dark tip. Upper wings and mantle are uniform light gray with no contrast in the primaries. Body is the same light gray, and vent and forked tail are white. Outer tail feathers are dark compared to the white outer tail feathers of a Forster's Tern. The tips of the underside of the primaries are dark. **NONBREEDING:** (Oct.–Mar.) Forehead is white, bill is dark. Dark carpal-bar on the shoulder. **FIRST YEAR:** Like nonbreeding but primaries are much darker, almost blackish. **JUVENILE:** (July–Aug.) Dark cap is brownish and incomplete. Mantle and upper wings are brownish with some dark and light patterns.

Bill and feet are orange. **VOICE:** Call is a high, slightly raspy *keeeyurr*. **BEHAVIORS:** Flight is buoyant. The downbeat of the wing often causes the bird to rise a little. **HABITAT:** Primarily coastal, rare on large reservoirs in migration. **STATUS:** Common fall and uncommon spring migrant on the coast and offshore. Rare migrant in the eastern half of the state. Casual winter visitor on the coast. Spring migration is early March–early June, fall mid-August–early November.

Nonbreeding

Breeding

Nonbreeding. Note pale wings compared to the back.

Forster's Tern

(Sterna forsteri)

IUCN RED LIST STATUS (2019): Least Concern
POPULATION TREND: Least Concern

Forster's Tern is the only tern that is exclusive to North America. It breeds both on the Gulf Coast and the northern Great Plains. **LENGTH:** 13 inches. **WINGSPAN:** 31 inches. **ADULT BREEDING:** (Mar.–Aug.) Black cap and orange bill with dark tip. Bill and legs are less red than in Common Tern. Body is white. Mantle and upper wings are pale gray. Primaries are lighter and contrast with upper wings. Deeply forked tail is pale gray with white outer tail feathers. **NONBREEDING:** (Aug.–Feb.) Like breeding but crown and nape are white and there is a dark face mask, a combination that distinguishes it from Common Tern. Bill is dark. Primaries are darker than breeding plumage. **FIRST YEAR:** Like nonbreeding but some dark on crown and nape. Outer primaries are dark. **JUVENILE:** (July–Dec.) Like first year but brownish above; that wears off by

August. Bill and legs are orange, but bill turns dark starting in August. **VOICE:** Raspy *KERR* or repeated *ker-ker-ker-ker*. **BEHAVIORS:** Flight is agile and graceful. Dives for prey from hovering over water. In courtship, parades with a fish, walking around the female before feeding her. **HABITAT:** Can use most wetland types in Texas. **STATUS:** Common coastal resident year-round. Common to uncommon migrant statewide. Winter resident south of the High Plains. Spring migration is early March–mid-May, fall late July–October.

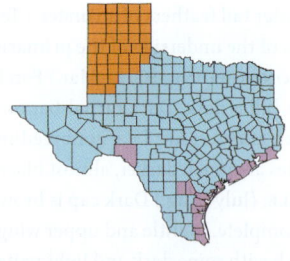

Breeding

Nonbreeding

Juvenile

Royal Tern

(Thalasseus maximus)

IUCN RED LIST STATUS (2018): Least Concern
POPULATION TREND: Stable

Royal Tern is a conspicuous and abundant resident in the warm waters of the Americas. It appears that none other than John James Audubon confused the Royal and Caspian Tern. In his opus work *The Birds of America*, he combined the two species into one called Cayanne Tern that was mostly a Royal Tern but with some Caspian Tern mixed in. LENGTH: 20 inches. WINGSPAN: 41 inches. ADULT BREEDING: (Mar.–June) Heavy bright orange to orange-red bill, never as red as in Caspian Tern. Dark cap, more ragged than a Caspian Tern. Upper wings and mantle are light gray. Tail is white. Legs are dark. NON-BREEDING: (June–Mar.) Like breeding but forehead is mostly white. Primaries are mostly dark. FIRST YEAR: Like nonbreeding. Primaries black. Upper wings are pale, almost white, with gray secondaries. JUVENILE: (July–Nov.)

Like first year. Legs and bill are yellow. Crown is rounded and not ragged. Dark gray primaries, and the leading and trailing edge of the wings are dark. VOICE: Call is a rolling *koor-rrrick*. BEHAVIORS: Strong flier; can usually outfly pursuers trying to steal food, including other Royal Terns. HABITAT: Entirely coastal, rarely away from the immediate coast. STATUS: Common resident on the coast. Accidental inland migrant, usually associated with the passage of a tropical cyclone.

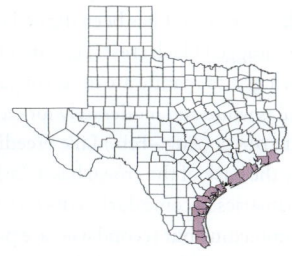

Breeding

Nonbreeding

Sandwich Tern

(Thalasseus sandvicensis)

IUCN RED LIST STATUS (2019): Least Concern
POPULATION TREND: Stable

Taxonomists have not reached consensus on what to do with Sandwich Tern. Currently there are three subspecies, two in the New World and one in the Old World. Some taxonomists believe the two New World subspecies should be elevated to full species, and the subspecies in North America would become Cabot's Tern.

LENGTH: 15 inches. **WINGSPAN:** 34 inches.
ADULT BREEDING: (Mar.–Aug.) Black bill with yellow tip, as if it has been dipped in mustard. Ragged black crest. Legs are black. Upper wings and mantle are uniform pale gray. Tail is white and moderately forked. **NONBREEDING:** (Aug.–Feb.) Like breeding but only the back of the crown has a dark crest. Primaries become dark. **FIRST YEAR:** Like nonbreeding but secondaries are pale gray. **JUVENILE:** (July–Sept.) Like first year but bill is initially orange. The centers of the upper wing and mantle feathers are darker, giving birds a light pattern. **VOICE:** Call is a high-pitched *kiiidiip.* **BEHAVIORS:** Flight is powerful and not delicate. Unpaired males in breeding season will approach a loafing flock with a fish to advertise to females their availability. **HABITAT:** Entirely coastal, nests on low, sandy islands near shore. **STATUS:** Common summer resident and rare to uncommon winter resident on the coast. Inland records are associated with the passage of a tropical cyclone.

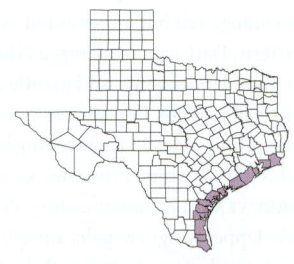

"Skimming" with lower mandible in the water

Black Skimmer

(Rynchops niger)

IUCN RED LIST STATUS (2016): Least Concern
POPULATION TREND: Decreasing

Black Skimmer is a fascinating species. The three species of skimmers worldwide are unique in that the lower mandible of the bill is longer than the upper mandible. The bill is also compressed and very knifelike. Both adaptations are for its unique feeding style—it skims along the surface with its lower mandible in the water. When it feels the touch of a fish, it snaps its bill shut.
LENGTH: 18 inches. **WINGSPAN:** 44 inches.
ADULT BREEDING: (Mar.–Sept.) Lower mandible is longer than the upper. It is bicolored, orange at the base and black at the tip, laterally compressed and knifelike. Black above and white below. Legs are orange. Secondaries are white-tipped. White tail has black central line.
NONBREEDING: (Sept.–Mar.) Like breeding but develops a white collar. **FIRST YEAR:** Like nonbreeding but the wing is gray. **JUVENILE:**

Like first year but boldly patterned above in black and white. Crown is finely streaked in black. **VOICE:** Call is a soft barking *yep*. **BEHAVIORS:** Feeds by flying over calm water with its lower knifelike mandible in the water. Often sleeps laying prone on the sand with its head flat on the ground. **HABITAT:** Exclusively coastal in open sandy or gravel areas with sparse vegetation. **STATUS:** Locally common resident on the coast. Very rare inland in late summer and early fall.

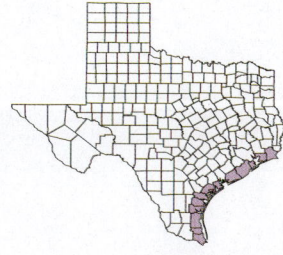

GAVIIDAE (LOONS)

The loon family is a small one consisting of just five species, four of which have been recorded regularly occurring in Texas and one, the Yellow-billed Loon, a rare visitor. The lineage of loons is an ancient one, with fossils of loon-like birds dating back to the late Cretaceous period 70 million years ago, contemporaneous with the dinosaurs. The family survived the great extinction event at the end of the Cretaceous period.

Loons have bodies highly evolved for swimming and diving. Their feet are way back on their torpedo-like bodies, ideally placed for swimming but making it impossible for them to move overland. They only leave the water to breed or nest, and never venture inland more than a body's length so they can slide back into the water on their bellies. Loons are stong fliers, but partially because they lack the hollow air-filled bones of other birds they must run across the water to get their heavier bodies airborne.

Loons were long considered the most primative of birds and so placed first in most taxonomies. Due to similarities, they were also once considered closely related to grebes. More recent taxonimies place the loons further up the evolutionary ladder and their closest relatives are now considered to be tropicbirds and penguins. While similar to grebes, the similarities are probably because of convergent evolution.

Winter plumage

Red-throated Loon

(Gavia stellata)

IUCN RED LIST STATUS (2018): Least Concern
POPULATION TREND: Decreasing

Red-throated Loon has historically been rare in Texas. Starting in the 1990 records, it increased in frequency and the Texas Bird Records Committee took the species off its review list in 2002. While this suggests it is becoming more frequent in the state, it is possible that greater awareness of the Red-throated Loon and the skill of Texas birders accounts for the increase in reports.
LENGTH: 25 inches. **WINGSPAN:** 36 inches.
ADULT BREEDING: (Apr.–Nov.) Rounded head and thin bill. Head and neck are gray with red throat patch. Back is dark. **NON-BREEDING:** (Oct.–Apr.) Dark crown and nape. Face and neck are white. Back is dark with white speckling. Sides are light. In flight, the white face and neck are diagnostic. **JUVENILE:** (Aug.–Jan.) Gray head and neck with white cheek. Dark on the back and never shows any

barring. **VOICE:** Generally silent in Texas. **BEHAVIORS:** Wing beats faster and deeper than other loons. Flies with head drooping more than other loons. **HABITAT:** Coastal near shore and freshwater lakes of all sizes. **STATUS:** Rare winter resident along the upper Texas coast and on reservoirs in the northeastern part of the state. Records exist from late October to early May.

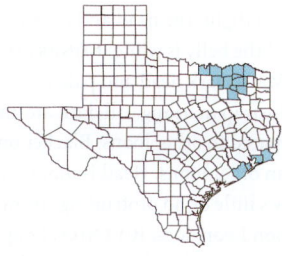

Winter plumage

Pacific Loon

(Gavia pacifica)

IUCN RED LIST STATUS (2018): Least Concern
POPULATION TREND: Increasing

Pacific Loon is likely the most abundant loon in North America, even though its breeding range is only about a third the size of the Common Loon. Not at all common in Texas, it is known for being an abundant and conspicuous migrant on the Pacific coast.
LENGTH: 25 inches. **WINGSPAN:** 36 inches.
ADULT BREEDING: (Feb.–Oct.) Small dark bill and dark face. Nape is pale gray and visible at a distance. Back is dark with small white patches. In flight, the head and neck look dark, and the belly is white. **NONBREEDING:** (Sept.–Mar.) Dark head and pale throat with dark chinstrap marking. All dark around the eye compared to other loons. Blacker on the back than other loons. Head is more rounded and shows little or no protruding "brow" like a Common Loon does. **JUVENILE:** (Aug.–Jan.) Like nonbreeding but face pattern is less

distinct. **VOICE:** Generally silent in Texas.
BEHAVIORS: Has to run on the surface of the water to take off; once in the air flies fast and direct. On the water, stretches its neck up fully erect then jumps upward before diving, compared to Common Loon, which simply slips underwater. **HABITAT:** Large lakes and coastal waters. **STATUS:** Rare winter resident along the upper Texas coast and on reservoirs in northeastern Texas. Fall birds arrive in Texas late October–early April.

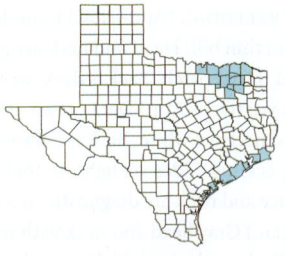

Nonbreeding

Common Loon

(Gavia immer)

IUCN RED LIST STATUS (2018): Least Concern
POPULATION TREND: Stable

Common Loon is the iconic symbol of the lake country of North America. It breeds in the lakes of the boreal forests of Canada and a small part of the United States. They are a long-lived species and have been recorded living 35 years. They typically only lay two eggs a year. Common Loons do not mate for life as is commonly quoted, but are serially monogamous—loyal to their mate until migration, then going their separate ways. **LENGTH:** 32 inches. **WINGSPAN:** 46 inches. **ADULT BREEDING:** (Mar.–Oct.) Dark head and heavy bill. Head has a prominent "brow." Extensive white checkering on back. White, vertically streaked collar. **NONBREEDING:** (Sept.–Mar.) Dark head with white throat and neck. Variable amounts of dark around the eye, though never completely dark. Back is dark. **JUVENILE:** (Aug.–Feb.) Like nonbreeding adults with partial neck collar. **VOICE:** Usually silent in Texas. **BEHAVIORS:** Usually takes off into the wind with a long run-up on the water. Flight is swift and straight. Dives by slipping smoothly into the water. **HABITAT:** Coastal bays and large freshwater lakes. **STATUS:** Uncommon to rare migrant throughout the state. Common winter resident on lakes and bays in the eastern part of the state. Uncommon in the west. Fall migrants arrive in late September. Spring migrants leave late March–early April.

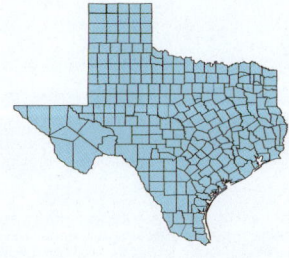

CICONIIDAE
(STORKS)

Storks are large, long-legged wading birds similar to herons and ibises. The Wood Stork was even formerly known as the Wood Ibis. They differ from herons and ibises by having a bare head and neck and no ornamental plumes. Storks often soar with neck and legs extended, while herons typically coil their necks and never soar. Early DNA studies placed the storks as being related to New World vultures. More recent studies with larger sample sizes show they are more closely related to frigatebirds, shearwaters, and petrels.

Most storks live in wetlands and hunt by feel for just about anything that swims in the water. They aren't opposed to scavaging dead fish and carrion. A typical feeding strategy is to stand with the bill open in the water and snap it shut when something is detected by touch.

There are eighteen species of storks worldwide. Two species have been recorded in Texas. One, the enourmous Jabiru, is one of the largest species of birds in the Americas, standing five feet tall with a wingspan of nine feet. Jabirus are very rare visitors to Texas.

Wood Stork

(Mycteria americana)

IUCN RED LIST STATUS (2016): Least Concern
POPULATION TREND: Decreasing

Wood Stork is the only stork that breeds in North America, mostly in Florida. The birds found in Texas are not from the Florida population but are presumed to be post-breeding visitors from Mexican colonies. **LENGTH:** 40 inches. **WINGSPAN:** 61 inches. **ADULT:** Head and neck are bare, dark gray with light crown. Bill is heavy, long, and down-curved at the tip. Body is white. Wings are white with black primaries and secondaries. Tail is short and black. Long legs are dark gray with yellow to pinkish feet. **SECOND YEAR:** Like adult but the neck up to the nape is feathered in white. Bill is lighter, a dirty yellowish. **JUVENILE:** Like immature but neck and head feature tan feathers. Bill is pale to yellowish. Legs are light gray to pinkish. **VOICE:** Silent except for bill-clattering and hissing. **BEHAVIORS:** Walks slowly. When feeding, head is held in the water with the bill open. Roosts in trees. Flies both by soaring and flapping. **HABITAT:** Natural and artificial wetlands where prey is available and water depth is 12 inches or less. At times in brackish water. **STATUS:** Uncommon to locally common post-breeding visitor to the coast and inland in the eastern third of Texas. During fall migration, groups of multiple thousands recorded at hawk watches on the central and upper coast. Historically bred in Jefferson, Chambers, and Harris counties.

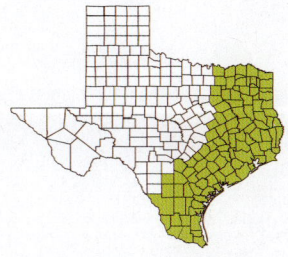

FREGATIDAE (FRIGATEBIRDS)

The frigatebirds are a small family of seabirds numbering only five species. They are mostly black with varying amounts of white on the body, mostly below. These swift masters of the air are kleptoparasites, meaning they steal a large portion of their food from other species. Sailors named these speedy raiders after a swift, lethal ship, calling them "man-o'-war birds."

Frigatebirds have a wingspan almost equal to that of a Bald Eagle, but they are one of the lightest birds for their size. They have an exceptionally light skeleton—so light that the bones of a frigatebird weigh about half as much as its feathers. Frigatebirds have poorly waterproofed feathers and can't rest on the water without becoming waterlogged and unable to fly. They spend their time in the air and only perch in trees when they come to land to roost or breed.

Male

Immature

Magnificent Frigatebird

(Fregata magnificens)

IUCN RED LIST STATUS (2020): Least Concern
POPULATION TREND: Decreasing

Magnificent Frigatebirds are striking prehistoric-looking birds. Their long wings and hooked beak make them seem almost reptilian as they soar overhead with little apparent effort. While known for stealing food on the wing from other birds, they are actually adept at catching their own prey.
LENGTH: 40 inches. **WINGSPAN:** 90 inches. **ADULT:** Male is all black with long, gray, hooked bill. Very long wings. Tail is long and deeply forked. Red throat pouch is not always visible unless inflated. Female is like adult male but white on breast. Upper wings have white pale stripe in the upper coverts. **IMMATURE:** White head and underside. Light white stripes on upper wing coverts. **VOICE:** Mostly silent away from breeding colonies. **BEHAVIORS:** Not able to walk but does perch on almost any structure with good visibility and that will aid takeoff. Almost never flies by flapping; once able to glide, can glide for extended periods of time. **HABITAT:** Warm coastal and offshore waters. **STATUS:** Uncommon summer and fall visitor along the coast. Can be locally common in the Gulf of Mexico and large bays. Birds in Texas are assumed to be post-breeding birds from colonies in Mexico. They are rare in Texas from late March until mid-April, when they increase in numbers. Most leave by the end of October, some linger into December.

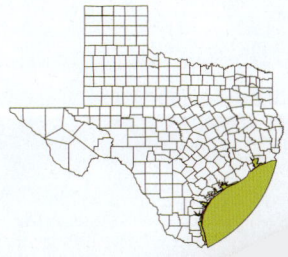

SULIDAE
(BOOBIES AND GANNETS)

Boobies and gannets are large seabirds with heavy pointed bills, long pointed wings, and a tapered tail that gives them a distinct profile in the air. They are well-known for their spectacular dives when feeding. They will often dive from great heights, over 100 feet in many cases, to shoals of fish or squid. In the dive they fold their wings back against their body and enter the water like a dart. Diving from great heights allows them to pursue fish other birds can't, as deep as 30 feet.

Boobies and gannets are pelagic, spending most of their lives on the open ocean. They nest mostly on isolated islands away from mammal predators. There are ten species worldwide. Five have been recorded in Texas. Two, Blue-footed and Red-footed Booby, are rare visitors to Texas.

Masked Booby

(Sula dactylatra)

IUCN RED LIST STATUS (2018): Least Concern
POPULATION TREND: Decreasing

The Masked Booby breeds throughout the tropical oceans and is locally abundant near hundreds of oceanic islands. They are exceptionally vulnerable to human development and to some introduced predators. In the eastern Pacific, the introduction of pigs to Clipperton Island caused nearly complete abandonment of a once-immense colony. Since the removal of the pigs, the island colony has recovered. Fortunately many colonies are small, remote, and protected.

In 2000 the eastern Pacific subspecies was recognized as a separate species, Nazca Booby, *Sula grant*. Much of the literature on Masked Boobies prior to the split refers to the Nazca Booby.

LENGTH: 32 inches. **WINGSPAN:** 62 inches.
ADULT: Mostly white with heavy conical, yellow bill. Face is black extending behind the eye. Primaries and the trailing edge of the wings are black. Tail is black. **FIRST YEAR:** Like adults but head is dark brown. Bill is pale grayish. Back is brown. **VOICE:** Generally silent away from breeding colonies. **BEHAVIORS:** Maintains altitude when flying. When possible, groups make up diagonal line formations in flight. Plunge-dives for fish from heights up to 100 feet. **HABITAT:** Primarily found in deep-blue water area of the ocean. **STATUS:** Uncommon to rare migrant and summer visitor offshore and rare to very rare on the coast. Birds found on the beaches are usually injured or sick. Most birds are found late March–early October.

Brown Booby

(Sula leucogaster)

IUCN RED LIST STATUS (2018): Least Concern
POPULATION TREND: Decreasing

The first record for Brown Booby in Texas wasn't until August 1967. The number of annual records steadily increased until it was removed from the Texas Bird Records Committee review list in 2017. In 2022 alone there were more than twenty records in the state.

While increasing dramatically in Texas, extensive human disturbance and introduced predators have dramtically reduced the worldwide population. Some estimates place the population at only 10% of its historic numbers. **LENGTH:** 30 inches. **WINGSPAN:** 57 inches. **ADULT:** Dark brown overall but clean white below, from mid-breast to vent. The demarcation between brown and white on the chest is always a clean straight line. Feet and bill are yellow. Males have blue-gray face mask. **FIRST YEAR:** Brown overall. Belly is paler, with a clean line separating the darker brown

of the upper breast and head from the lighter belly. Feet are dull yellow. Bill is gray. **VOICE:** Generally silent away from breeding colonies. **BEHAVIORS:** Flies with strong flapping interspersed with gliding. Plunge-dives for prey, often at very shallow angles. Good swimmer and well waterproofed, able to spend long hours on the water. **HABITAT:** Mainly coastal and pelagic waters. **STATUS:** Rare to uncommon visitor to the coast. Rare to very rare visitor to large inland reservoirs. There are records from almost every month of the year, and birds often linger for many months.

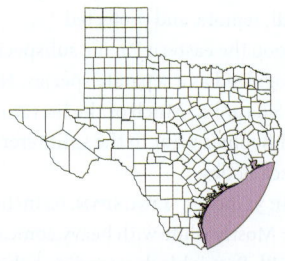

Adult

Immature

Northern Gannet

(Morus bassanus)

IUCN RED LIST STATUS (2018): Least Concern
POPULATION TREND: Increasing

The Northern Gannet is the largest member of the booby family. In size, it falls between Brown Pelican and Double-crested Cormorant. It is known for breeding in large colonies of sometimes thousands of birds, though there are only six breeding colonies in North America.

LENGTH: 37 inches. **WINGSPAN:** 72 inches.
ADULT: Mostly white with pale-yellow head. Primaries are black. Heavy conical bill is light gray. Feet are gray. **THIRD YEAR:** Like adults but back is mottled with dark. Secondaries are dark. Tail is black. **SECOND YEAR:** Head is pale. Mostly dark above. Tail is dark with white upper tail coverts. **FIRST YEAR:** Mostly dark gray with white upper tail coverts. **VOICE:** Generally silent away from breeding colonies. **BEHAVIORS:** Flies strong and direct. Dives from heights with little splash. Often feeds in groups. **HABITAT:** Remains mostly offshore. **STATUS:** Uncommon to common migrant and winter visitor offshore in the Gulf of Mexico. Most records in Texas are nonadults. No records inland from the coast. Most birds are found between early November and early May.

ANHINGIDAE (ANHINGAS)

The anhinga family is made up of four species. One for the Americas and one each for Asia, Africa, and Australia. These species are very similar and some have suggested they are one worldwide species. Anhingas are closely related to cormorants and share similar habitats. They have specially shaped vertebrae that allow them to coil their necks back in an S-shape and spear fish with their dagger-like bill.

Both cormorants and anhingas perch with their wings spread when they emerge from the water. While cormorants need to dry their poorly waterproofed feathers, anhingas do this for thermoregulation. Anhingas have much less body insulation and very slow metabolisms, so time in the sun helps them bring their body temperature back up.

The microstructure of anhinga feathers lets water into the feather structure and soaks the feathers quickly. They can control their buoyancy, allowing them to swim with just their head above water and then submerge without diving.

Migrating Anhingas

Anhinga

(Anhinga anhinga)

IUCN RED LIST STATUS (2018): Least Concern
POPULATION TREND: Decreasing.

Anhinga is one of Texas's most distinct birds, with its snakelike neck, spear-like bill, and turkey-like tail. Its wetable plumage allows it to be less bouyant underwater and stalk and hunt fish to spear with its bill, but also results in a great loss of body heat. This is why the Anhinga spends so much time sunning. **LENGTH:** 35 inches. **WINGSPAN:** 45 inches. **ADULT:** Male is black with straight, yellow, dagger-like bill. Eye is red. In breeding season, the bare skin around the eye is blue. Silvery-streaked mantle and upper wings. Tan terminal band on the end of a long tail. Female is like males but head and long neck are tan. **JUVENILE:** Like adult female but little or no silvery streaking on back and upper wings. **VOICE:** Series of descending croaks: *krr-kr-krr-kr-krrr*. **BEHAVIORS:** Typically perches over water. Perches with wings spread to dry and thermoregulate. Often swims with just the head above water. Flight is strong, flapping alternating with gliding. In migration moves in large flocks soaring on thermals. **HABITAT:** Shallow, slow-moving fresh water with available perches and banks for drying and sunning. **STATUS:** Uncommon to locally common summer resident from South Texas through the Pineywoods. Very rare to rare winter resident on the coast and in the Lower Rio Grande Valley. Spring migrants arrive as early as February; most have left by the end of October.

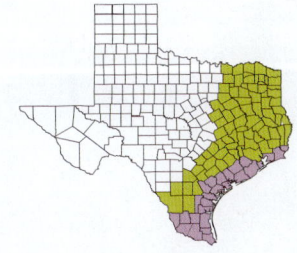

PHALACROCORACIDAE (CORMORANTS AND SHAGS)

Cormorants and shags are large waterbirds with webbed feet, long necks, and long hooked bills. They use the bill to catch aquatic prey, usually fish underwater. There are many Asian paintings of domesticated cormorants being used to catch fish, prevented from swallowing the fish by a metal ring around their neck. Most cormorants have blue eyes but they focus poorly underwater; scientists speculate that this is why they often catch larger fish.

Cormorants and shags are foot-propelled in water. Their outer feathers are not waterproof. While this seems like an odd adaptation for a waterbird, it makes them less bouyant and better able to chase fish underwater. After they emerge they often perch with wings spread to dry off.

There are forty species of cormorants and shags worldwide, inhabiting both fresh- and saltwater environments. Two species occur in Texas. At least fourteen species of cormorants are in decline. One, the Spectacled Cormorant of the Bering Sea, went extinct shortly after it was described in 1741.

Double-crested Cormorant

(Nannopterum auritum)

IUCN RED LIST STATUS (2018): Least Concern
POPULATION TREND: Increasing.

Double-crested Cormorant is the most abundant cormorant species in North America. They have been blamed for impacts on game and commercial fish species, and because cormorant populations are growing, the blame has been increasing. In nature, Double-crested Cormorants have not been shown to take game fish in large numbers, though in fish farms they do take enough to have a significant impact.

LENGTH: 33 inches. **WINGSPAN:** 52 inches.
ADULT BREEDING: (Mar.–May) All-black glossy-green overall. Glossy black margins can give a scaly appearance. Yellow facial skin extends to above the eye. Plumes on either side of the head vary from black in the east to almost white in the west. **NONBREEDING:** (June–Feb.) Like breeding but lacks plumes on sides of the head. **FIRST YEAR:** Dark gray with a pale neck and breast. **VOICE:** Usually silent away from the nest. Bullfrog-like grunt: *yaaa yaa ya.* **BEHAVIORS:** Flies quickly on short wings. Swims for long distances when fishing but leaves the water when done fishing. **HABITAT:** Breeds on ponds, lakes, slow-moving rivers, and open coastlines. Winters in large numbers on the coast and in large reservoirs. **STATUS:** Uncommon to abundant migrant and winter resident along the coast and uncommon to locally abundant inland winter resident. Begins arriving in the state in early September and begins to leave in late March, with most gone by early May.

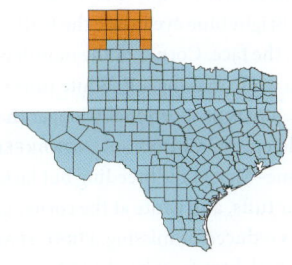

Neotropic Cormorant

(Nannopterum brasilianum)

IUCN RED LIST STATUS (2018): Least Concern
POPULATION TREND Increasing

Distinguishing Neotropic Cormorant from Double-crested Cormorant is one of the great challenges for Texas birders. On a perched bird one can usually see that the bare facial skin on a Double-crested Cormorant extends above the eye, while on a Neotropic Cormorant the eye is always in the dark feathers on the face.

LENGTH: 33 inches. **WINGSPAN:** 52 inches.
ADULT BREEDING: (Apr.–May) Glossy black overall. Bright blue eye inside the feathering on the face. Corner of the mouth is edged in white making a V. White tufts over the ears. Tail is proportionally longer than a Double-crested Cormorant. **NONBREEDING:** (June–Mar.) Like breeding but lacks the white ear tufts, and white at the corner of the mouth is reduced or missing. **FIRST YEAR:** Usually dark brown on head and breast. Small amount of white at corner of mouth. **VOICE:** Mostly silent away from breeding colonies. Low frog-like grunts and croaks. **BEHAVIORS:** Perches often in trees and on structures. Flight is swift and often low over the water. Flocks often fly in V formation. **HABITAT:** Coastal marshes and swamps, inland reservoirs, and lakes. Needs water deep enough for diving and elevated perches. **STATUS:** Uncommon to common resident throughout the coastal prairies and south to the Lower Rio Grande Valley. In summer, large numbers can be found on inland reservoirs in the eastern half of Texas.

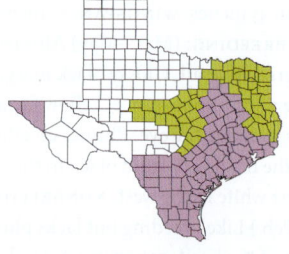

PELECANIDAE (PELICANS)

"A wonderful bird is the pelican, his bill will hold more than his belly can," remarked the poet Dixon Merritt. This is not just poetic license. Depending on the species, the pelican's distinctive pouch can hold up to three times the volume of its stomach. It was once erroneously thought that pelicans used this enormous pouch to store fish for transport back to the nest. Instead it is used as a dip net to capture fish; the water is squeezed out and the fish is swallowed immediately. Young pelicans are then fed a soup of half-digested fish.

A pelican's pouch is a remarkable adaptation. It consists of the two bones of the lower jaw fused at the tip, and the bones are very flexible. When the bill is dipped into the water they flex and form a hoop and the flexible pouch expands. Some species can hold up to 3 gallons of water in the pouch. The pouch is actually very sensitive and some species of pelicans can feed at night by feel. In order to keep the pouch flexible, most pelicans do stretching exersizes.

There are eight species of pelican worldwide and two occur in Texas. Two species are brown in color and are mostly marine species and live in coastal areas. The other species are mostly white with black accents and breed in inland bodies of water. The two brown-colored species forage by diving from great heights after fish, while the white-colored species paddle and dip-net fish from the water, often cooperatively feeding in groups.

American White Pelican

(Pelecanus erythrorhynchos)

IUCN RED LIST STATUS (2016): Least Concern
POPULATION TREND: Increasing

The American White Pelican is unmistakable, large and white with an enormous bill and pouch. There are two breeding colonies on the central Texas coast, but most winter in Texas and breed in the northwestern parts of North America. In migration, huge flocks can be seem migrating on thermals.
LENGTH: 62 inches. **WINGSPAN:** 108 inches.
ADULT BREEDING: (Feb.–June) Large white bird with large bill and pouch. Laterally flattened horn or ridge on the bill. In flight, dark primaries and outer secondaries. **NON-BREEDING:** (June–Feb.) Like breeding but dusky-colored on head and the bill ridge has been lost. **JUVENILE:** (July–Mar.) Like non-breeding but dusky white. **VOICE:** Silent away from breeding grounds. **BEHAVIORS:** Flies and soars well. Migrating flocks can number in the thousands, soaring on thermals. Does not dive. Seines small prey with the bill, often in small groups of other pelicans. **HABITAT:** Shallow coastal bays, inlets, and estuaries with fish and loafing sites. Inland below dams. **STATUS:** Uncommon to common migrant in the eastern half of Texas. Less common in the west. Common winter resident in the southern half of Texas. Locally common in the northern part of Texas on reservoirs and in the eastern Trans-Pecos. Uncommon summer resident at numerous locations statewide. Known to breed at only two sites in Texas, in Nueces and Kleberg counties.

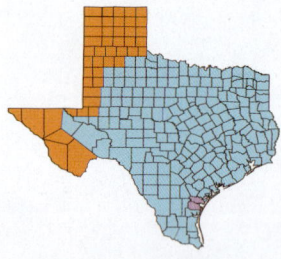

Brown Pelican

(Pelecanus occidentalis)

IUCN RED LIST STATUS (2018): Least Concern
POPULATION TREND: Increasing

Brown Pelicans were nearly extirpated from Texas by pesticides, DDT in particular. The population in Texas dropped to 12 to 15 individuals by the 1970s. Bans on some chemicals and controls on others started to have positive effects, and today's levels match historic numbers.

LENGTH: 51 inches. **WINGSPAN:** 79 inches.
ADULT BREEDING: (Dec.–Aug.) Head is creamy-yellow-colored. Large bill is grayish with a dark pouch. Nape is brown. Dark brown below. Gray to silvery-gray above.
NONBREEDING: (Aug.–Jan.) Like breeding but head and nape are whitish. **FIRST YEAR:** Head, bill, and neck are a dark gray-brown. Light below. Wings and mantle are dark gray-brown.
VOICE: Generally silent. **BEHAVIORS:** Glides low over the waves, barely above the water. Makes spectacular plunging dives from 100 feet at times. **HABITAT:** Mainly the Gulf Coast near shore. Usually within 15 miles of breeding sites in breeding season and up to 45 miles off shore post breeding. Occasional in large bodies of water post-breeding. Offshore distance is limited by the need to return to land to roost. Breeds on barrier islands, natural estuary islands, and dredge spoil islands that are predator-free. **STATUS:** Common to uncommon resident along the Texas coast. Does not breed on the lower Texas coast. In late summer and fall, accidental inland up to 150 miles from the coast.

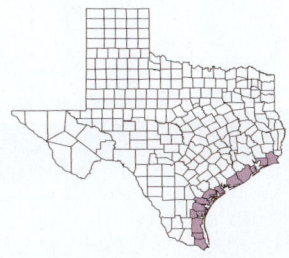

ARDEIDAE (HERONS, EGRETS, AND BITTERNS)

Herons, Egrets, and Bitterns are medium-size wading birds with long legs, long necks and bills, and long broad wings. They range in size from the tiny Least Bittern at 13 inches to the African Goliath Heron that can stand up to 5 feet high. They vary in color from pure white to dark blue, brown, gray, and even black. All members of the family have rump patches and powder-down breasts, on which the down disintegrates to a powder that they use to preen their feathers with a serrated or pectinate toe.

Most species of herons and egrets live near or in wetlands. Most are colonial nesters and form large, often mixed-species colonies, favoring isolated islands as protection from mammal predators. During the breeding season, many species acquire ornate nuptial plumes. These plumes became the target of commerical hunters for the millinery trade. The plumes at one time sold at a price per ounce twice that of gold, and millions of birds were killed just for the couple of feathers that had commerical value. This wanton slaughter led to the founding of the Massachusetts and Pennsylvania Audubon Societies. Other like-minded societies were soon founded and the network eventually became the National Audubon Society in North America.

There are seventy-one species of herons, egrets, and bitterns worldwide. Thirteen species have been recorded in Texas. One, the Bare-throated Tiger-Heron is a very rare visitor to Texas.

Least Bittern

(Botaurus exilis)

IUCN RED LIST STATUS (2016): Least Concern
POPULATION TREND: Stable

One of the most inconspicuous marsh birds, the Least Bittern is the smallest member of the heron family. In suitable habitat it can be common. Often late in the day they will perch up high in vegetation, sunning themselves. They forage along the water's edge, moving through vegetation with amazing dexterity on their long toes.

LENGTH: 13 inches. **WINGSPAN:** 17 inches.
ADULT BREEDING: Male has dark crown and dark back. Rich brown on wings and neck. Streaked with bold brown and white on throat and breast. Belly is pale. In flight, primaries are dark. Very long grasping toes. Female's back is brown and not dark. **JUVENILE:** (July–Oct.) Like female but lacks dark crown. **VOICE:** Song is a croaking *hork-hork-hork-hork*. Call is a high-pitched screeching: *skree-skree-skree-skree*. **BEHAVIORS:** Clambers through vegetation, grasping with its toes as it moves. Can run and burrow into dense vegetation. Flight is fluttering and short back into vegetation. **HABITAT:** Freshwater and brackish marshes with tall emergent vegetation, reeds, and cane species. **STATUS:** Locally common summer resident on the coastal prairies and in the eastern Lower Rio Grande Valley. Uncommon summer resident in the eastern part of the state east of the Balcones Escarpment. Spring migrants arrive in late March. Fall migration begins in early September and most have departed by late October. Very rare winter resident on the coastal prairies.

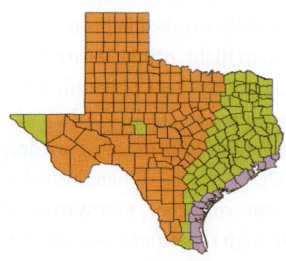

American Bittern

(Botaurus lentiginosus)

IUCN RED LIST STATUS (2016): Least Concern
POPULATION TREND: Decreasing

Solitary and cryptic, a motionless American Bittern is easy to pass right by. Its bold veritcal stripes blend right into the vertically striated habitat it prefers. Often a bird will freeze head up and motionless when exposed in the open, doing its best to attact no attention to itself. **LENGTH:** 28 inches. **WINGSPAN:** 42 inches. **ADULT:** Heavy body with long neck and pointed bill. Dark mustache stripe on neck. Mottled brown on back. Neck, breast, and nape are boldly streaked in brown. Legs are yellow. In flight, primaries are dark. **IMMATURE:** (July–Sept.) Like adult but crown is duller-colored and lacks the dark whisker mark on the neck. **VOICE:** Song is a deep resonant *GLOONK-bloynk*. When flushed, makes a hoarse *craw-craw-craw*. **BEHAVIORS:** Moves slowly through reeds and grass. Often remains very still for long periods of time. At times will sway with the grass to blend in. **HABITAT:** Brackish and freshwater wetlands with tall grasses and reeds. Sometimes found in wet fields, where the grass is tall enough to conceal it. **STATUS:** Rare to uncommon migrant in the eastern third of Texas. Rare to uncommon winter resident on the coastal prairies. There are a handful of breeding records for Chamber, Galveston, and Wilbarger counties. Birds arrive in fall starting in late September and can be encountered until early May.

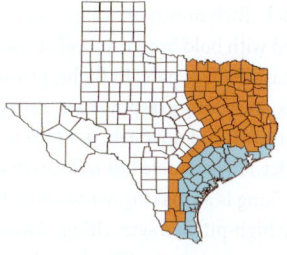

Tricolored Heron

(Egretta tricolor)

IUCN RED LIST STATUS (2016): Least Concern
POPULATION TREND: Stable

Formerly known as the Louisiana Heron, the Tricolored Heron is the most coastal of herons in Texas except for the Reddish Egret. **LENGTH:** 26 inches. **WINGSPAN:** 36 inches. **ADULT BREEDING:** (Feb.–July) Blue-gray head and neck and yellow throat. Bill is bicolored, pale blue-gray at base and dark-tipped. Short white plumes on head. Back has pale rust plumes on lower portion. Belly is white; the combination of a dark chest and dark breast is unique in all plumage. In flight, underwing is pale with dark flight feathers. **NONBREEDING:** (Aug.–Jan.) Like breeding but lower mandible of bill is yellow with dark upper mandible. White plumes on head are absent. **JUVENILE:** (July–Feb.) Bill is yellow. Head, neck, upper wings, and top part of the mantle are rusty-red-colored. Underside is like adults. **VOICE:** Nasal *craw*. **BEHAVIORS:** Very active while feeding, dashing from place to place. In flight draws the head back and extends feet behind. Flies at a steady pace. **HABITAT:** Breeds mostly in coastal brackish habitats but will use fresh water also. Nonbreeding birds will use a variety of wetland habitats. **STATUS:** Common summer resident along the coast and locally common on the coastal prairies. Uncommon to rare in the eastern third of the state in late summer and fall. Uncommon to rare winter resident on the coastal prairies.

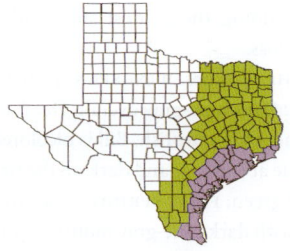

Adult

Little Blue Heron

(Egretta caerulea)

IUCN Red List Status (2017): Least Concern
Population Trend: Decreasing

Little Blue Heron, because it is dark-colored and doesn't have any showy plumes, was spared the ravages of the commerical plume hunters of the early 20th century. It has been a difficult bird population to assess, though, since it is solitary when foraging and its habit of nesting under canopy make it difficult to spot in the aerial surveys used to count colonial waterbirds. The population is thought to be declining; the greatest cause is likely habitat loss.

LENGTH: 24 inches. **WINGSPAN:** 36 inches. **ADULT:** Dark blue-gray overall. Head and upper neck are purplish. Bill is bicolored, pale blue at the base and dark on the tip. Legs are dull green. **FIRST SPRING:** (Apr.–June) White with dark blue-gray mottling. Bill is the same bicolored bill of adults. **JUVENILE:** (June–Apr.) White overall. Bill is pale gray with dark tip. Legs are greenish. **VOICE:** Raspy SKRAAA. **BEHAVIORS:** Typically forages leaning forward. **HABITAT:** Freshwater and marine-estuarine wetlands. **STATUS:** Common summer resident in the eastern third of the state. Locally abundant on the coastal prairies to the Lower Rio Grande Valley. Uncommon winter resident on the coastal prairies. Spring migration is early March–early April, fall early September–early November.

Immature

Dark morph

White morph

Reddish Egret

(Egretta rufescens)

IUCN RED LIST STATUS (2020): Near Threatened
POPULATION TREND: Decreasing

The Reddish Egret is well-known for its "drunken dance" while it feeds. It seems to stagger around rapidly chasing small prey in shallow water, and hops into the air with raised wings to create shadows to attract prey, a process known as canopy feeding. One of the least-numerous herons in North America, the worldwide population likely doesn't exceed 11,000.

LENGTH: 30 inches. **WINGSPAN:** 46 inches.
ADULT BREEDING: Dark morph has stout bicolored bill with dark tip and lower two-thirds are pink. Bare skin around eye is light blue. Head and neck are shaggy rusty-red color. Pale slate-gray upper and lower parts. Legs and feet are dark gray. White morph is like dark morph but all feathers are white. **NONBREEDING:** Like breeding but the bare facial skin is pale brown to pale violet. Lower two-thirds of bill is pale brown to pale violet. **JUVENILE:** Dark morphs is a pale chalky version of adults with an all-dark bill. White morph is like nonbreeding with an all-dark bill. **VOICE:** Normally quiet. Raspy *craw* like a Tricolored Heron. **BEHAVIORS:** Very active, frenetic forager. Dances, or moves about rapidly holding wings up to create shade to attract prey. **HABITAT:** Nests on natural and artificial islands with low vegetation. Feeds in shallow bays and lagoons. **STATUS:** Uncommon to locally common resident along the coast. More numerous south of Matagorda Bay.

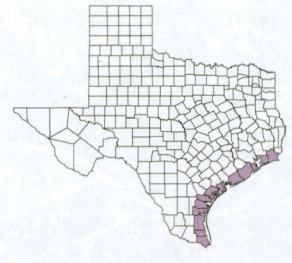

Snowy Egret

(Egretta thula)

IUCN RED LIST STATUS (2016): Least Concern
POPULATION TREND: Increasing

Snowy Egret was one of the most sought-after birds for the hat trade. Its delicate back plumes were worth an astounding $32 per ounce in 1886. At the time, that was almost twice the value of gold! The hunting was banned in 1910, and Snowy Egrets have shown a remarkable comeback.
LENGTH: 24 inches. **WINGSPAN:** 36 inches.
ADULT BREEDING: (Feb.–July) White overall with yellow lores and dark bill. At times the lores can be orange-yellow. Short plumes on back of the head. Legs are black with yellow feet. **NONBREEDING:** (Aug.–June) Plumes on back of the head are absent. Lower mandible pale at the base. Backs of the legs become greenish, but black is usually present on front of the legs. **JUVENILE:** (July–Apr.) Like nonbreeding but base of upper mandible can be pale also. **VOICE:** Raspy *kraaa*. **BEHAVIORS:**
Very active feeder; often chases prey around in the shallows. Sometimes forages from a crouched position. **HABITAT:** Nests in colonies. Habitats vary but usually shallow sites including salt marshes, tidal channels, shallow bays, freshwater marshes, lakeshores. Nests in trees, shrubs, even prickly pear cactus. **STATUS:** Uncommon to common summer resident to the eastern part of the Rolling Plains and south to the Lower Rio Grande Valley. Rare summer visitor to the Panhandle. Uncommon winter resident on the coastal prairies, becoming rare as one moves inland.

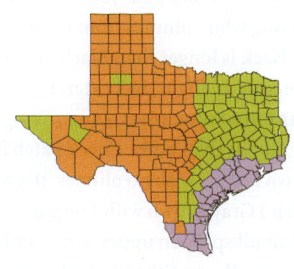

Adult

Juvenile

Yellow-crowned Night-Heron

(Nyctanassa violacea)

IUCN RED LIST STATUS (2018): Least Concern
POPULATION TREND: Stable

Even though Yellow-crowned Night-Herons do have long plumes, they were spared the devastation of early-20th-century plume hunters. The current population is considered stable and has even expanded its range in the last century. They are known to consume most species of available crabs and crayfish. **LENGTH:** 24 inches. **WINGSPAN:** 42 inches. **ADULT:** Pale slate-gray overall. Dark head with white cheek and pale-yellow crown patch. Long white plumes from the head. Bill is dark. Neck is longer than Black-crowned Night-Heron. In flight, feet extend beyond the tail. **FIRST SUMMER:** (Mar.–Aug.) Wings are slightly darker than body. Cheek patch is gray and crown is dark with no plumes. **JUVENILE:** (July–Feb.) Gray-brown with fine pale streaking and small spots on upper wings. Bill is all dark. **VOICE:** Raspy *CROAW*, higher-pitched

than a Black-crowned Night-Heron. **BEHAVIORS:** Feeds both day and night; tides influence when it feeds in coastal areas. More stationary when feeding than other herons. **HABITAT:** Breeds in a wide variety of habitats, including in wooded neighborhoods. Often seen foraging at night on lawns in urban areas. More coastal in winter. **STATUS:** Uncommon to locally common summer resident on the coastal prairies and the eastern third of Texas and west to the Rolling Plains. Locally common winter resident on the coastal prairies south of Matagorda Bay. Spring migration is mid-March–early May, fall mid-August–October.

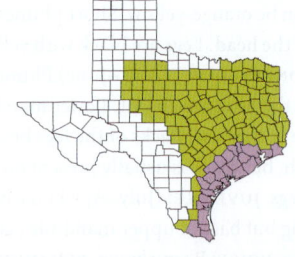

Adult

Juvenile

Black-crowned Night-Heron

(Nycticorax nycticorax)

IUCN RED LIST STATUS (2019): Least Concern
POPULATION TREND: Decreasing

Black-crowned Night-Heron breeds on every continent except Antarctica and Australia. It is widespread and common in North America and Texas, but many birders have never seen one because its coloration and habit of feeding at night and twilight make it a less-conspicuous species than most other herons.

LENGTH: 25 inches. **WINGSPAN:** 44 inches. **ADULT:** Pale gray with dark crown and mantle. Bright red eye. Legs are yellow. Two long plumes on head. Bill is dark. In flight, feet barely extend beyond the tail. **FIRST SUMMER:** (Feb.–Aug.) Like adult but less contrast; back and crown are lighter, body is darker gray. Lower mandible of the bill is pale. **JUVENILE:** Slate-gray overall with pale streaking and spots on upper wings. Eye is yellow-orange. Lower mandible of the bill is yellow. **VOICE:** Call is a *QUOP*. **BEHAVIORS:** Mostly nocturnal, often roosts in groups during the day. Forages from a crouched position. **HABITAT:** Wide variety of wetlands. **STATUS:** Common resident on the coastal prairies. Locally common to uncommon summer resident west of the Pineywoods. Rare to uncommon summer resident in the Pineywoods. Rare to uncommon winter resident in all but the Panhandle.

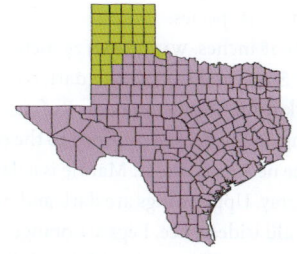

Green Heron

(Butorides virescens)

IUCN RED LIST STATUS (2022): Least Concern
POPULATION TREND: Stable

The relationship between Green Heron and Striated Heron has been much debated. In the 1970s they were combined into one worldwide species. In the 1990s they were again split into two species.

Small and solitary, the Green Heron is a bird of thicketed wetlands. It is most often seen flying away when disturbed. Recreational use of water channels will drive it away. Threats and changes to wetlands are the main threats to this species.

LENGTH: 18 inches. **WINGSPAN** 26 inches.
ADULT: Small and stocky with dark crown and dark rufous head and neck. Rufous streaking on the neck is confined to the center of the neck and breast. Mantle is a dark rufous-gray. Upper wings are dark and have an emerald iridescence. Legs are orange.
FIRST SUMMER: (Mar.–Aug.) Like adult but streaking on neck and breast more extensive. Legs are greenish-yellow. **JUVENILE:** (July–Mar.) Like first summer but more extensive streaking. Upper wings spotted with buff.
VOICE: Sharp *skow skow skow*. **BEHAVIORS:** Forages from a crouch, waiting patiently, staring into the water. Known to use bait to attract fish. **HABITAT:** Feeds on the edges of water from cover. **STATUS:** Common summer resident in the eastern two-thirds of Texas. Uncommon westward. Rare to uncommon on the lower coast and the Lower Rio Grande Valley. Spring migration is from late March, fall mid-August–late October.

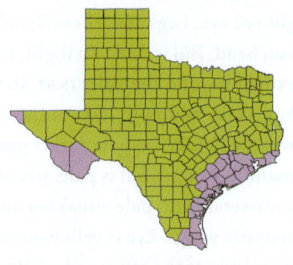

Great Egret

(Ardea alba)

IUCN RED LIST STATUS (2019): Least Concern
POPULATION TREND: Unknown

The Great Egrets has worldwide distribution on all continents except Antarctica. Its relationships in the heron family have been revised several times, and over time it has been placed in several different genera and had many English names: American Egret, Common Egret, Great White Egret, and even Great White Heron.

Its cosmopolitan distribution and generalist adaptability have contributed to its rebound from population decimation by overhunting in the 20th century. Its decline and that of other wading birds kindled the formation of conservation societies. Even today the National Audubon Society uses the Great Egret as its symbol.

LENGTH: 39 inches. **WINGSPAN:** 51 inches.
ADULT BREEDING: (Feb.–July) All white with yellow bill. Lores become bright lime-green for a short time. Back is adorned with long shaggy plumes. Legs and feet are black. **NON-BREEDING:** (Aug.–Jan.) Like breeding but lacks the green lores and plumes on the back. **FIRST YEAR:** Identical to nonbreeding adult. **VOICE:** Deep gravely *KRAAW*. **BEHAVIORS:** Flight is buoyant with the head and neck retracted, long legs extended behind. Walks with head and neck extended. **HABITAT:** Nests in colonies with other waterbirds. Nests often over water with a preference for islands. Feeds in a variety of wetlands. **STATUS:** Common resident on the coastal prairies. Uncommon to locally common in eastern and central Texas. Rare to common post-breeding visitors in the rest of Texas.

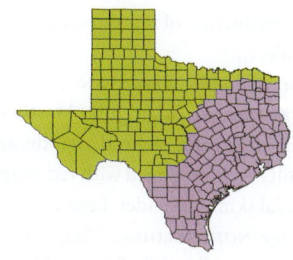

Cattle Egret

(Bubulcus ibis)

IUCN RED LIST STATUS (2019): Least Concern
POPULATION TREND: Increasing

Cattle Egrets were originally native to Africa, the Asian tropics, and Northern Australia. After 1900 Cattle Egrets exploded worldwide. It is now known on all continents, including a record from the South Shetland Islands in Antarctica. It did not reach North America until 1941, in Florida, and was not known in Texas until 1955. It was breeding in Texas by 1959. The amazing expansion was probably aided by land-clearing for agriculture and the production of more grazing animals worldwide.
LENGTH: 20 inches. **WINGSPAN:** 36 inches.
ADULT BREEDING: (Mar.–July) Mostly white with buffy rust color on head, mantle, and chest. Bill is yellow-tipped with red-orange base; facial skin is lavender. Legs are bright red-orange. **NONBREEDING:** (Aug.–Feb.) All white with yellow bill. Legs are black.

JUVENILE: (July–Oct.) Like nonbreeding but bill is dark with yellow base. **VOICE:** Mostly silent away from breeding grounds. Guttural *ack-ack-ack* or quiet *crup-crup*. **BEHAVIORS:** Walks with an exaggerated head-pumping motion. Rarely forages in or near water, usually in fields, often with livestock. Flies with rapid wingbeats. **HABITAT:** Breeds with other colonial waterbirds in heronries. Forages in upland grassland habitats. **STATUS:** Common to abundant summer resident in Texas. In winter, rare to uncommon on the coastal prairies south to the Lower Rio Grande Valley.

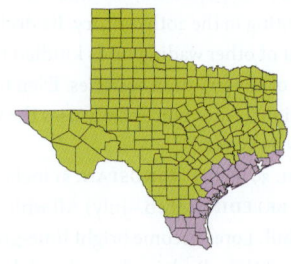

Adult

Juvenile

Great Blue Heron

(Ardea herodias)

IUCN RED LIST STATUS (2020): Least Concern
POPULATION TREND: Increasing

The Great Blue Heron is at home in almost any wetland. Extremely adaptable, their diet can include almost anything they can subdue. All varieties of rodents, snakes, fish of many sizes, small alligators, a Pied-billed Grebe, and a Clapper Rail have been recording being consumed by Great Blue Heron.

Even though it was heavily hunted for plumes in the 20th century, populations have recovered well. Current threats to Great Blue Heron are contaminants in the environment and loss of wetland habitats. **LENGTH:** 54 inches. **WINGSPAN:** 72 inches. **ADULT:** Large dagger-like yellow bill. Head is pale with dark plumes. Neck is light gray with shaggy plumes on breast. Upper wings and mantle are blue-gray. Wings are two-toned with darker flight feathers. **JUVENILE:** Gray overall with dark crown. Bill is two-toned with dark upper mandible and yellowish lower mandible. Belly is paler with dark streaking on breast. **VOICE:** Deep hoarse *KRAWK*. **BEHAVIORS:** Flies with head pulled back and legs dangling. Walks upright with long strides. Often up to the belly in the water. Nests in colonies. **HABITAT:** Widespread and adaptable. Will use wetlands of all types: saltwater, brackish, and freshwater. Nests in trees, bushes, and on the ground. Will use artificial structures. Prefers islands and swamps for nesting. **STATUS:** Common to uncommon migrant and summer resident statewide. Less common in winter away from the coastal prairies.

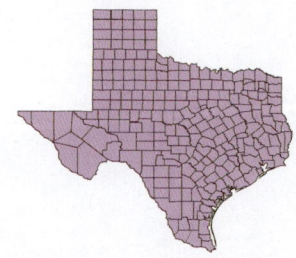

THRESKIORNITHIDAE (IBISES AND SPOONBILLS)

Ibises and spoonbills are long-legged and heron-like but differ from the herons and egrets notably by their strangely shaped bills. Ibises have long down-curved bills and spoonbills have flat spatulate bills. Both groups feed by feel. The ibises probe for food and the spoonbills sweep back and forth through the water.

Different cultures have treated this family differently. The Sacred Ibis was deified as the god Toth by the ancient Egyptians, and the Japanese Ibis has been declared a national treasure in Japan. In contrast, in North America, Roseate Spoonbill and White Ibis were hunted by the thousands to make ladies' hats.

Worldwide there are thirty-six species of ibises and spoonbills. Four species have been recorded in Texas and none are considered rare here. Eleven are at risk worldwide. One, the Dwarf Ibis of São Tomé, was thought extinct until it was rediscovered in 1990.

Immature

Adult

White Ibis

(Eudocimus albus)

IUCN RED LIST STATUS (2018): Least Concern
POPULATION TREND: Stable

Striking white with a long, red, decurved bill, White Ibis is an abundant wading bird on the Gulf of Mexico.
LENGTH: 25 inches. **WINGSPAN:** 38 inches.
ADULT: Red bill and face. Iris is brilliant sky blue. Rest of body is bright white with black wingtips. Legs and feet are red. **JUVENILE:** (July–Dec.) Orange bill and face. Iris is sky blue. Head and neck are brown with white streaking. Back and upper wings are brown. Underparts and underwings are white. Underwing flight feathers are dark. Rump and tail are white with black terminal band. Legs are orange. **FIRST SUMMER:** (Mar. –Aug.) Gradually molts from brown to white, brown blotchy at times as the white adult feathers molt in. **VOICE:** Deep low *urnk urnk urnk*. **BEHAVIORS:** Colonial nester. Colonies can be large; records exist of colonies with more than 20,000 pairs. Feeds often in large flocks. Migrates in large flocks in fall. **HABITAT:** Nests on barrier, marsh, and spoil islands. Salt stress from saltwater crustaceans prevent normal development of chicks, so shallow freshwater habitat for foraging is needed. In winter, shifts to mostly coastal marshes. **STATUS:** Common and abundant on the coast and coastal prairies. Uncommon and local inland in the eastern third of the state. Migrants begin to arrive in the breeding areas away from the coast in mid-March and depart in October and November.

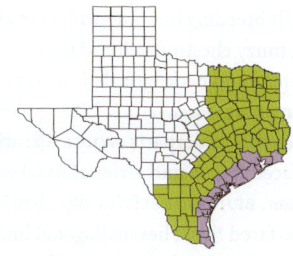

Nonbreeding

Breeding

Likely hybrid male with female on nest

Glossy Ibis

(Plegadis falcinellus)

IUCN RED LIST STATUS (2018): Least Concern
POPULATION TREND: Decreasing

Glossy Ibis is a worldwide species undergoing a slow expansion.
LENGTH: 23 inches. **WINGSPAN:** 36 inches.
ADULT BREEDING: (Mar.–Aug.) Bill is brownish, face is lead-gray. Facial skin is bordered in blue-white skin that does not extend behind the eye. Eye is dark. Front of face is dark, almost black. Bronzy chestnut on nape, neck, chest, and belly. Wings are darker and show a greenish gloss. **NONBREEDING:** (Sept.–Feb.) Like adult breeding but blue-white facial skin duller, bronzy chestnut colors darker and less glossy. Head and neck "grizzled" with fine white streaking. **IMMATURE:** Like nonbreeding but darker underparts. **VOICE:** Identical to White-faced Ibis. Rapid series of nasal *waaa waaa waaa*. **BEHAVIORS:** Basically identical to White-faced Ibis. Flies in diagonal line or V and is indistinguishable from White-faced in flight. Will soar high and ride thermals. **HABITAT:** Mainly freshwater marshes, but will use brackish and saltwater marshes, mudflats, ponds. Rice fields are important habitat when present. Often shares habitat with White-faced Ibis. **STATUS:** Increasingly regular on the upper and central coast. There are scattered records across the state. Has become a regular nester on the upper Texas coast and often present in White-faced Ibis colonies. Hybrids are common and may outnumber pure birds. The status may be heading to few pure birds in the contact zone in Texas.

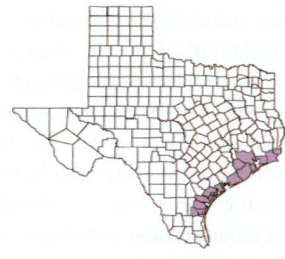

Breeding

Winter

White-faced Ibis

(Plegadis chihi)

IUCN RED LIST STATUS (2018): Least Concern
POPULATION TREND: Increasing

The White-faced Ibis is the most common dark ibis in Texas.
LENGTH: 23 inches. **WINGSPAN:** 36 inches.
ADULT BREEDING: (Mar.–Aug.) Gray bill with reddish face bordered by creamy white feathers. Border extends completely behind the eye. Eye is bright blood-red. Head, neck, and chest are rich bronzy chestnut. Wings and back are dark, often iridescent purplish, but iridescent green can also be present. Legs are bright red throughout. **NONBREEDING:** (Sept.–Mar.) Bill is gray, face is dull red with a red eye. Breeding-plumage white face border is absent. Head and neck are grizzled with fine white streaking. Legs are olive-gray. **IMMATURE:** (Sept.–Mar.) Similar to winter plumage but bill and face are the same gray. Eye is brown, becoming red usually in December or early January. **VOICE:** Rapid series of nasal *waaa*

waaa waaa. **BEHAVIORS:** Often feeds in flocks, probing deep with its long bill. Often plunges its head completely underwater. Flocks fly in loose V formation. **HABITAT:** Nests mostly in coastal prairie wetlands. Colonial nester, frequently in stands of common cane and California bulrush. Feeds in shallow flooded wetlands and prefers fresh water. Frequently uses rice fields with water. **STATUS:** Common to uncommon resident on the coast. Migrant and wintering birds boost the population outside of breeding season. Rare to casual in winter away from the coast in most of the state.

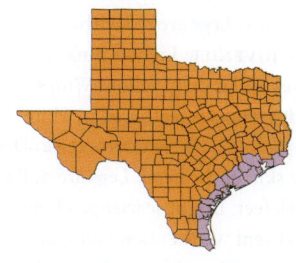

Adult

Second year

Roseate Spoonbill

(Platalea ajaja)

IUCN RED LIST STATUS (2018): Least Concern
POPULATION TREND: Stable

Roseate Spoonbill is one of the six species of spoonbills worldwide. It is the only spoonbill species in the Americas and is the only pink one; the others are various shades of white. **LENGTH:** 32 inches. **WINGSPAN:** 50 inches. **ADULT:** Distinctive spoon-shaped bill, light gray in color. Head is bare and often greenish-tinged with dark band of skin on nape. Body, flight, and belly feathers are light pink. Upper wing and eye are bright scarlet. Neck is white. Legs are red with dark joints and feet. **JUVENILE:** Lighter pink than other plumages. Bill is dirty yellow. Wings can have dusky tips. Head is feathered in white. Legs may be pink to dark. **SECOND YEAR:** Develops the bare skin of the head. Legs are dull red with dark feet. Bright scarlet patch on upper wing is absent. **VOICE:** Low *huh-huh-huh-huh.* **BEHAVIORS:** Flies without gliding. Feeds in shallow water sweeping its bill. Colonial nester and shares colonies with other species of colonial waterbirds. **HABITAT:** Variety of salt, brackish, and freshwater habitats. Needs shallow water to feed. Most nesting sites are on islands or over standing water. **STATUS:** Locally common along the coast until about August. Post-breeding birds can disperse inland. Birds return to breeding sites in mid-March and breeding begins in earnest in mid-April. Rare in summer and fall in east Texas. Accidental west to about Abilene.

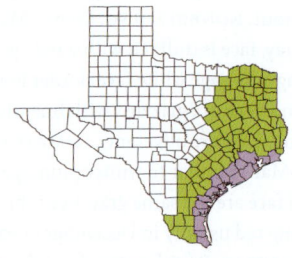

CATHARTIDAE
(NEW WORLD VULTURES)

The New World Vultures are the low-energy lifestyle masters. Each species can travel hundreds of miles without ever flapping their wings, gracefully using rising air currents to gain altitude and soar across the landscape. New World vultures have one of the lowest weight-to-wing-area ratios of any bird.

Most forage by sight, drifting along on air currents, on the alert for carrion. Man has made life easy for them with an abundance of roadkill. The Turkey Vulture is an exception and has a sensitive sense of smell for sniffing out a meal.

Old World and New World Vultures have a strong resemblance, but they actually don't share any close ancestors. Instead, their resemblance is a case of convergent evolution. Because they share similar ecologies, they have come to resemble each other.

Worldwide there are seven species of vultures. Three have been recorded in Texas. One is known only from a single semi-fossilized skeleton discovered in Big Bend National Park.

Black Vulture

(Coragyps atratus)

IUCN RED LIST STATUS (2016): Least Concern
POPULATION TREND: Increasing

Black Vultures are often seen in large communal roosts. These roosts may serve to communicate the location of food. Black Vultures are almost exclusively carrion eaters and spend much of their soaring in search of carcasses. Unlike Turkey Vultures, they don't use smell to locate carrion but may follow a Turkey Vulture to carrion and force the rival away.

Black Vultures do not build nests, instead laying eggs directly on the ground in a dark cavity like cave, hollow log, or abandoned building. Pairs are monogamous and will stay together year-round. Pairs my feed young for up to eight months. Black Vultures have strong social bonds with kin and will drive unrelated Black Vultures away from a carcass. **LENGTH:** 25 inches. **WINGSPAN:** 59 inches. **ADULT:** All black with black, featherless, wrinkled head. Legs and feet are pale. Tail is short and fan-shaped and only extends to the feet, both perched and in flight. Light primaries. **JUVENILE:** (July–Nov.) All black with black feathered head. **VOICE:** Usually silent. **BEHAVIORS:** When taking off, wingbeats are short and snappy. Soars with nearly flat wings. **HABITAT:** Woodlands and forested wetlands. Forages more open spaces. Avoids urban areas, croplands, and barren lands. **STATUS:** Common to locally abundant resident in the eastern two-thirds of the state. Uncommon on the Rio Grande from Val Verde County to Presidio County.

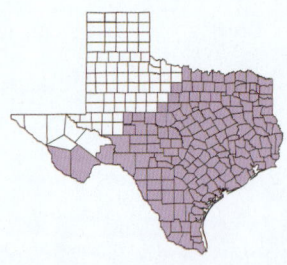

Turkey Vulture

(Cathartes aura)

IUCN RED LIST STATUS (2016): Least Concern
POPULATION TREND: Increasing

Turkey Vultures are scavengers and eat almost exclusively carrion. The genus name *Cathartes* means "purifier." They have a sharp sense of smell and can locate a carcass under a forest canopy. Black Vultures, who lack this sense of smell, will follow Turkey Vultures to a carcass and displace them. Turkey Vultures specialize in smaller food items they can gulp down quickly before a competitor arrives. **LENGTH:** 26 inches. **WINGSPAN:** 67 inches. **ADULT:** Black with brownish color to the back. Head is naked and bright red and legs are light-colored. In flight and perched, wings and tail extend well beyond the feet. Flight feathers are silvery, contrasting with dark underwing coverts. **JUVENILE:** (July–Nov.) Like adults but the naked head is ash-gray. **VOICE:** Usually silent. **BEHAVIORS:** Wingbeats are clumsy and slow in flight. Soars with wings in a dihedral. While soaring, it seems unstable and rocks side to side. **HABITAT:** Mixed forests and farmland. Seems to avoid areas of row crops. **STATUS:** Common to abundant summer resident statewide. In winter, they withdraw from the western half of the state. Abundant migrant in fall in the eastern half of the state. Spring migration begins in late February, and most nonresident birds are gone by mid-October.

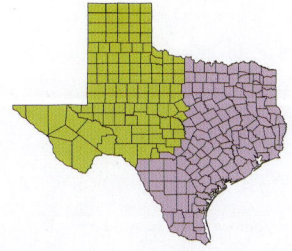

PANDIONIDAE (OSPREY)

The osprey family has long been considered a subfamily of the hawks and eagles. Molecular phylogenetic evidence is strong that the families are very closely related. The two sister families together form a clade that is sister to the Secretary Bird of Africa.

Ospreys feed exclusively on fish. Agricultural chemicals like DDT concentrate in fish and the Ospreys who eat them, causing low nesting success because of eggshell thinning. Since the banning of DDT in 1972, Osprey populations have recovered.

Osprey

(Pandion haliaetus)

IUCN RED LIST STATUS (2021): Least Concern
POPULATION TREND: Increasing

Ospreys are one of thirty-four species of birds in a monotypic family. In the osprey family, there is just Osprey. They are worldwide in distribution, essentially only absent from Antarctica and southern South America.

Ospreys have a network of raised ridges on their feet, and one of their toes is reversible—both adaptations for carrying slippery fish. **LENGTH:** 23 inches. **WINGSPAN:** 63 inches. **ADULT:** White head with black eye stripe. Forehead and nape are dark, but crown is white. Uppers are all dark. White below with spotted breast band. Underwings are light with dark "wrists" and secondaries. Primary tips are dark. Tail has narrow bands. **JUVENILE:** (July–Jan.) Like adults but the upper parts have pale scaling. Breast band is absent, and breast is buffy. The buffy color fades quickly. **VOICE:** Call is a shrill, whistling *twep twep twep*. **BEHAVIORS:** Steady flight like it's rowing through the air. Often hovers before diving. When flying with a fish, always maneuvers it headfirst to reduce drag. Perches high to eat. **HABITAT:** Rivers, lakes, and bays, always near shallow water for fishing. **STATUS:** Uncommon to rare migrant statewide. Common to uncommon winter resident on the coast and in the eastern third of the state. Rare and local breeder along the coast and in the Pineywoods near large reservoirs. Spring migration is mid-March–late May, fall early September–mid-November.

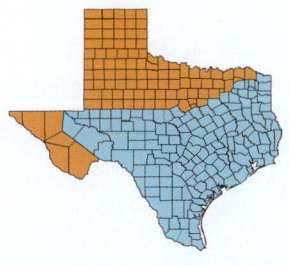

ACCIPITRIDAE
(HAWKS, EAGLES, AND KITES)

Hawks, eagles, and kites are collectively known as the diurnal raptors. Most have stongly hooked bills for tearing flesh, and long, strong talons for grasping and killing prey. They range in size from huge eagles to tiny hawks in the genus *Accipiter*. Most species have distinct immature and adult plumage. Raptors occupy virtually all terrestrial habitats. While most have a diet of a variety of smaller vertebrates, many species have very specialized diets of monkeys, snails, wasps, palm fruit, and even bone marrow.

Hawks, eagles, and kites have large eyes facing forward on their heads, giving them binocular vision and very good depth perception. Their eyes are packed with a much higher density of receptors than humans', allowing them to make out details at much greater distances. Their eyesight has been described as being able to read a page of text from across a large room. This allows a soaring bird to detect prey from hundreds of feet in the air.

There are 250 species of hawks, eagles, and kites worldwide. Twenty-seven species have been recorded in Texas. Five of these are very rare visitors to Texas: American Goshawk, Snail Kite, Roadside Hawk, Crane Hawk, and Short-tailed Hawk. Double-toothed Kite, Steller's Sea-Eagle, and Great Black Hawk are known from just a single record.

White-tailed Kite

(Elanus leucurus)

IUCN RED LIST STATUS (2016): Least Concern
POPULATION TREND: Increasing

From 1981 to 1994, White-tailed Kite was considered a subspecies of the Eurasian Black-shouldered Kite, but the American Ornithologists Union elevated it back to a full species. In the early 20th century, White-tailed Kite was threatened with extinction in North America due to egg collection and hunting but has since recovered.

It is perhaps one of the few species that may have benefited from human modification of the landscape to agriculture, as they favor open landscape that supports the small mammals that make up most of their diet. **LENGTH:** 15 inches. **WINGSPAN:** 39 inches. **ADULT:** White head with dark eye patch. Back and upper wings are pale gray. Prominent dark wing patches on shoulders. Primaries are dark. From below, wings have dark wrist patches. Tail is white. **FIRST YEAR:** Like adult but with a buffy wash on the breast, head, and upper back. The buffy wash fades as the bird ages. Back has a scalloped pattern. **VOICE:** Quick, high-pitched *choot*. **BEHAVIORS:** Flies with fast, shallow wingbeats followed by a glide. Hovers high over prey then drops quickly with wings high. **HABITAT:** Low-elevation grasslands, agricultural lands, and savannah-type habitats. **STATUS:** Uncommon to common resident in the coastal prairies, the southern part of the Post Oak Savannah, and the South Texas Brush Country. Rare to uncommon summer resident in the northern part of the Post Oak Savannah.

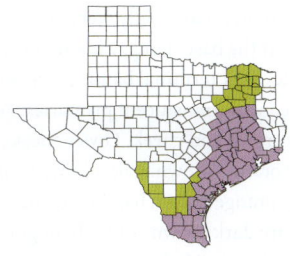

Hook-billed Kite

(Chondrohierax uncinatus)

IUCN RED LIST STATUS (2020): Least Concern
POPULATION TREND: Decreasing

The Hook-billed Kite was not recorded in Texas or the United States until 1964, when a pair nested in Santa Ana National Wildlife Refuge. Abundance of this speecies is likely tied to the abundance of tree snails, its main food item. During long droughts the number of snails drops, so the number of kites found in Texas drops.

LENGTH: 18 inches. **WINGSPAN:** 36 inches.
ADULT: Male is slate gray overall, with white barring on breast and belly. Very large bill that is yellow at the base. Long, wide-banded tail with wide, dark terminal band. Underwings are barred with translucent primaries. Female has gray crown and cheek. Lower cheeks and neck are orange-brown. Chest is light with the same orange-brown barring. Upper wings and tail are dark brown. Tail is light gray with wide dark bands. Underwings are barred with translucent primaries. **FIRST YEAR:** Like females but lower cheek and neck are white. Underside is white with narrow brown barring. **VOICE:** Rapid *kekekekekekek*, woodpecker-like. **BEHAVIORS:** In flight, glides on bowed wings, usually low over canopy. When occasionally riding thermals, periods of soaring are interspersed with slow, floppy wingbeats. **HABITAT:** Mesquite woodlands and riparian floodplains. **STATUS:** Rare and local resident in Hidalgo and Starr counties of the Lower Rio Grande Valley. The Texas population is extremely variable and tied to the abundance of tree snails.

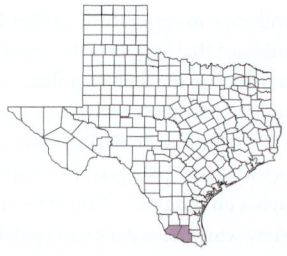

Swallow-tailed Kite

(Elanoides forficatus)

IUCN RED LIST STATUS (2016): Least Concern
POPULATION TREND: Increasing

The Swallow-tailed Kite is one of the most distinct and easily identifiable birds in Texas. They once nested in most of the southeastern United States. Historically, they nested in southeastern Texas until about 1915. From 1915 to 1989 they were only occasionally seen as migrants in the state. In 1989, a pair was found nesting in Jefferson County. Today they are uncommon but regular nesting birds in most of their former range in Texas.
LENGTH: 22 inches. **WINGSPAN:** 51 inches. **ADULT:** White head, breast, and underside. Upper parts all dark. Flight feathers under the wing are dark. Tail is long and deeply forked. **JUVENILE:** (July–Feb.) Like adult but with a buffy wash that fades in a few weeks. Forked tail is about half the length of an adult's tail. **VOICE:** Short, high-pitched *weet-weeet-weet-weet*. **BEHAVIORS:** In flight

most of the time. Often perches early and late, when there are few thermals. Graceful and buoyant in flight. When foraging, soars low over treetops. **HABITAT:** Forested habitats with tall trees for nesting; stands of trees can be small. Often near a stream or river. **STATUS:** Rare to uncommon migrant through the coastal prairies and the eastern third of Texas. Rare to uncommon summer resident in the southern part of east Texas west to Jackson County. Spring migration is early March–early May, fall mid-August–mid-October.

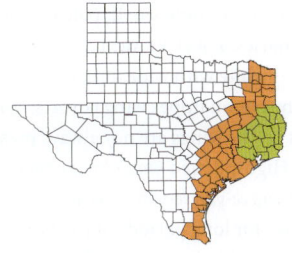

Golden Eagle

(Aquila chrysaetos)

IUCN RED LIST STATUS (2021): Least Concern
POPULATION TREND: Stable

The Golden Eagle is one of, and perhaps *the* most, formidable bird of prey in Texas. In a full swoop after prey it can attain speeds greater than 200 miles per hour.
LENGTH: 30 inches. **WINGSPAN:** 79 inches.
ADULT: Dark brown with golden-brown head. Heavy bill merges smoothly into the head, unlike the high forehead of a Bald Eagle. Pale bar across the upper wing coverts. **SECOND YEAR:** Like adult but upper tail is white. There are barely discernible small, pale patches in the primaries. **FIRST YEAR:** Like second year but tail is white with wide dark band at the tip. Lacks the bar on upper wing coverts. White patches on the wings are not always present. **VOICE:** High-pitched *kroup-kroup-kroup* or shrill *ee-yup ee-yup ee-yup*. **BEHAVIORS:** Soars and glides for long periods of time and wings are held in a slight V. When stooping after prey, the wind in the wings makes a loud rushing sound. Regularly perches on telephone poles. **HABITAT:** Usually forages in open grassland habitat, nests in nearby cliffs and sometimes trees. **STATUS:** Rare to locally uncommon resident in the Panhandle and western part of the Trans-Pecos. Uncommon winter resident from the Panhandle south through the Rolling Plains and western Edwards Plateau. Very rare winter resident on the upper Texas coast. Migrants can be found statewide late October–late March.

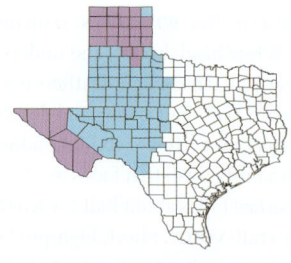

Female

Male

Female

Northern Harrier

(Circus hudsonius)

IUCN RED LIST STATUS (2016): Least Concern
POPULATION TREND: Decreasing

The Northern Harrier is found throughout Texas. Its bold white rump and flight with wings in a V shape make it easily recognizable. The genus name *Circus* comes from the Greek *kirkos*, or "circle," for its habit of flying in circles while foraging. Formerly cospecific with the Eurasian Hen Harrier, it has since been elevated to a separate species.

LENGTH: 18 inches. **WINGSPAN:** 43 inches.
ADULT: Male's head and upper parts are pale gray. Face has distinct facial disk, much like an owl. Underparts are white with brown spots. Outer primaries and secondaries are dark. Rump is white. Long tail is gray with dark bands. Female has dark brown head and upper parts. Obvious owl-like facial disk. Light underparts are heavily and boldly streaked in brown. Rump is white and long brown tail has several dark bands. **FIRST YEAR:** Like female but underwings, breast, and belly are unmarked buffy-brown. **VOICE:** Dry, barking *chef-chef-chef-chef*. **BEHAVIORS:** Often perches on fence posts. When foraging, glides low with wings held in a V. Tips side to side as it hunts. **HABITAT:** Open wetlands, wet prairies, old fields, upland grasslands. **STATUS:** Common to uncommon migrant and winter resident in all parts of the state. Uncommon breeder on the coastal prairies and in the Panhandle. In fall, migrants begin arriving in late August. In spring, birds depart late March–early May.

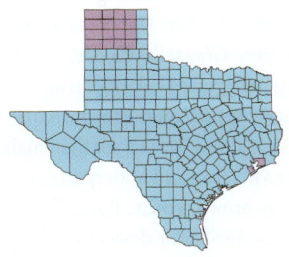

Sharp-shinned Hawk

(Accipiter striatus)

IUCN RED LIST STATUS (2016): Least Concern
POPULATION TREND: Increasing

Even though small mammals and insects do appear in the Sharp-shinned Hawk's diet, it feeds mostly on small birds. Those who maintain winter bird feeders in Texas often fear the arrival of a Sharp-shinned Hawk in their neighborhood, where they might be seen snatching songbirds.
LENGTH: 11 inches. **WINGSPAN:** 23 inches.
ADULT: Slate-gray crown and nape. Rufous cheeks. Heavy rufous barring to the vent. Undertail is bright white. Upper wings and back are slate-gray. Gray tail with dark band. Underwings are light with dark barring. Female upper parts appear more brownish than gray. Females are almost 50% larger than males.
FIRST YEAR: Head and underparts are coarsely streaked in brown. **VOICE:** Rapid, high-pitched *kiw-kiw-kiw-kiw-kiw* or slower *keeep keeep keeep*.
BEHAVIORS: In flight, wingbeats rapid and snappier than similar Cooper's Hawk. Makes three to six shallow wingbeats followed by a glide. When gliding, wings are forward compared to Cooper's Hawk and head rarely extends beyond the wings. **HABITAT:** Wide variety of forested habitats. Frequents feeders in urban and suburban areas to prey on birds visiting the feeder. **STATUS:** Uncommon to common migrant and uncommon winter resident statewide. Very rare summer resident in the Pineywoods and higher elevations of the Chisos, Davis, and Guadalupe mountains. Fall migrants arrive in Texas beginning in mid-August; spring migrants depart late March–early May.

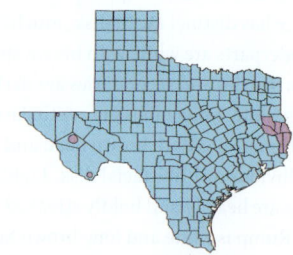

Immature

Adult

Cooper's Hawk

(Accipiter cooperii)

IUCN RED LIST STATUS (2016): Least Concern
POPULATION TREND: Increasing

Cooper's Hawk was named for William Cooper, a noted naturalist of New York. It is regularly seen in urban and suburban areas. **LENGTH:** 16.5 inches. **WINGSPAN:** 31 inches. **ADULT:** Slate-gray crown; nape is paler. Slate-gray on back and upper wings. Heavily barred rufous on breast and belly. Underwing coverts are rufous. Bright undertail coverts. Long tail is gray with dark bands. White terminal band. Males have light-gray cheek; females have rufous cheek. Females are larger than males. **FIRST YEAR:** Dark-brownish head. Throat, breast, and belly have thin dark-brown streaks. Dark brown on back. Underwings are like adults but with white spotted underwing coverts. **VOICE:** Nasal *kek-kek-kek-kek-kek* or *KHEEE kek-kek-kek-kek-kek*. **BEHAVIORS:** In flight, alternates several rapid wingbeats with glides. Often low to the ground or under tree canopy. In flight, holds wings further back than similar Sharp-shinned Hawk so the head projects out in front of the wings. **HABITAT:** Wide variety of forested habitats. Very tolerant of human disturbance and uses urban or suburban habitat at higher density than natural habitat because of preferred prey availability. **STATUS:** Uncommon to rare migrant and winter resident statewide. Rare to locally uncommon summer resident in all areas of the state but the High Plains, western Trans-Pecos, and the coastal prairies south to Matagorda County. Fall migrants begin to arrive in late August; spring migrants depart mid-March–late April.

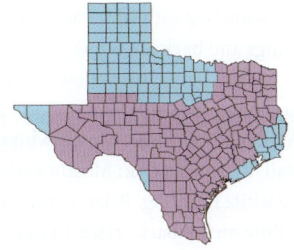

Adult

Immature

Adult

Bald Eagle

(Haliaeetus leucocephalus)

IUCN RED LIST STATUS (2016): Least Concern
POPULATION TREND: Increasing

The Bald Eagle was selected by the Continental Congress to be the iconic symbol of the United States in 1782. By the mid-1900s Bald Eagles were rare in the continental United States because of persecution and pesticides that dramatically reduced breeding success. Listed as endangered in 1978, they had recovered enough to be removed from the endangered species list by 2007.
LENGTH: 31 inches. **WINGSPAN:** 80 inches.
ADULT: Distinctive with white head and tail. Upper wings and back are brown. Underparts are brown. Very large yellow bill. **FIRST YEAR:** Dark overall. Underwings mottled white. Axillaries are white. Mostly white in base of tail. **SECOND YEAR:** Mostly dark with extensive white mottling. Paler on the crown. Mostly white on the back. **THIRD YEAR:** Mostly dark overall, mottled white below.

Head is white with heavy streaking and obvious dark eye stripe. **VOICE:** Rapid, chirping *choot-choot-choot-choot*. **BEHAVIORS:** Soars extensively with wings held straight and flat. Capable of powerful flight when pursuing prey. **HABITAT:** Usually near a body of water with suitable roosting trees nearby. **STATUS:** Rare to uncommon resident in the eastern third of Texas. Uncommon to locally common winter resident in the eastern half of the state south to Nueces County. Uncommon to rare winter resident in the High Plains, Panhandle, and Davis mountains. Fall migrants arrive in Texas starting in early October and depart by mid-March.

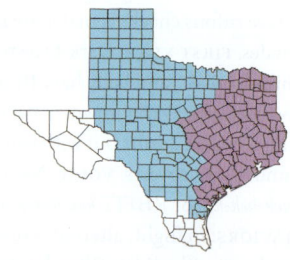

Adult

Immature

Mississippi Kite

(Ictinia mississippiensis)

IUCN RED LIST STATUS (2016): Least Concern
POPULATION TREND: Increasing

Mississippi Kites are one of the most graceful and recognizable raptors in Texas. Primarily insect eaters, their long, straight, and pointed wings and dark fanlike tail form one of the most distinctive profiles of the raptor family. A flock will climb in a swirling, chaotic kettle on a rising column of warmer air until reaching the top of the invisible elevator, where they all stream off in the same direction for the next rising air column.

LENGTH: 14 inches. **WINGSPAN:** 31 inches.
ADULT: Pale gray head with dark eye patch. Red eye is often obvious. Plain, unpatterned gray overall. In flight, the upper sides of the secondaries are white. Primaries are dark. Tail of males is uniformly black. Female's tail appears lighter because of the lighter inner webs on the tail feathers. **JUVENILE:** (Aug.–Mar.) Pale, gray-streaked head with white patch above eye. Heavily streaked below in brown. Feathers on upper parts are edged in white. **VOICE:** Thin, whistled *pee-teew*. **BEHAVIORS:** Flight is graceful and powerful. Often captures and consumes prey on the wing. **HABITAT:** Mature bottomland forest. West of the Pineywoods, nests in many urban neighborhoods. **STATUS:** Common to uncommon migrant statewide. Common summer resident on the High Plains and Rolling Plains. Uncommon summer resident in the Oaks and Prairies region to the upper coast. Spring migration is early April–mid-May, fall late August–mid-October.

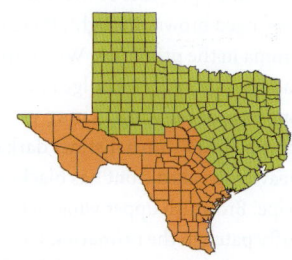

Common Black Hawk

(Buteogallus anthracinus)

IUCN RED LIST STATUS (2020): Least Concern
POPULATION TREND: Decreasing

The Common Black Hawk's secretive habits, small numbers, and inaccessible distribution north of Mexico make it a challanging species to study. It feeds on a wide range of prey, especially fish, frogs, crustaceans, snakes, and lizards.
LENGTH: 21 inches. **WINGSPAN:** 46 inches.
ADULT: All dark with yellow face and lores. Tail is white with one dark band and white terminal band. Upper sides of the wings are faintly tinged brown. In flight, there is a pale comma in the primaries. Wings are very broad, and secondaries bulge out on the trailing edge of the wings. **FIRST YEAR:** Bold patterned head with a distinct dark eye stripe. Heavily streaked front has blackish malar stripe. Brownish upper wings have a light buffy patch in the primaries. Underwings are pale. Tail is white with black terminal band and wavy bars. **VOICE:** Piercing *weet-weet-weet-wi-wi-wee*. **BEHAVIORS:** In flight, wingbeats are slow and strong. Soars with wings flat and tail completely fanned. **HABITAT:** Riparian mature forests associated with perennial streams. **STATUS:** Rare and local summer resident in riparian areas of the Davis Mountains and along the Rio Grande in Brewster and Presidio counties. There is a small nesting population along Devils River in Val Verde County. Spring migrants arrive in mid-March and depart September–early October. Rare winter visitor to the Lower Rio Grande Valley from November to mid-March.

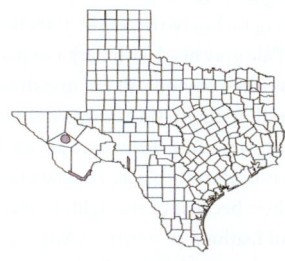

Harris's Hawk

(Parabuteo unicinctus)

IUCN RED LIST STATUS (2016): Least Concern
POPULATION TREND: Decreasing

Harris's Hawk was named for Edward Harris, a New Jersey gentleman farmer and naturalist who was a friend of John James Audubon and who accompanied the famous illustrator on some of his expeditions. Harris's Hawks are some of the most social hawks, breed in groups of up to seven birds, and practice both monogamy and polyandry, and sometimes polygyny.

LENGTH: 20 inches. **WINGSPAN:** 42 inches.
ADULT: Dark brown overall, with distinct bay or rufous-colored wing patches. Underwing coverts are also rufous. Tail has bold white upper tail, one wide dark band, and a narrow white terminal band. Fluffy white undertail coverts are easily visible on a perched bird.
FIRST YEAR: Like adult but breast and belly are boldly streaked in dark. Tail and secondaries have narrow dark bars on gray. **VOICE:**
Harsh, low *craaaaaaa*, sometimes up to three seconds in length. **BEHAVIORS:** Most flights are a flap-flap-glide type. In Texas, most often hunts from a perch. **HABITAT:** Semi-open desert scrub, savannah, grasslands, and wetlands. Availability of perches is important.
STATUS: Common to uncommon resident in South Texas Brush Country north to Edwards Plateau and east to Victoria and Goliad counties. Rare resident on the southern High Plains. Rare from Refugio County to the upper Texas coast. Locally uncommon to rare along the Rio Grande from Big Bend National Park to El Paso County.

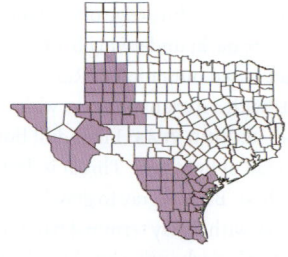

Adult   Immature

White-tailed Hawk

(Geranoaetus albicaudatus)

IUCN RED LIST STATUS (2020): Least Concern
POPULATION TREND: Unknown

For birders, the White-tailed Hawk is a specialty of Texas—if you expect to see it in the United States, you must visit Texas. In its grassland habitat, this beautiful hawk is attracted to fires, which may attract up to a dozen that hang around the burn for several days foraging.

LENGTH: 20 inches. **WINGSPAN:** 51 inches.
ADULT: Pale gray head, back, and upper wings. Clean white below. Shoulders have rusty patches. Trailing edge of the long, pointed wings are dark; underwing is white. Wings are broad-based but pointed. Rump and tail are white with a dark band and a white terminal band. **SECOND YEAR:** Like adult but dark gray instead of light gray. Throat is dark with a white chest. Belly is gray to gray-brown. Tail is pale gray with a gray terminal band. **FIRST YEAR:** Mostly dark with white breast patch.

Most birds, but not all, show a pale cheek patch and supercilium. Tail is light gray and faintly barred. **VOICE:** High-pitched *aiiii-yup ee-yup ee-yup ee-yup ee-yup.* **BEHAVIORS:** Goes to ground in inclement weather. In flight, the wingbeats are slow. Soars with wings in a dihedral. **HABITAT:** Open and semi-open humid to arid grasslands. **STATUS:** Uncommon to local common resident of the coastal prairies and South Texas Brush Country.

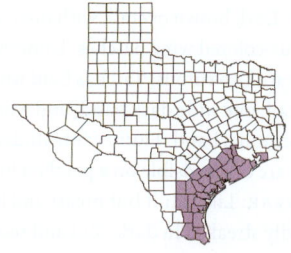

Immature

Adult

Gray Hawk

(Buteo plagiatus)

IUCN RED LIST STATUS (2020): Least Concern
POPULATION TREND: Stable

This tropical hawk is found in the United States only, in Texas and Arizona. There are thought to be fewer than 100 pairs of Gray Hawk in both states combined. Despite this, the population seems stable and is perhaps increasing.

LENGTH: 17 inches. **WINGSPAN:** 34 inches.
ADULT: Smooth gray above. Breast and belly are finely barred in gray. Tail is long, black, and white banded with a narrow white terminal band. Wingtips are dark with a narrow dark trailing edge. **FIRST YEAR:** Crown is dark brown. White supercilium, dark eye stripe, white cheeks, and dark malar mark the face. Breast and belly are boldly spotted in dark brown. Upper parts are dark brown. Upper tail has white band like a Northern Harrier. Tail is long and pale gray-brown with dark narrow bands. **VOICE:** High-pitched, plaintive

peeooooooooooo, descending slightly in pitch.
BEHAVIORS: Flight is like a Cooper's Hawk, a rapid series of wingbeats followed by a glide. Wings are held flat when gliding and soaring. When soaring, tail is fanned. **HABITAT:** Usually mesquite woodlands with taller trees for nesting. **STATUS:** Rare to locally uncommon resident along the Rio Grande from Hidalgo County to Webb County. Has nested in Val Verde County. Rare summer resident in the Davis Mountains and along the Rio Grande in Brewster and Presidio counties, where birds arrive in mid-March and depart by mid-October.

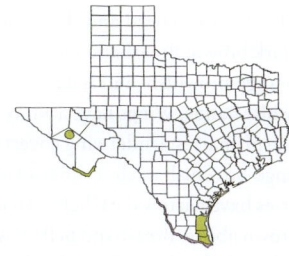

Adult

Immature

Red-shouldered Hawk

(Buteo lineatus)

IUCN RED LIST STATUS (2016): Least Concern
POPULATION TREND: Increasing

The raucous call of a pair of Red-shouldered Hawks is common in its range in Texas. Our suburban neighborhoods actually mimic its preferred natural habitat, and they can do very well nesting in them. Most of their diet consists of small mammals they swoop down on from a perch. They will readily consume reptiles and amphibians if mammals are not available.
LENGTH: 17 inches. **WINGSPAN:** 40 inches.
ADULT: Rusty brown head and shoulders. Back is dark brown. Breast and belly are barred orange-brownish. Tail is dark with narrow white bands. Wings have a white crescent in the primaries. Underwing coverts have orange-brownish wash. Primaries and secondaries have narrow dark bars. **FIRST YEAR:** Brown above. Breast and belly have even, bold dark streaks. Tail is reddish-brown with even dark bands. Wings have white crescents in the primaries like adults but lack the orange-brownish wash on underwing coverts. **VOICE:** Call is a loud, screeching *crah-crah-crah-crah*, often given at length. **BEHAVIORS:** Soars with wings flat and tail spread. When hunting, flies slow and directly at prey. **HABITAT:** Mature forest with open parklike understory. Common in suburban areas with mature trees. **STATUS:** Common to uncommon resident in the eastern two-thirds of Texas south to the Nueces River. Uncommon winter resident south of the Nueces River. In South Texas, birds arrive in late September and depart by late March.

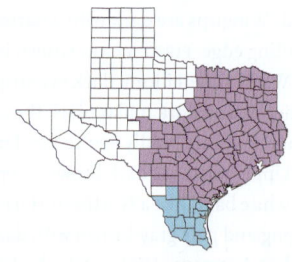

Immature

Adult

Migrating

Broad-winged Hawk

(Buteo platypterus)

IUCN RED LIST STATUS (2016): Least Concern
POPULATION TREND: Increasing

Broad-winged Hawks are the most numerous hawk in North America but are secretive and seldom seen while breeding. In migration they become conspicuous, moving silently across the sky in flocks of thousands.
LENGTH: 15 inches. **WINGSPAN:** 4 inches.
ADULT: All brown above. Belly and breast heavily barred in brown; on breast the bars merge almost into a hood. Light underwings have narrow dark border. Dark tail has one white band. **FIRST YEAR:** Brown above and heavily streaked with brown below. Underwings are pale with dark tips. Tail is pale gray with narrow dark bands and a wider dark terminal band. **DARK MORPH:** Essentially the same as light morphs previously described, but underwing coverts, belly, and breast a dark chocolate brown. **VOICE:** High-pitched, piercing *ti-teeeeeeeeeee*. **BEHAVIORS:** Makes

short flights through the trees, moving from branch to branch. Soars over the forest during breeding season. In migration, forms flocks that number into the thousands, traveling almost exclusively by soaring. **HABITAT:** Nests in continuous or mixed-deciduous forest with openings and water nearby. **STATUS:** Common to abundant migrant throughout the eastern half of the state. Uncommon to very rare going west from the eastern edge of Edwards Plateau. Uncommon to common summer resident in the Pineywoods and uncommon west to north-central Texas and Edwards Plateau. Spring migration is mid-March–mid-May, fall late August–mid-October.

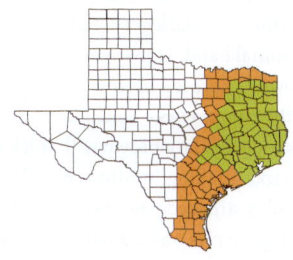

Swainson's Hawk

(Buteo swainsoni)

IUCN RED LIST STATUS (2016): Least Concern
POPULATION TREND: Stable

Swainson's Hawks, named for illlustrator William Swainson, largely leave North America for the winter, and some birds make a one-way migration of up to 6000 miles from Canada to southern South America. During this migration, they often form large flocks. **LENGTH:** 19 inches. **WINGSPAN:** 51 inches. **ADULT:** Gray head with white face and throat. Breast is dark rufous-brown. Belly is white. Underwing coverts are creamy white and flight feathers are dark gray. Tail is gray with dark terminal band. Upper parts are dark gray-brown. There is a narrow pale U at the base of the rump. **FIRST YEAR:** Like adults but breast is mottled dark brown with thick dark malar marks, making "mutton chops" on the face. Head is streaked pale and can look white. Underwing coverts washed reddish-brown. **VOICE:** Raspy, piecing *chaaaaawwww*.

BEHAVIORS: Strong graceful flapping flight. Soars with wings in a dihedral. "Kites," or hangs motionless in the wind, when foraging. **HABITAT:** Grasslands, sparse shrubland, and open woodland. Can use agricultural areas if crops are not higher than native grasses and hiding prey. **STATUS:** Common and locally abundant migrant statewide except the Pineywoods. Uncommon to common summer resident from the Panhandle south to Edwards Plateau. Rare to uncommon summer resident in the South Texas Brush Country and coastal prairies. Spring migration is mid-March–mid-May, fall early August–early November.

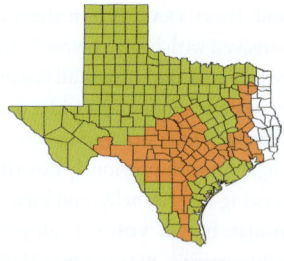

Zone-tailed Hawk

(Buteo albonotatus)

IUCN RED LIST STATUS (2020): Least Concern
POPULATION TREND: Stable

Zone-tailed Hawks could be thought of as always undercover. They mimic the common Turkey Vulture in pattern and flight style and may even loosely associate with a kettle of Turkey Vultures. But unlike the carrion-foraging Turkey Vulture, the Zone-tailed Hawk is seeking live prey. In flight, it can be distinguished from a Turkey Vulture by its much larger feathered head and the long banded tail. The bright yellow legs often also stand out. **LENGTH:** 20 inches. **WINGSPAN:** 51 inches. **ADULT:** All dark charcoal-gray. Tail is long and has gray bands. Underwings are dark with gray-checkered flight feathers. In flight, the bright yellow legs stand out against the dark undertail. **FIRST YEAR:** Like adult but tail is narrowly banded in gray. **VOICE:** Long drawn-out *kwaaaaaaaaaaaaaaaa*.

BEHAVIORS: Soars like a Turkey Vulture. **HABITAT:** Riparian forest, desert uplands, and mixed-conifer forests. **STATUS:** Uncommon and local summer resident in the mountains of the Trans-Pecos east through the southern Edwards Plateau. Rare in summer in the Guadalupe Mountains. Rare winter resident in the Lower Rio Grande Valley. Occurs between late March to mid-October in the west. Winter records fall between early November and mid-April.

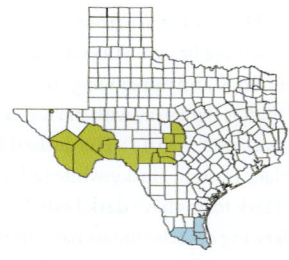

Harlen's form, juvenile

Red-tailed Hawk

(Buteo jamaicensis)

IUCN RED LIST STATUS (2016): Least Concern
POPULATION TREND: Increasing

In winter in Texas, any hawk perched on the roadside is probably a Red-tailed Hawk until proven otherwise. Its darkest subspecies is known as a Harlen's Hawk.
LENGTH: 19 inches. **WINGSPAN:** 49 inches.
ADULT: Extremely variable. All forms have dark bars on the leading edge of the wing close to the body. Underwing flight feathers are light with dark tips. Most forms have a bright red to pink-washed tail. Underwing coverts and body can vary from almost white to dark chocolate brown. **FIRST YEAR:** Like adults but body and upper wings tend to be mottled. Tail is light gray with narrow dark bands. **HARLEN'S, ADULT:** The darkest form, almost black overall. Dusky-white tail lacks any red. Dark forms have dark body feathers and underwing coverts. Breast has variable white streaking. The light form is clear white below with no buffy wash. **HARLEN'S, FIRST YEAR:** Like adult dark form but underwings mottled white. Tail is gray with narrow dark bands. **VOICE:** Raspy, screaming *skrawwwww-wwwww*. **BEHAVIORS:** Flapping flight is slow and ponderous. Soars and kites on updrafts. Most hunting is from a perch. **HABITAT:** Open-canopy forest and woodlands, grasslands with perches for hunting. **STATUS:** Common resident virtually statewide. Common to uncommon summer resident statewide, except uncommon in the South Texas Brush Country. Absent as breeders in the Lower Rio Grande Valley. Common migrant and winter resident statewide.

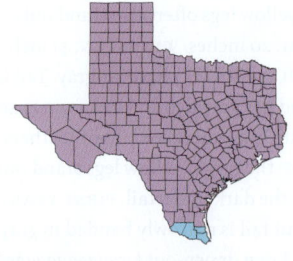

Eastern form

Eastern form

Krider's form

Harlen's form, adult

Western form

Rough-legged Hawk

(Buteo lagopus)

IUCN RED LIST STATUS (2021): Least Concern
POPULATION TREND: Stable

The Rough-legged Hawk is named for the feathers on its legs, which extend all the way to its feet.
LENGTH: 21 inches. **WINGSPAN:** 53 inches.
ADULT: Light morph is streaky gray-brown above with a distinctly paler head. Tail has single dark band at the end. Underwings have bold, square dark patches at the "wrists." Flight feathers are white with dark tips. Breast and underwing coverts are streaked gray-brown, and belly is dark. Dark morph is rare in Texas. Almost black overall. White undertail with dark band at the end. Flight feathers are white with dark tips. **FIRST YEAR:** Light morph is like adult but tail has a single, wide gray band. Upper wings show a pale patch in the primaries. Dark morph is like adult but underwing coverts are brownish as a darker square "wrist" patch is visible. Pale upper wing patch is visible. Tail band is gray and not blackish as an adult. **VOICE:** A piercing *skweeerrrrrrrr*. **BEHAVIORS:** Leisurely and flexible wingbeats alternate with glides. Often kites, or hovers, high in the air before pouncing on prey. **HABITAT:** Mostly open, treeless areas. **STATUS:** Uncommon to rare winter resident in the Panhandle and the South Plains. Fall birds begin to arrive in mid-October, and most have departed by early March.

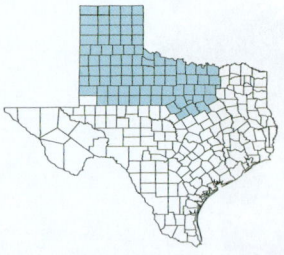

Ferruginous Hawk

(Buteo regalis)

IUCN RED LIST STATUS (2021): Least Concern
POPULATION TREND: Increasing

The largest of North America's hawks has two color morphs, light and dark.
LENGTH: 23 inches. **WINGSPAN:** 56 inches.
ADULT: Light morph's head is light with gray cheeks. Snowy white below. Legs are feathered in rusty brown and make a rusty brown V in flight. Underwings are light with a rusty wash on the underwing coverts. Upper wings are dark with rusty shoulders and back. White patch on upper wing. Undertail is white; upper side a rusty color. Dark morph is rich chocolate brown overall with long yellow gape. White patch on upper wing. Undertail is white; upper side of tail is light gray. **FIRST YEAR:** Light morph has pale head. Brown back and upper wings. White patch on upper wing. Underwing including the coverts is mostly clean white. Tail is pale gray below, gray above. Dark morph is like adult but lacks any rusty

tones and upper tail is a slate color. **VOICE:** Screeching *wraaaaaaaah*. **BEHAVIORS:** Flies low to the ground when approaching prey. **HABITAT:** Open terrain from grassland to desert, especially where prairie dogs are present. **STATUS:** Common to locally uncommon winter resident in the Panhandle, South Plains, and Trans-Pecos. Rare to casual migrant and winter resident in the rest of the state. Rare summer resident in the western and central Panhandle. Fall birds arrive in early October and depart mid-February–mid-March.

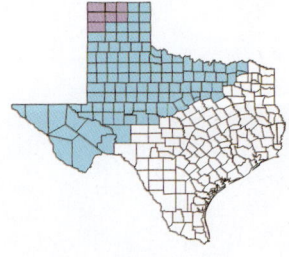

TYTONIDAE
(BARN OWLS)

Barn Owls are sometimes known as Monkey-faced Owls because of ther prominent facial disks. They are generally slim, medium-sized owls with long, sparsely feathered legs. They seem more suited to warm climates than the typical owls. Many Barn Owl species are endemic to tropical islands.

The Barns Owl's facial disk funnels sound to its asymmetric ear openings. Having ears on both sides of the head allows it to locate sounds from right to left, while the asymmetric placement of the ear openings allows it to pinpoint sounds on the vertical axis too. Barn Owls have the most sensitive hearing of any animal tested. They can hunt in total darkness and locate prey under snow or leaf litter.

Worldwide there are eighteen species of Barn Owls, though only one is known to Texas. Many have very limited ranges, and 31% of Barn Owl species are of conservation concern. While the North American Barn Owl has been extensively studied, most of the other species have not.

American Barn Owl

(Tyto furcata)

IUCN RED LIST STATUS (2019): Least Concern
POPULATION TREND: Stable

The Barn Owl is one of the most cosmopolitan of land bird species, with a range on all continents except for Antarctica. While well-studied in Europe and North America, most of its twenty-eight subspecies have yet to be studied in detail. Highly adaptable, it can exist close to humans and, as its name implies, readily use man-made structures for nesting. Its range is only limited by the availability of small mammals for prey (mostly rodents) and extreme cold temperatures. When prey is abundant, American Barn Owls may nest in any month of the year, and typically monogamous pairs may raise two or three broods a year.

LENGTH: 16 inches. **WINGSPAN:** 42 inches.
ADULT: Male is very pale below. Face is white, flattish, and heart-shaped. A tawny ruff surrounds the facial disk. Upper parts are light tawny with gray mottling and gray and black spots. In flight the long legs extend slightly beyond the tail. Female is like male but underparts tawny with more prominent spotting. **VOICE:** Raspy, breathy, drawn-out *skraaaaaaaaaaaaah*. **BEHAVIORS:** In flight has deep, buoyant, and slow wingbeats. **HABITAT:** Able to use most open habitats with available prey. **STATUS:** Rare to locally uncommon resident throughout most of the state. Very rare in the Pineywoods and avoids closed canopy forest. Locally common on the coastal prairies.

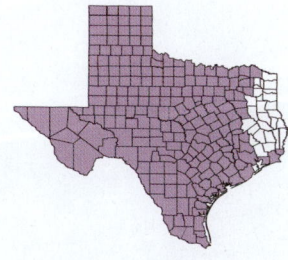

STRIGIDAE (OWLS)

The typical owls are small to large, mostly nocturnal predatory birds. Owls generally have round heads and flat faces with large eyes. Many species have earlike or hornlike feather tufts on the head that aid in camouflage and displays. Like the closely related barn owl family, they have asymmetric ear openings that aid in locating prey by sound.

Soft feathers allow owls to fly almost silently. The leading edge of the forward primary feather on the wing is serrated. This disrupts the airflow to reduce wingtip noise. Owls are able to silently pounce on prey from above. Small food items are swallowed whole, but larger prey is ripped apart with their powerful beaks.

DNA evidence shows that nighthawks and nightjars are the closest relatives to owls. Similarities to hawks are the result of convergent evolution due to similar feeding methods; they are not closely related. Worldwide there are 229 species of typical owls. Fifteen species have been recorded in Texas. Five of these are considered rare visitors to Texas: Snowy Owl, Northern Pygmy-Owl, Mottled Owl, Stygian Owl, and Northern Saw-whet Owl.

Flammulated Owl

(Psiloscops flammeolus)

IUCN RED LIST STATUS (2016): Least Concern
POPULATION TREND: Decreasing

These tiny owls have an unusually deep voice for a bird of their size. The males have a specialized muscle in the syrinx, the bird equivalent of the larynx, to produce these surprisingly deep sounds. Flammulated Owls may benefit from the illusion that they are a much larger owl much farther away.

LENGTH: 6.75 inches. **WINGSPAN:** 16 inches.
ADULT: The only small owl with dark eyes. Short, rounded ear tufts, usually flattened. Facial disk is incomplete, only showing from the eat tufts to the "mustache." Vermiculated gray overall. Red form has flashes of rufous color.
VOICE: Deep, resonant *whoot* repeated every 2 to 3 seconds. Also, a three-note *who-who whooo*.
BEHAVIORS: Flies quicker than other owls while foraging. **HABITAT:** Open and near-open conifer forest. **STATUS:** Uncommon to rare summer resident in the Chisos, Davis, and Guadalupe mountains above 6000 feet. Rare migrant in the western third of the state.

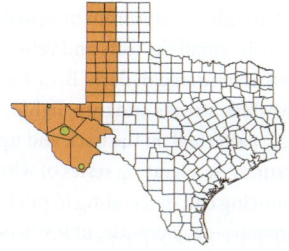

Western Screech-Owl

(Megascops kennicottii)

IUCN RED LIST STATUS (2016): Least Concern
POPULATION TREND: Decreasing

Western Screech-Owl was considered cospecific with the Eastern Screech-Owl until 1983, when the American Ornithological Union recognized it as a separate species. Despite being an abundant owl, this relatively recent elevation to a full species means it has not yet been studied on its own in great detail. Much of its basic biology is still unknown, such as breeding success, habitat use, and home-range size. **LENGTH:** 8.5 inches. **WINGSPAN:** 20 inches. **ADULT:** Variable in color from brownish to plain grayish. Small ear tufts and yellow eyes. Facial disk is bordered in dark. Breast and belly have dark streaks and thin, lighter bars. Intricately patterned on the back and upper wings. **VOICE:** Accelerating series of whistles, like a bouncing ball descending in pitch: *pwep pwep pwep-pwep-pwep-pep-pep.* **BEHAVIORS:** A sit-and-wait predator; perches on a projecting twig and captures prey on the ground. Will also hawk insects from the air. **HABITAT:** Woodland habitats in the Trans-Pecos and live oak–mesquite savannah on the western Edwards Plateau. **STATUS:** Common to uncommon resident in the southern Trans-Pecos west to western Kerr County and north to the Concho Valley.

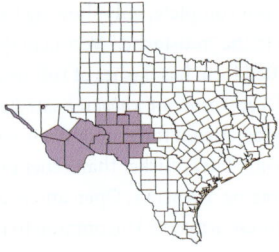

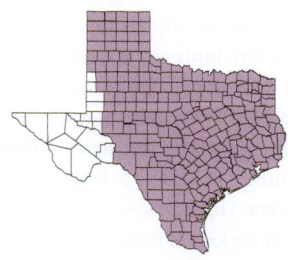

Eastern Screech-Owl

(Megascops asio)

IUCN RED LIST STATUS (2019): Least Concern
POPULATION TREND: Decreasing

Eastern Screech-Owl is often the only small owl species present in its range. It is a non-specialist when it comes to habitat and can use almost any wooded habitat with an open understory. It adapts well to being around humans and will readily take to nest boxes when natural cavities are in short supply. It can be secretive but common in urban neighborhoods with enough trees to support it.

The Mexican subspecies *Megascops asio mccallii* presents in the Lower Rio Grande Valley and the South Texas Brush Country, and it differs from other Eastern Screech-Owls by its smaller size. Some authorities have proposed that it be called a separate species. **LENGTH:** 8.5 inches. **WINGSPAN:** 20 inches. **ADULT:** Variable in color, from rufous to brown to gray. Small ear tuffs and facial disk outlined in dark. Breast and belly have dark streaks with thin, lighter bars. Intricately marked on the back and upper wings. Bright yellow eyes. **VOICE:** Call is a descending whinny. Also, a tremolo song of a long trill on one patch up to 3 seconds in length. **BEHAVIORS:** Hunts from a perch, usually right below the canopy. Nests in cavities and can often be found roosting in a tree cavity. **HABITAT:** Suburban and forest landscapes with an open understory. **STATUS:** Common resident in the eastern three-quarters of Texas, roughly east of the Pecos River.

Great Horned Owl

(Bubo virginianus)

IUCN RED LIST STATUS (2018): Least Concern
POPULATION TREND: Stable

The Great Horned Owl is nearly the same size as a Red-tailed Hawk and outweighs it. With extremley powerful talons that can sever the spine of prey and a powerful beak for tearing it apart, Great Horned Owls are arguably one of the most powerful birds in North America. They have one of the most varied diets and are in all North American habitats except the arctic tundra.
LENGTH: 22 inches. **WINGSPAN:** 44 inches.
ADULT: Large ear tufts are unmistakable. The face of eastern birds is tawny-colored, and gray in western birds. White throat. Densely barred below. Underwings are pale with bold dark comma marking the "wrist." **VOICE:** Deep, muffled hooting: *hoo hoodoo hoooo hoo.*
BEHAVIORS: Flight is powerful and straight, with short periods of flapping followed by gliding. Wings are held level in a glide. Hunts from a perch and grabs prey with strong talons. **HABITAT:** Wide variety of habitats, but mostly open and secondary growth. Usually the home range includes some open fields. **STATUS:** Common resident statewide except for the Pineywoods, where it is uncommon.

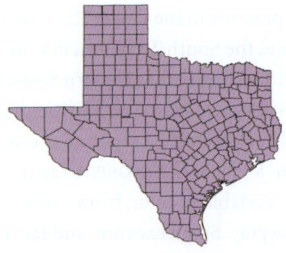

Ferruginous Pygmy-Owl

(Glaucidium brasilianum)

IUCN RED LIST STATUS (2022): Least Concern
POPULATION TREND: Decreasing

Ferruginous Pygmy-Owl is a common tropical owl whose range barely includes deep southern Texas (and a small part of Arizona). Formerly found in several publicly accessible locations along the Rio Grande River in South Texas, the population seems likely to have been extirpated by flooding in 2010. Presently, most of the Texas population is found on a couple of very large ranches in Kenedy and Brooks counties, where a small number of tours allow lucky birders to record this rare North American owl on their life list.
LENGTH: 6.75 inches. **WINGSPAN:** 12 inches.
ADULT: Pale orange-brown overall. Crown is streaked. Sides are unmarked brown with heavy brown streaks. Dark patches on nape mimic eyes in back of the head. Tail is orangish with broad pale bars. Underwing is pale.
JUVENILE: (Apr.–Aug.) Like adult but less rusty-colored; browner with almost no crown streaking. **VOICE:** High-pitched, monotonous series of whistled notes: *pwip pwip pwip pwip*.
BEHAVIORS: Short, direct flights with rapid wingbeats. **HABITAT:** In Texas, found in undisturbed live oak–mesquite forest and mesquite brush, ebony, and riparian areas. **STATUS:** Uncommon and local resident on the Coastal Sand Plains in Kenedy County and eastern Brooks County. Formerly a rare resident in the riparian areas of the Lower Rio Grande Valley, until the floods of 2010. It is unknown if any of that population still exists.

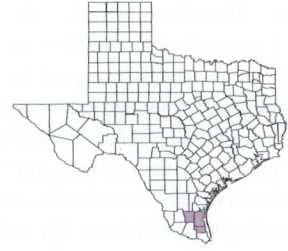

Elf Owl

(Micrathene whitneyi)

IUCN RED LIST STATUS (2020): Least Concern
POPULATION TREND: Decreasing

Elf Owl is the smallest owl in the world and can be found in two populations along the Texas–Mexico border. Fiercely territorial, nesting pairs plus neighboring ones will mob and attack the head of predators many times their size. When handled, female Elf Owls are known to feign death, perhaps in an attempt to lure a predator into loosening its grip. **LENGTH:** 5.75 inches. **WINGSPAN:** 13 inches. **ADULT:** Brown facial disk with bold white "eyebrows." Blurry brown streaks on breast. Back and upper wings are speckled gray. **JUVENILE:** (May–Sept.) Gray facial disk with gray barred breast. Speckled gray above. **VOICE:** High-pitched *hup-hup-hup-hup-pa*. **BEHAVIORS:** Flies with uniformly rapid wingbeats in a straight line when hunting. Occasionally glides and hovers. Climbs like a parrot in trees. Nests and roosts in old woodpecker holes. Roosts in dense mistletoe and especially dense evergreen trees, like alligator junipers. May form flocks in migration. **HABITAT:** Riparian forest and riparian canyon forest, desert-wash woodlands, upland desert. **STATUS:** Uncommon to locally common summer resident, from the southern Trans-Pecos eastward to the western edge of the Edwards Plateau and the Lower Rio Grande Valley. Spring birds arrive in early March to early April. In fall, birds leave between mid-September and mid-October.

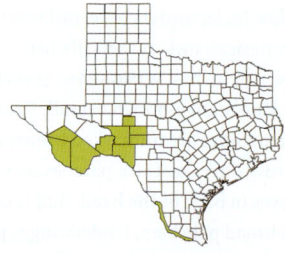

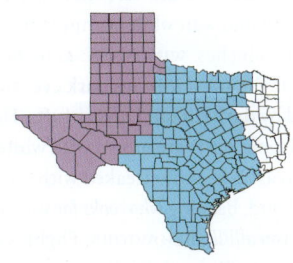

Burrowing Owl

(Athene cunicularia)

IUCN RED LIST STATUS (2016): Least Concern
POPULATION TREND: Decreasing

The charismatic Burrowing Owl is unique among North America's owls by being active both day and night. It nests in underground burrows, typically in small groups. In Texas, it's often found in prairie dog towns but also uses the burrows of many other species or, when natural burrows aren't available, artificial ones. Numbers have rapidly declined, likely caused by conversion of habitat to agriculture.
LENGTH: 9.5 inches. **WINGSPAN:** 21 inches.
ADULT: Pale brown with white spots above. Brown face with white throat and white "eyebrows." Spotted, almost barred brown on breast. Legs are long and extend beyond the tail in flight. **JUVENILE:** (Apr.–Sept.) Unmarked buffy belly and buffy throat. Upper wing coverts are pale tawny, making a tawny patch on upper wing in flight. **VOICE:** Nasal *huk-coooo*. **BEHAVIORS:** Flies low to the ground using slow wingbeats. Hunts by flying, most often by hover-hunting. **HABITAT:** Breeding habitat is typically open, treeless areas with low, sparse vegetation. Often closely associated with prairie dog towns. In winter, will use culverts, pipes, and rock piles instead of burrows for roosts. **STATUS:** Uncommon to common summer resident and uncommon rare winter resident in the western half of the state. Rare to locally uncommon migrant and winter resident in the Oaks and Prairies, Lower Rio Grande prairie, and coastal prairies. Fall migration is mid-September–mid-October, spring late March–late April.

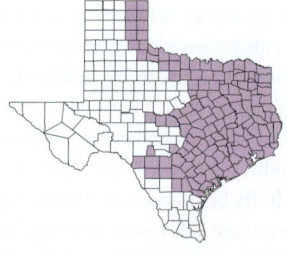

Barred Owl

(Strix varia)

IUCN RED LIST STATUS (2016): Least Concern
POPULATION TREND: Increasing

If you've ever been in the East Texas bottom-land forests at night, chances are you've heard the echoing call of Barred Owls: *who cooks for you, who cooks for you allllll.* They also give an impressive array of bark, squawks, and hoots that can raise goose bumps when one finds themself close to a calling owl.

Barred Owls are generalist feeders and will take small mammals, birds, reptiles, amphibians, fish, and invertebrates. A nocturnal hunter, it will sit and wait while scanning for prey. **LENGTH:** 21 inches. **WINGSPAN:** 42 inches. **ADULT:** Light gray face with dark eyes surrounded by dark ruff and no ear tufts. Head is pale. Upper parts are brown with white spots. Breast is heavily streaked with brown. **VOICE:** Loud, barking *who cooks for you, who cooks for you allllll.* **BEHAVIORS:** Flight is light and buoyant. Glides skillfully among tree branches. **HABITAT:** Large, unfragmented blocks of mature forest. **STATUS:** Uncommon to common resident in the eastern two-thirds of Texas south to Nueces County.

Long-eared Owl

(Asio otus)

IUCN RED LIST STATUS (2021): Least Concern
POPULATION TREND: Decreasing.

The secretive Long-eared Owl is seldom observed. By day it forms communal roosts. Most roosts are numbered 2 to 20 birds, but roosts of over 100 have been recorded. Long wings and light bodyweight make Long-eared Owls active search-hunters. Its main prey are small rodents that it locates by sound. **LENGTH:** 15 inches. **WINGSPAN:** 36 inches. **ADULT:** Tawny-orange face with dark vertical stripes through the eyes. Long ear tufts that can be raised and lowered. Dark streaking and barring in a pattern like tree bark. In flight, underwing is pale with a dark comma mark at the "wrist." Upper wing has tawny patches in the primaries and pale gray coverts. **VOICE:** Soft, hooting *woooo* given about every 3 seconds. **BEHAVIORS:** Agile flight through dense vegetation. Long glides on level wings. Wingbeats are deep. **HABITAT:** Often forms communal roosts in dense vegetation adjacent to open areas for foraging. **STATUS:** Uncommon migrant and winter resident in the Panhandle, and rare to locally uncommon migrant and winter resident on the South Plains. Fall birds arrive between late September and early December. Spring birds depart early March through mid-April.

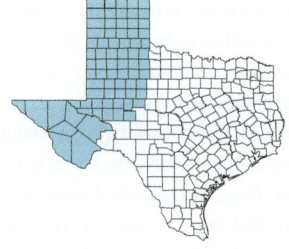

Short-eared Owl

(Asio flammeus)

IUCN RED LIST STATUS (2021): Least Concern
POPULATION TREND: Decreasing

Short-eared Owls are one of the most widely distributed owls in the world. This bird of open country spends its time in open grasslands, roosting and nesting on the ground. Often active early and late in the day while the sun is up, it hunts low over the ground with slightly dihedral wings. In recent decades, the North American population has declined, with loss of open grassland habitat appearing to be the major cause.

LENGTH: 15 inches. **WINGSPAN:** 38 inches.
ADULT: Male has light grayish facial disk with dark "mascara" around the eyes. Short ear tuffs. Pale below with grayish-brown streaking. Boldly barred wingtips with tawny wing patches in primaries. Slightly checkered gray-brown on upper wings and mantle. Female is like males but more tawny-colored than gray. **VOICE:** Short, nasal *rawl* like a hoarse kitten. Drawn-out screeching *craaaa-ooooow*.
BEHAVIORS: While foraging, flight is mothlike and buoyant. Wingbeats are slow and deliberate. **HABITAT:** Open areas with dense grass.
STATUS: Rare to locally uncommon migrant and winter resident across the northern third of the state, the Oaks and Prairies region, and the coastal prairies. Fall birds begin to arrive in mid-October. In spring, birds depart between early March and mid-April.

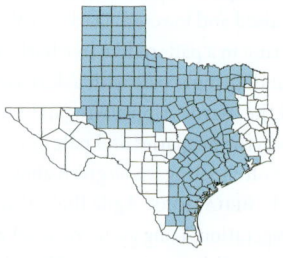

ALCEDINIDAE (KINGFISHERS)

Kingfishers range in size from tiny pygmy kingfishers to the crow-sized kookaburras of Austrialia. Kingfishers are found in waterside and wooded habitats worldwide. As the name suggests, many species are skilled at catching small fish, but other species eat no fish and are skilled hunters in the forests, diving to snatch lizards and insects from the trees and ground. Those species that do live an aquatic life have a special off-axis, cam-shaped lens in the eye. This allows them to switch instantly between focal points as they dive into the water.

Worldwide there are 117 species of kingfisher. Of those, 42 are of conservation concern. One, the Guam Kingfisher, is extinct in the wild but being bred in captivity. Many species of concern are endemic to small island groups. In Texas, four species have been recorded. One, the Amazon Kingfisher, is a very rare visitor to Texas.

Ringed Kingfisher

(Megaceryle torquata)

IUCN RED LIST STATUS (2020): Least Concern
POPULATION TREND: Stable

Ringed Kingfisher is the largest of North America's kingfishers, about 25% larger than the more common Belted Kingfisher. It is a relative newcomer to Texas, first recorded in 1888, and the first nest wasn't recorded until 1970. Ringed Kingfisher continues to expand its range. There are records from the upper Texas coast, Dallas County, and even Oklahoma and Louisiana, and it has the potential to occur almost anywhere in Texas.
LENGTH: 16 inches. **WINGSPAN:** 25 inches.
ADULT: Male is brick-red below and blue on the back. White throat and blue head. Massive bill is dark with pale base. There is a small dark spot in front of the eye. Underwings and undertail coverts are white. Female is like male but breast has a blue band. Underwing and undertail coverts are the same brick-red as the breast and belly. **VOICE:** Loud *clack* similar to a Great-tailed Grackle, often repeated in long rapid series in flight. **BEHAVIORS:** Strong slow wingbeats, often so low their wingtips touch the water. Flies over land more often than other kingfishers. **HABITAT:** Water that supports fish and near-vertical banks for nesting. Avoids water that is overgrown with vegetation. **STATUS:** Locally common resident in the Lower Rio Grande Valley west to Webb County. Uncommon and local in the South Texas Brush Country.

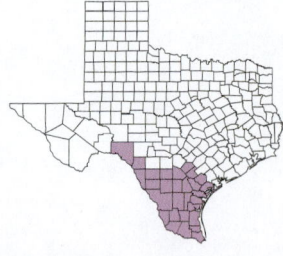

Belted Kingfisher

(Megaceryle alcyon)

IUCN RED LIST STATUS (2022): Least Concern
POPULATION TREND: Stable

Belted Kingfisher is the most widely distributed kingfisher in Texas. If there is enough water to support it, there is a good chance one will winter there. It's not uncommon to see one miles from a body of water, intently staring down into a wet ditch from a telephone line. Like the phalaropes, females are more colorful than the males, sporting dashes of bright rufous.

LENGTH: 13 inches. **WINGSPAN:** 20 inches. **ADULT:** Male has blue head and upper parts with a wide white collar. Shaggy crest and small white dot in front of the eye. Dark bill is pale gray at the base. Dark blue-gray breast band. In flight, white wing patch in primaries. Female is like male but with rusty breast band and sides. **JUVENILE:** (May–Sept.) Like female but blue breast band is mostly rufous or blue with rufous mottling. **VOICE:** Long uneven rattle, like woodpeckers. **BEHAVIORS:** Hunts from a perch over water. Flight is strong and direct, with alternating flapping and gliding. Will hover while hunting. **HABITAT:** Needs clear water with near-vertical banks for nesting. In winter, almost any body of water with perches and prey. **STATUS:** Uncommon to locally common winter resident throughout the state. Uncommon and local summer resident in the northern third of the state.

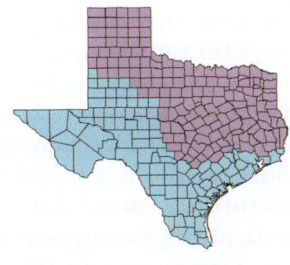

Green Kingfisher

(Chloroceryle americana)

IUCN RED LIST STATUS (2020): Least Concern
POPULATION TREND: Decreasing

Green Kingfisher is the smallest kingfisher in Texas with perhaps the proportionally largest bill. About the size of a Northern Cardinal, it is still a fish eater, chasing after the small minnows that venture too close to its hunting perch. With their barely audible clicking calls and habit of sticking close to the bank and overhanging shrubs, they are inconspicuous along clear rivers and streams in Texas. **LENGTH:** 8.75 inches. **WINGSPAN:** 11 inches. **ADULT:** Male's head is green with a proportionally large dark bill. White collar and green back. Upper wings are blackish with white dots. Underparts are white with rufous breast band. White outer tail feathers are conspicuous in flight. Female is like male, with two greenish breast bands. **VOICE:** Soft click like two stones tapping together. **BEHAVIORS:** Flight is rapid and low over water. Often perches under overhanging branches. **HABITAT:** Fresh clear water surrounded by bushy vegetation for perching. **STATUS:** Uncommon resident from the Edwards Plateau south to the Lower Rio Grande Valley and north to Victoria County on the coastal prairie.

PICIDAE (WOODPECKERS)

Woodpeckers are known for clinging upright to trees, supported by their stiff tails. Most species have zygodactyl feet, or two toes pointed forward and two pointed backward, and short legs for clinging to a tree. They have sharp bills that are reinforced at the skull for drilling into wood for food and to excavate nest holes. Woodpeckers live in every tree habitat, some non-treed habits above the treeline, and some grasslands.

Most woodpeckers excavate a cavity in wood. These nest cavities are seldom reused. Many species of cavity-nesting birds, from tiny chickadees to Wood Ducks, depend on woodpecker cavities for nest sites.

There are 235 species of woodpeckers worldwide. These are divided into three subfamilies: the wrynecks of the Old World, the piculets found in both the New World and Old World tropics, and the typical woodpeckers found worldwide. Fifteen species have been recored in Texas. One, the Ivory-billed Woodpecker, is widely considered extinct. The Red-breasted Sapsucker is a very rare visitor to Texas.

Lewis's Woodpecker

(Melanerpes lewis)

IUCN RED LIST STATUS (2016): Least Concern
POPULATION TREND: Decreasing

The Lewis's Woodpecker is named for Meriwether Lewis of the Lewis and Clark expedition, where the first specimens were collected. Unusual for woodpeckers, they don't peck into wood for wood-boring insects but instead hunt them by gleaning them off tree trunks or hawking them out of the air. In winter, their diet switches to nuts, acorns in native forests, but also pecans according to many records in Texas. **LENGTH:** 10.75 inches. **WINGSPAN:** 21 inches. **ADULT:** Dark red face. Dark glossy-green crown and pale gray collar. Pale gray below with pink belly. Dark glossy-green on back. Long wings are dark above and below. **JUVENILE:** (July–Nov.) Gray-green head and dark glossy-green on back. Gray-green below, paler on belly. **VOICE:** Rattling *kreeiiip kreeeiiip kreeeiiip*. **BEHAVIORS:** Flight is crow-like, like rowing through the air. Spends considerable time gliding and hawking insects. Gleans insects off trunks of trees. **HABITAT:** Oak woodland, commercial nut orchards, pecan bottomland forests. **STATUS:** Irregular and very rare migrant and winter visitor in the eastern two-thirds of the state. They are irruptive in their pattern of occurrence in Texas. Fall migrants generally arrive around mid-October through November, and birds usually depart by late March.

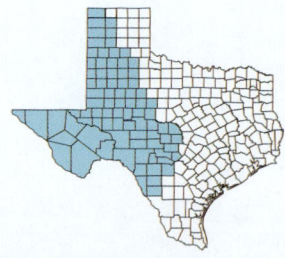

Red-headed Woodpecker

(Melanerpes erythrocephalus)

IUCN RED LIST STATUS (2018): Least Concern
POPULATION TREND: Decreasing

One of the most recognizable of North American birds, its all-red head is unique among woodpeckers. They are also one of the few woodpeckers that cache insects and nuts for later consumption while foraging. They are skilled flycatchers and will perch high on an open perch, sally out for insects, and return to their perch with their prize.
LENGTH: 9.25 inches. **WINGSPAN:** 17 inches.
ADULT: Bright red head and a dark back with a wide white rump. White secondaries. Underparts are all snowy white. **JUVENILE:** (July–Feb.) Like adult but head is brown. White secondaries have two dark bars across them. **VOICE:** Wheezy *quuuur*. **BEHAVIORS:** Flight is less undulatory than other woodpeckers. Often fly-catches from a perch. **HABITAT:** Wide variety of treed habitat, usually with dead limbs or trees for nesting. **STATUS:** Locally common to rare in riparian habitat from the western Panhandle to north-central Texas. Uncommon to locally common resident in the Pineywoods. Rare to locally uncommon in the Post Oak Savannah.

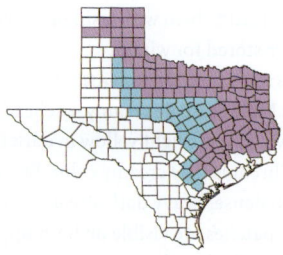

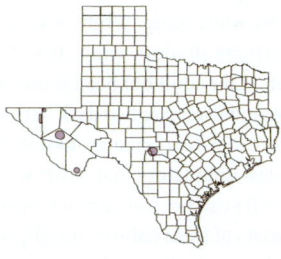

Acorn Woodpecker

(Melanerpes formicivorus)

IUCN RED LIST STATUS (2016): Least Concern
POPULATION TREND: Increasing

The Acorn Woodpecker's bold harlequin facial pattern, with a bright yellow eye in a dark mask that always looks startled, is distinctive. In Texas, they are restricted to the oak and oak-pine forests of some of the western sky islands. Even if not seen, their presence can often be detected by one of their larder trees, telephone poles, or even the sides of wooden buildings. Acorn Woodpeckers will sometimes make hundreds of small holes, and a family group will stuff them with a single acorn each, which are stored for winter.
LENGTH: 9 inches. **WINGSPAN:** 17.5 inches.
ADULT: Male has pale creamy forehead and red crown. Eye is bright yellow in dark face mask. Throat is pale creamy color. Breast is pale with dense, heavy dark streaks. White primary patches are visible on both upper and underwing. Back and upper wings are glossy black with wide white rump. Female is like male but forecrown is dark. **VOICE:** Breathy *heka heka heka*. **BEHAVIORS:** Typical undulating flight of woodpeckers. Stores acorns in numbers in small holes it makes in trees and telephone poles. **HABITAT:** Oak and pine-oak woodlands. **STATUS:** Common resident in the Chisos, Davis, and Del Norte mountains. Uncommon and local in the Sierra Diablo and Guadalupe mountains. Small population on private property in Bandera, Kerr, and Real counties.

Golden-fronted Woodpecker

(Melanerpes aurifrons)

IUCN RED LIST STATUS (2016): Least Concern
POPULATION TREND: Stable

The Golden-fronted Woodpecker ranges south to Nicaragua and north to Oklahoma through a good part of the center of Texas. Golden-fronted Woodpeckers consume about equal parts fruit and nuts as they do insects. Summer birds have been noted with their faces stained purple from prickly pear fruit and the wild grapes that grow abundantly in central Texas, which are a favorite food source. **LENGTH:** 9.5 inches. **WINGSPAN:** 17 inches. **ADULT:** Yellow forehead, red crown, and orange nape. Head, neck, and underparts are light gray. Back is black and white-barred. Rump is white and tail is unmarked black. In flight, primaries are dark. Female is like males but lacks the red crown patch, and nape is more yellow than orange. **JUVENILE:** (June– July) Like female but bill not as long. **VOICE:** Clear *cheerp cheerp cheerp*. Also, a trilling *ker-r-r-r-p*. **BEHAVIORS:** Typical undulating flight of a woodpecker. **HABITAT:** Brushlands and riparian woodlands, usually with mesquite. **STATUS:** Common resident in the South Texas Brushlands, Edwards Plateau, and western Rolling Plains to the south-central Panhandle. Found in the riparian habitat of the Rio Grande west to Presidio County.

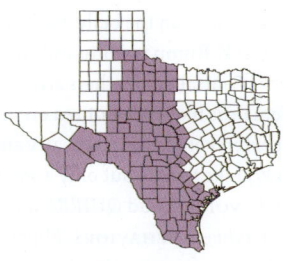

Female

Male

Red-bellied Woodpecker

(Melanerpes carolinus)

IUCN RED LIST STATUS (2016): Least Concern
POPULATION TREND: Stable

The Red-bellied Woodpecker is an extraordinary generalist. It can use almost any forest, from dry pine forests, to mangroves, to northern hardwood forests. Its diet is even more varied, switching from fruit to nuts to insects as availability changes.

LENGTH: 9.25 inches. **WINGSPAN:** 16 inches.
ADULT: Red nape extending over the crown. Forehead is washed with red. Cheeks are gray. Gray on breast and belly. Many birds show a faint reddish wash on the belly. Back is barred gray and black. Rump is white and speckled with black. Central tail feathers are mostly white. Female is like male but red on the nape does not extend onto the crown. **JUVENILE:** (June–July) Like female but only a small red nape patch. **VOICE:** Loud *QUIRRR* and harsh *chig-a chig-a chig-a.* **BEHAVIORS:** Flight typically undulating for a woodpecker. **HABITAT:** Almost any forested habitat. **STATUS:** Common to abundant resident in the eastern third of the state and eastern edge of the Panhandle.

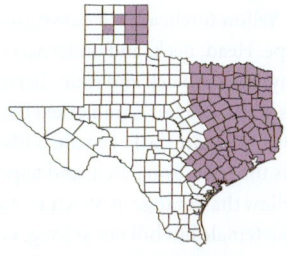

Male

Female

Williamson's Sapsucker

(Sphyrapicus thyroideus)

IUCN RED LIST STATUS (2016): Least Concern
POPULATION TREND: Stable.

The striking Williamson's Sapsucker's diet consists of a large portion of pine sap, and one can often detect their presence by looking for fresh sap wells on pine trees. Fruits, particularly madrone berries in Texas, are a major calorie source during migration. They also glean ants from trees.

LENGTH: 9 inches. **WINGSPAN:** 17 inches.
ADULT: Male has black head with white supercilium and a white line from the lore almost back to the nape. Black above with white wing patch and white rump. Tail is black. Red throat patch. Black breast and yellow belly. When perched, the wing patch has prominent white bar. Female has gray head. Upper parts are dark with thin white bars. White rump and dark tail with white, dark-barred central tail feathers. Breast is black, belly is yellow.
JUVENILE: (July–Aug.) Male is like adult male but lacks yellow belly and red throat. Female has gray head and is grayish overall with dark bars. **VOICE:** Hoarse *quaaaa* or *queeah queeah queeah*. **BEHAVIORS:** Undulating flight. Often flies from the crown to the base of the next tree. **HABITAT:** Low to mid-level oak-juniper and pine-oak forests. **STATUS:** Rare to locally uncommon migrant and rare winter resident in the upper elevations of the Davis and Guadalupe mountains. Fall migrants begin passing through in mid-September; spring migrants move between late February and early April.

Female

Juvenile

Yellow-bellied Sapsucker

(Sphyrapicus varius)

IUCN RED LIST STATUS (2016): Least Concern
POPULATION TREND: Decreasing

Yellow-bellied Sapsuckers maintain a system of sap wells on trees that they attend to daily. These wells are important to other bird species that feed on the insects the sap attracts or the sap itself. Ruby-throated Hummingbirds nest near these wells and may even time their migration to coincide with sapsuckers.
LENGTH: 8.5 inches. **WINGSPAN:** 16 inches.
ADULT: Male has red forecrown and black hind crown. White eye stripe extends to back of head. White stripe runs from lore to neck and down onto breast. Limited red throat with complete black border. Upper wings are dark, with a bold white wing patch visible on both a perched bird and one in flight. Back is mottled gray and black. Rump is pale and outer tail feathers are dark. Belly is pale yellow. Female is like male but throat is white. **JUVENILE:** (July–Mar.) Pale-yellow-washed head with dark spotted crown and fainter version of the adult facial pattern. Pale throat without the bold black-and-white stripes of adults. **VOICE:** Mewing *QUEEah*. **BEHAVIORS:** Hitches up and down trunks. Typical woodpecker undulating flight. **HABITAT:** Almost any forested habitat. **STATUS:** Uncommon to locally common migrant and winter resident statewide except west of the central mountains of the Trans-Pecos, where it is rare. Fall migrants arrive starting in mid-September; spring migration begins in early February and most have departed by early April.

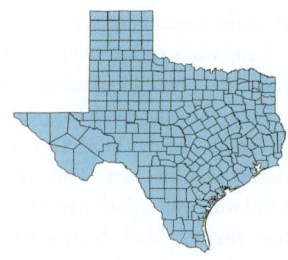

Adult

Immature

Red-naped Sapsucker

(Sphyrapicus nuchalis)

IUCN RED LIST STATUS (2018): Least Concern
POPULATION TREND: Decreasing

Red-naped Sapsuckers, like other sapsuckers in the genus *Sphyrapicus,* create sap wells in the bark of trees and feed on the sap that collects in them. This system of wells takes mainte-nance, and sapsuckers defend the wells from other species. Like the association between Yellow-bellied Sapsuckers and Ruby-throated Hummingbirds, Rufous Hummingbirds will associate with Red-naped Sapsuckers to take advantage of the sap wells.
LENGTH: 8.5 inches. **WINGSPAN:** 16 inches.
ADULT: Male has red crown with a black hind crown. Red patch at top of nape. White line around red throat, which is not completely bordered in black. Dark wing with white patch visible in flight and perched. Back is mottled in white in two wide bands. Belly is pale yel-low. Female is almost identical to male but chin is usually white. **JUVENILE:** (July–Oct.)

Gray-headed with faint version of the adult fa-cial pattern. Throat and breast are plain gray.
VOICE: Almost identical to Yellow-bellied Sapsucker, a mewing *QUEEah.* **BEHAVIORS:** Hitches up and down trunks. Typical wood-pecker undulating flight. **HABITAT:** Found in diverse woodlands. **STATUS:** Rare to locally common migrant and winter resident in the Trans-Pecos. Rare migrant in the High Plains and western Rolling Plains. In fall, the first migrants may arrive in late August. Spring migration is early February–mid-March.

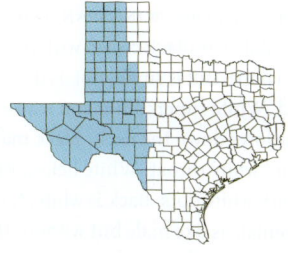

Male   Female

Downy Woodpecker

(Dryobates pubescens)

IUCN RED LIST STATUS (2016): Least Concern
POPULATION TREND: Stable

Downy Woodpecker is the smallest of North America's woodpeckers. The species was named by none other than the father of modern taxonomy, Swedish zoologist Carl Linnaeus, based on the work of colonial naturalist Mark Catesby. Catesby gave it the name Downy, reportedly due to the soft white feathers on its lower back. Those on the back of the very similar Hairy Woodpecker are longer and more hairlike.

LENGTH: 6.75 inches. **WINGSPAN:** 12 inches.
ADULT: Male has black crown with red patch on nape. Bill is small, not as deep as the head. White supercilium and white line from the lores down the neck. Black malar stripe and white throat. White below. Black wings with white dots. Back is white, tail is black. Female is like male but without the red nape. **JUVENILE:** (May–Aug.) Like female but forecrown is red. **VOICE:** High-pitched trill: *kikikikikikikik.* **BEHAVIORS:** Typical hitching up branches and undulating flight of a woodpecker. **HABITAT:** Generally present in open deciduous forest. **STATUS:** Common to uncommon resident of the eastern half of the state, from the eastern edge of the Rolling Plains south to the Guadalupe River Delta. Rare to uncommon resident in the western Panhandle.

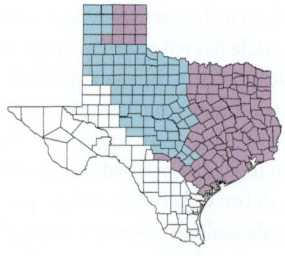

Male

Female

Ladder-backed Woodpecker

(Dryobates scalaris)

IUCN RED LIST STATUS (2016): Least Concern
POPULATION TREND: Stable

The Ladder-backed woodpecker is the Southwestern cousin of the Downy Woodpecker. It almost completely replaces the Downy Woodpecker as one moves west in the state. There is some overlap in range, in the eastern Panhandle and on the southern part of the coastal prairies.
LENGTH: 7.25 inches. **WINGSPAN:** 13 inches.
ADULT: Male has buffy forehead and a red crown that extends from the back to above the eye. Face is white with dark eye stripe that extends down the neck and merges with a dark malar stripe. Dirty white below with dark streaking on the flanks. Even black-and-white bars on back and upper wings make a ladder pattern. Tail is all dark. Female is like male but crown is dark instead of red. **JUVENILE:** (July–Aug.) Like female but with small red patch at top of crown. **VOICE:** Sharp, squeaky *CHEP*.

A rattling *chechechechechechechurchurchur*.
BEHAVIORS: Hitches vertically up tree trunks. Flight is typical for woodpeckers, swift and undulating. **HABITAT:** Deserts, desert scrub, thorn forests, live oak savannahs. **STATUS:** Common to uncommon resident in the western two-thirds of Texas.

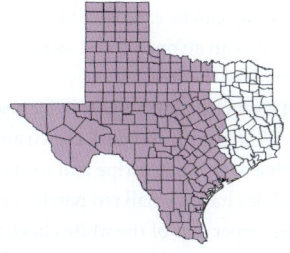

Female

Red-cockaded Woodpecker

(Dryobates borealis)

IUCN RED LIST STATUS (2020): Near
Threatened
POPULATION TREND: Decreasing

The US Fish and Wildlife Service lists
Red-cockaded Woodpecker as an endangered
species. They are unusual in that they nest in
colonies in living trees that have been infected
with red heart fungus, which naturally enters
trees from a broken branch. As protection
to the nest cavity, the woodpeckers excavate
small holes to encourage the flow of sap, which
forms a sticky barrier around the cavity. The
amount of sap can be extensive and visible for
some distance in an open forest as a gray wash
on the trunk of the tree.
LENGTH 8.5 inches. **WINGSPAN** 14 inches.
ADULT: White cheek and dark crown and
nape with a dark malar stripe and white
throat. Males have a small red patch or cock-
ade at the upper rear of the white cheek patch
that is virtually invisible in the field. White

below with spotted flanks. Back and upper
wings are barred black and white. Tail is
dark. **JUVENILE:** (July–Aug.) Like adult but
small red patch on forehead. **VOICE:** Sharp,
high-pitched *PEK PEK*. Rapid, twittering *pep-
peppeppep*. **BEHAVIORS:** Typical undulating
flight of woodpeckers, usually below the canopy. Hitches up the trunk. **HABITAT:** Mature
open pine forest maintained by fire. **STATUS:**
Rare to locally uncommon resident in the
open pine forests of the southern Pineywoods.
Listed as endangered by the US Fish and
Wildlife Service.

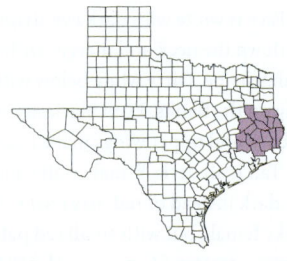

Hairy Woodpecker

(Dryobates villosus)

IUCN RED LIST STATUS (2020): Least Concern
POPULATION TREND: Increasing

The Hairy Woodpecker is very similar to the Downy Woodpecker but has a much larger and heavier bill and is about 50% larger overall. The Hairy Woodpecker gets its name from the long feathers on its back that are hairy compared to the soft, short back feathers of the Downy Woodpecker.

LENGTH: 9.25 inches. **WINGSPAN:** 15 inches. **ADULT:** Eastern male has dark head with white supercilium and red patch on nape. White chin stripe merges into the back. White below. Back is white with dark rump and tail. Wings are dark and spotted. Eastern female is like male but lacks the red on the nape. Western adult is like eastern adult female but white stripes in the face are thinner. Below is a dingy white. Spotting on upper wings is limited and restricted to primaries. **JUVENILE:** (May–Aug.) Like female but with red forehead

patch. **VOICE:** Sharp *PEEK*. **BEHAVIORS:** Typical of woodpeckers, hitches up trees. Flight is undulating. **HABITAT:** Forest habitats, usually mature woodlands. Less common in small woodlots, parks, cemeteries, and other urban areas. **STATUS:** Uncommon resident in the Pineywoods. Rare and local resident west to Tarrant and Bastrop counties. Uncommon resident in the eastern Panhandle. The western form is an uncommon resident in the high elevations of the Guadalupe Mountains.

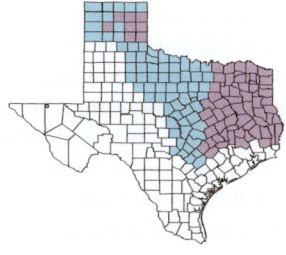

Yellow-shafted

Red-shafted

Northern Flicker

(Colaptes auratus)

IUCN RED LIST STATUS (2016): Least Concern
POPULATION TREND: Decreasing

Northern Flicker was formerly considered two separate species, the Red-shafted Flicker in the west and the Yellow-shafted Flicker in the east. Intergrades between the two forms exist and are found in central Texas in winter. **LENGTH:** 12.5 inches. **WINGSPAN:** 20 inches. **ADULT:** Yellow-shafted male has brown face and throat. Black malar stripe. Crown and nape are gray. Red crescent on nape. Back and upper wings are brown with dark bars. Breast has black crescent and bold black spots. Rump is white with dark tail. Underwing and under-tail are yellow. Yellow-shafted female is like male but lacks dark malar stripe. Red-shafted male is like yellow-shafted male but head is gray with only brown on the lores. Malar strip is red. Red crescent on nape is absent. Underwings are red. Red-shafted female is like male but malar stripe is brown. **VOICE:** Piercing *keewp*. Long, drawn-out, rattling *kep-kep-kep-kep-kep-kep-kep*. **BEHAVIORS:** Forages on the ground. Flight is undulating like other woodpeckers. **HABITAT:** Woodlands with dead or dying trees for nesting and open areas for foraging for ants. **STATUS:** Common to uncommon summer resident in the eastern Panhandle, northwestern Rolling Plains, and the Pineywoods. Uncommon summer resident in the mountains of the Trans-Pecos. Common winter resident in the northern two-thirds of Texas, uncommon in the South Texas Brush Country, and rare in the Lower Rio Grande Valley.

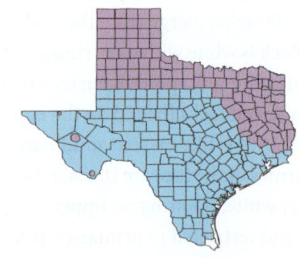

Pileated Woodpecker

(Dryocopus pileatus)

IUCN RED LIST STATUS (2016): Least Concern
POPULATION TREND: Increasing

Pileated Woodpecker is the largest surviving species of woodpecker in North America; only the likely extinct Ivory-billed Woodpecker was larger. Pileated Woodpeckers are conspicuous when present. Its loud call carries for some distance in the woods. Pairs defend their territories year-round. If one member of a pair dies, the remaining member may continue to defend the territory.

LENGTH: 16.5 inches. **WINGSPAN:** 29 inches. **ADULT:** Male is all black. Head is white with black eye stripe and black throat. Red malar stripe. Crest is bright red. Underwing coverts are white. On upper wing, bases of the primaries are white. Female is like male but red malar is lacking. Forecrown is dark and only the rear of the crest is red. **VOICE:** Rapid *KeeKeeKeeKeeKeee* and a slower, deeper *kuk kuk kuk kuk*. **BEHAVIORS:** Strong flier with slightly undulating flight. **HABITAT:** Mature forests and younger forests with scattered dead trees. **STATUS:** Uncommon to locally common resident in the Pineywoods and Post Oaks Savannah.

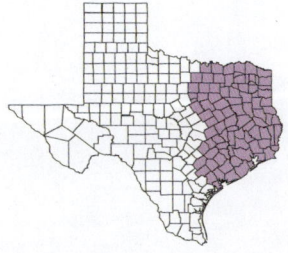

FALCONIDAE (FALCONS AND CARACARAS)

Falcons have often been placed close to the hawks and eagles because superficially they seem very closely related. But DNA analysis has shown that falcons are more closely related to South American Seriemas and New Zealand Parrots. Generally, falcons have long, pointed wings and long tails. Caracaras have more rounded wings but share the notched cutting edge of the upper bill, called a tomial tooth.

The falcons and caracaras are made up of three groups. The falcons are birds of open country and speedily capture prey, often on the wing. Caracaras are also open-country birds but aren't opposed to scavaging carrion. They have bare facial skin, longer legs, and feet adapted to walking to better take advantage of slow-moving or dead meals. The third group, forest-falcons, are birds of deep forests and jungles in the tropics.

Falcons have been used since ancient times as hunting animals, and the sport of falconry has long been regarded as the sport of kings, especially in the Middle East. Falconry is still practiced today, and Texas Parks and Wildlife still issues falconry licences. Illegal trapping of falcons for falconry is a concern, and the widespread Saker Falcon is endangered in part because of the black market for birds.

Worldwide there are sixty-five species of falcons and caracaras. Fifteen are of conservation concern. The Reunion Kestrel probably went extinct because of human persecution. The Guadalupe Caracara was also persecuted by humans and was extinct by 1900. Texas has recorded nine species of falcons and caracaras. Two, the Collared Forest-Falcon and Gyrfalcon, are considered rare visitors to the state. The Bat Falcon has only one record for Texas.

Adult

Juvenile

Crested Caracara

(Caracara plancus)

IUCN RED LIST STATUS (2016): Least Concern
POPULATION TREND: Increasing

Crested Caracara can be described as a falcon with a vulture's tastes. It's not fair to call them scavengers, though. Crested Caracara do take advantage of carrion, but they are adept and versatile hunters. They will hunt on foot, walking with ease in search of prey, hunt from a perch and pounce on prey, and hunt from the air.

Crested Caracara can "blush" in reverse. Normally the bare facial skin of a caracara is red, but when agitated, the blood flow to the skin is restricted and the face can turn yellow or even gray in a matter of seconds.
LENGTH: 23 inches. **WINGSPAN:** 49 inches.
ADULT: Heavy grayish bill with bare facial skin that varies from red to orange to pale yellow. Black crest and white head. Throat is also white. Back and upper wings are dark. Underwings are dark. Primaries are white with dark tips. Tail is white with black terminal band. **FIRST YEAR:** Like adult but more brown than black. **VOICE:** Staccato *kek-kek-kek-kek.* **BEHAVIORS:** Often can be seen at carrion. Hunts on foot by walking, and also hunts from the air and from a perch. **HABITAT:** Varied open habitats. **STATUS:** Uncommon to common resident in the southern third of Texas, including barrier islands. Uncommon on the upper Texas coast. Locally common to uncommon on the Oaks and Prairies, Edwards Plateau, and the eastern half of the Rolling Plains.

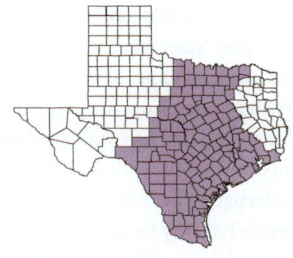

American Kestrel

(Falco sparverius)

IUCN RED LIST STATUS (2016): Least Concern
POPULATION TREND: Stable

American Kestrel is the smallest of North America's falcons. It is only one of the seventeen kestrel species found regularly in North America. They are cavity nesters and use old woodpecker cavities, rock crevices, or nooks in buildings. Nest sites appear to be the limiting factor in where they breed. They readily take to nest boxes, and there are now nest box programs for American Kestrels.

LENGTH: 9 inches. **WINGSPAN:** 22 inches.
ADULT: Male has boldly patterned head with "mustache" and "sideburns" stripe. Crown is gray. Dark patch on nape. Back is rufous with dark bars and upper wings are gray with dark spots. Underwings are pale with dark spots. Tail is unbarred rufous with dark terminal band. Underparts have rufous wash and dark spots. Female has gray head with gray "mustache" and "sideburns." Back and upper wings are rufous with dark bars. Tail is rufous with dark bars and no terminal bands. Underparts are pale with brown streaking. **VOICE:** Repeating *teep-teep-teep-teep*. **BEHAVIORS:** Hunts from a perch; often seen perched on wires. Bobs tail while perched. **HABITAT:** Wide variety of open to semi-open habitats. **STATUS:** Common to abundant migrant and winter resident statewide. Uncommon and local summer resident in the High Plains, Trans-Pecos, Rolling Plains, Pineywoods, and South Texas Brush Country. Fall migrants arrive in late August and most have departed by late April.

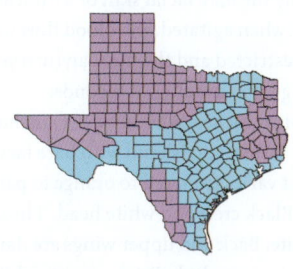

Merlin

(Falco columbarius)

IUCN RED LIST STATUS (2021): Least Concern
POPULATION TREND: Stable

Merlins breed all across nothern North America, Europe, and Asia. These small, agile, and swift falcons feed almost exclusively on other birds. They can capture birds almost as big as themselves out of the air.

Because of their small size, they were never used by falconers for large quarry, but they became known as a "lady's hawk." Noblewomen flew them against skylarks in duels; the skylark would fly up vertically and the Merlin would try to overtake it. Catherine the Great was an enthusiast of the sport.
LENGTH: 10 inches. **WINGSPAN:** 24 inches.
ADULT: Male has gray head with pale supercilium and pale cheek patch. Back and upper wings are gray. Tail is gray with narrow bands and a wider band at the end. Underparts are heavily streaked in brown. Female is like male but upper parts are more brown than

gray. **VOICE:** Rapid twittering *ti-ti-ti-ti-ti* or a slower-paced *tee-tee-tee-tee.* **BEHAVIORS:** In flight, gives an impression of power. Often flies low to the ground in straight flight. **HABITAT:** Open forest and grasslands. Often winters in cities and seen perched on buildings and power poles. Regularly seen hunting prey on tidal flats. **STATUS** Rare to uncommon migrant and winter resident statewide, most common on the coastal prairies. Fall migrants start arriving between mid-August and mid-September. In spring, they depart between early February and late April.

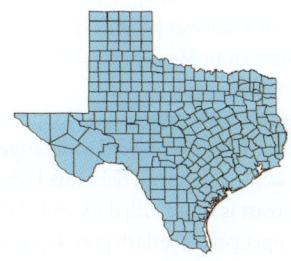

Aplomado Falcon

(Falco femoralis)

IUCN RED LIST STATUS (2018): Least Concern
POPULATION TREND: Decreasing

The colorful Aplomado Falcon was once a resident from the central Trans-Pecos east to Midland and in the South Texas Brush Country. There were records of them in Texas until the 1950s, when the species was extirpated from the state. In 1989, a reintroduction program was started in South Texas. Hundreds of captive-raised falcons were released, and in 1995, the first pair nested in the wild in Cameron County. A West Texas introduction program was not successful, but very rare visitors from Mexico are seen from time to time in that area.

LENGTH: 16 inches. **WINGSPAN:** 35 inches.
ADULT: Dark head with bold white supercilium. White cheek with dark mustache stripe. Breast is white with dark wide belly band. Upper parts are dark gray. Long tail is dark gray with narrow white bands. Orange undertail coverts. **FIRST YEAR:** Like adult but face, breast, and supercilium are buffy-orange. Breast is streaked dark. **VOICE:** Harsh repeating *kek kek kek kek kek*. **BEHAVIORS:** Agile on the ground, chases grounded prey swiftly. Flight is powerful. Able to make sharp turns and hover while pursuing birds. Flaps continuously while flying. **HABITAT:** Coastal prairies and desert grasslands with scattered yuccas and mesquite. Small woodlands in desert grasslands. **STATUS:** Rare resident from Matagorda County south to Cameron County on the coastal prairies. Very rare visitor to the mid-level grasslands in the Trans-Pecos.

Adult

Juvenile

Peregrine Falcon

(Falco peregrinus)

IUCN RED LIST STATUS (2021): Least Concern
POPULATION TREND: Increasing

Peregrine Falcon populations declined in the mid-20th century to endangered levels, mostly due to pesticides. Once these were banned, populations started to recover, and today populations are still increasing.

Peregrine Falcons are frequently billed as the fastest animal in the world. Theoretical studies suggest that they might be able to achieve speeds greater than 220 mph in a full free-fall stoop or dive from heights. In the wild, their speeds have been difficult to document.
LENGTH: 16 inches. **WINGSPAN:** 41 inches.
ADULT: Tundra form has gray crown and nape, white cheeks, and dark mustache stripe on face. Gray above with a dark tail with thin gray bars. Breast is white with dark-barred belly. *Anatum* form: Differs from the tundra form by having a very wide mustache stripe that essentially forms a hood. Breast has a buffy wash. **FIRST YEAR:** Like adult but brownish and not gray above. Breast and belly are heavily streaked. **VOICE:** Drawn-out *reeeehhhkkk*. **BEHAVIORS:** Flight is powerful and swift. Dives or stoops on prey from great height. **HABITAT:** Variety of habitats, including urban. Needs cliffs or analog-like building for nesting. **STATUS:** Uncommon to rare migrant throughout Texas. Locally uncommon to common winter resident on the coastal prairies and the immediate coast. Rare winter resident inland. The *Anatum* form is a very local breeding resident in the Trans-Pecos Mountains.

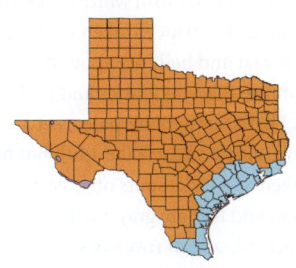

Prairie Falcon

(Falco mexicanus)

IUCN RED LIST STATUS (2016): Least Concern
POPULATION TREND: Increasing

The Prairie Falcon is more dependent on small mammal prey than most other falcons. This fact allowed them to escape most of the ravages of pesticide contamination that other species suffered from. Currently, the main threats Prairie Falcons face are harvesting for falconry and conversion of foraging lands to large-scale agriculture.

LENGTH: 16 inches. **WINGSPAN:** 40 inches.
ADULT: Brown head with white cheek and brown mustache stripe. Broken white supercilium. Breast and belly are spotted brown. Brown above. Black axillaries and underwing coverts. Flight feathers are pale. Feet and legs are yellow. **FIRST YEAR:** Like adult but heavily streaked and lacks spots on breast and belly. Legs and feet are gray. **VOICE:** Raspy *ack-ack-ack-ack-ack.* **BEHAVIORS:** Flapping flight is direct and swift. Soars on flat wings with tail slightly fanned and glides on flat wings. **HABITAT:** Breeds in open habitat at all elevations where cliffs or bluffs are present for nesting. Winters in grassland habitats. **STATUS:** Rare and local summer resident in the mountains of the Trans-Pecos. Rare to uncommon migrant and winter resident in the High Plains, western Rolling Plains, and the Trans-Pecos. Rare and local winter visitor from the eastern Rolling Plains to the South Texas Brush Country. Fall migrants begin arriving in early September and most depart by early May.

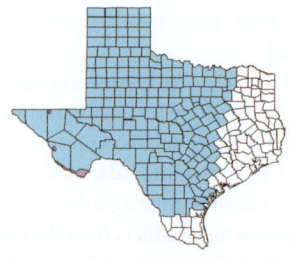

PSITTACIDAE (NEW WORLD AND AFRICAN PARROTS)

The New World and African parrot family is easily recognized. Most are green-colored, have large heads and short necks, hooked bills, and short legs. Many have colorful accents of red, blue, yellow, black, or gray. They range in size from the tiny sparrow-sized parrotlets to the huge macaws. The past few decades have seen a number of taxonomic revisions; all the parrots and cockatoos were once one big family, but they are now split into four families: New Zealand parrots, cockatoos, Old World parrots, and New World and African parrots. The New World and Old World parrots are closely related sister families.

Worldwide there are 177 species of New World and African parrots. In Africa there are two genera of New World parrots, overlapping in range with some Old World parrots. In Texas, four species are on the Texas Bird Records Committee checklist. Only the now-extinct Carolina Parakeet occurred in Texas historically. The South American Monk Parakeet is now introduced and established in several Texas cities, and in South Texas there are now established populations of Red-crowned Parrots and Green Parakeets whose orgins are still unclear and may never be known. Were they introduced as escaped pets? Did they arrive on their own from their core range in Mexico? Or was it some combination? In addition to these three, there are many other parrot species found free-flying in Texas, all considered to be feral and unestablished species.

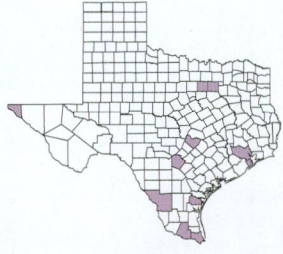

Monk Parakeet

(Myiopsitta monachus)

IUCN RED LIST STATUS (2018): Least Concern
POPULATION TREND: Increasing

This drab parakeet is a native of southern South America and has been introduced to Texas via the pet trade. The climate of their native range is similar to Texas, and these hardy birds have done well and seem to still be expanding.

Monk Parakeets are the only member of the parrot family that do not nest in cavities. Instead, they construct large stick nests that can host a single pair or up to a dozen pairs in a colony. These nests are often found on power-line towers and poles where they have become established.
LENGTH: 11.5 inches. **WINGSPAN:** 19 inches.
ADULT: Bright green with pale gray forehead and breast. Tail is long, pointed, and dark green. Flight feathers are dark cobalt-blue.
VOICE: Rolling throaty *graaa*. **BEHAVIORS:** Swift flight, often in small flocks. Wings are held level and never above the body. **HABITAT:** Largely urban and suburban environments. Nests in tall trees and structures with good visibility. **STATUS:** Introduced to North America and Texas from South America. Locally common resident in several metropolitan areas, including Austin, Corpus Christi, Dallas, Galveston, Houston, and San Antonio.

Green Parakeet

(Psittacara holochlorus)

IUCN RED LIST STATUS (2018): Least Concern
POPULATION TREND: Decreasing

Green Parakeets were first reported in the Lower Rio Grande Valley in 1960. There are several well-established populations, even though the provenance of these birds is debated. There are arguments that they were displaced from northern Mexico by agricultural conversion of habitat, and there are those who argue they escaped from the pet trade, though these birds are reportedly not popular in the pet trade because they are noisy. Regardless of origin, an evening pre-roost gathering of Green Parakeets is one of the great avian spectacles in Texas.
LENGTH: 13 inches. **WINGSPAN:** 21 inches.
ADULTS: Uniformly bright green. Relatively short tail compared to other parakeets. Thick gray orbital ring. **VOICE:** Shrill *rep-rep-rep*.
BEHAVIORS: Before going to roost, flocks gather at a pre-roost site, then travel in a group to the actual roost to sleep. **HABITAT:** Mostly urban areas in Texas. **STATUS:** Uncommon to locally common resident of the Lower Rio Grande Valley.

Red-crowned Parrot

(Amazona viridigenalis)

IUCN RED LIST STATUS (2021): Endangered
POPULATION TREND: Decreasing

Like the Green Parakeet, the origin of the Red-crowned Parrot in Texas is debated. It began breeding in Brownsville in the early 1980s. Loss of habitat and continued harvest of nestlings for the pet trade have made this species one of the most endangered in Texas. The IUCN in 2021 estimated that there were only 2000–4300 individuals in the wild. The Texas population has now become critical to this species' survival in the wild.
LENGTH: 12 inches. **WINGSPAN:** 25 inches.
ADULT: Green face with white orbital ring. Crown is red with blue nape. Wings have blue flight feathers and red patches in secondaries. Tail is short and square with broad yellow tip. **FIRST YEAR:** Like adult but only the forehead is red and most of the crown is blue. **VOICE:** Raspy, screeching *skee-oow-oww*. **BEHAVIORS:** Flight is labored with shallow wingbeats. Quiet and sluggish moving about trees midday. Uses its bill as a "third leg" while walking. **HABITAT:** In Texas it is largely urban, in areas with large trees. **STATUS:** First appeared in the Lower Rio Grande Valley in the early 1970s. Now common in urban areas, primarily in Brownsville, Harlingen, McAllen, and Weslaco.

TYRANNIDAE (TYRANT FLYCATCHERS)

Tyrant flycatchers are part of the huge order Passeriformes, or the songbirds. This order is divided into the oscines and suboscines. The suboscines are a New World group from mainly Central and South America. The tyrant flycatchers are the only family of the suboscines found in North America.

They are generally very small to medium-sized birds with upright posture. They have large heads and broad flattened bills. Most have rictal bristles around the mouth that aid in funneling prey into their mouths. Most have medium-length tails, short legs, and pointed wings. Most are drably colored but some are conspicuous, like the vibrant Vermilion Flycatcher.

Most are insectivores. The name tyrant fits, as most rule over their territory like a cruel king. The group known as kingbirds will attack and chase off birds as large as a Great Egret or Red-tailed Hawk. Insects are always on the menu, but almost all will eat fruits when they are available.

The family is made up of more then 400 species. It contains many cryptic species that are best differentiated by their voice. Species like Willow and Alder Flycatchers are nearly identical and even in the hand are not reliably identified, yet they have different voices and rarely hybridized even when breeding together. Forty-one species of tyrant flycatchers have been recorded in Texas. Nine of these are rare visitors to Texas: Social Flycatcher, Sulphur-bellied Flycatcher, Piratic Flycatcher, Thick-billed Kingbird, Gray Kingbird, Fork-tailed Flycatcher, Tufted Flycatcher, Greater Pewee, and Buff-breasted Flycatcher. Five species have only been recorded once in Texas: Greenish Elaenia, Small-billed Elaenia, White-crested Elaenia, Nutting's Flycatcher, and Variegated Flycatcher.

Northern Beardless-Tyrannulet

(Camptostoma imberbe)

IUCN RED LIST STATUS (2021): Least Concern
POPULATION TREND: Decreasing

The tiny Northern Beardless-Tyrannulet is found only in the southern tip of Texas. With its habits and bill shape, it resembles vireos or kinglets, traveling through the trees with deliberate movements that mimic those of a vireo, gleaning insects from the bark of trees. Unlike most flycatchers, it rarely fly-catches and lacks bristles at the base of its bill—hence the name "Beardless." *Tyrannulet* is a diminutive for "tyrant" because this tiny bird is a fierce defender of its nest.

LENGTH: 4.5 inches. **WINGSPAN:** 7 inches.
ADULT: Bushy crest on gray head and upper body. Breast is gray with pale-yellow belly. Bill is short, blunt, and has an orange base. The wings have pale grayish wingbars. Wings are short, barely extending past base of tail. **JU-VENILE:** (May–Sept.) Like adult but slightly browner and less green. May show pale lore

and short, indistinct supercilium. **VOICE:** High-pitched, descending *peer-peer-peer-peer*. **BEHAVIORS:** Moves through the trees in a kinglet- or vireo-like manner, gleaning food. Makes short upward flights after insects, usually from its perch under the canopy of a tree. **HABITAT:** Thorn, cedar-elm, and live oak forests with epiphytic moss. **STATUS:** Rare to locally uncommon resident in the Lower Rio Grande Valley north to Zapata County and on the Coastal Sand Plains to Kleberg County.

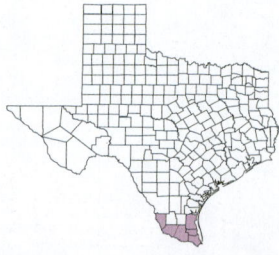

Dusky-capped Flycatcher

(Myiarchus tuberculifer)

IUCN RED LIST STATUS (2016): Least Concern
POPULATION TREND: Decreasing

Prior to 1991, there were only three records of Dusky-capped Flycatcher in Texas. The first breeding record for Texas was in 2000 in the Chisos Mountains, and since 2003 they have bred annually in the Davis Mountains. **LENGTH:** 7.25 inches. **WINGSPAN:** 10 inches. **ADULT:** Brown-headed with small, bushy crest. Bill is thin and dark. Throat and upper breast are gray. Belly is yellow. There is little or no rufous in the long tail. Secondary feathers have rufous edges only. Wingbars are inconspicuous. Unlike other flycatchers in the genus *Myiarchus*, there is little to no yellow on the underwing coverts. **JUVENILE:** (June–Sept.) Like adult but rufous edges to the tail feathers, which are more conspicuous. **VOICE:** Hoarse, drawn-out *whereeee* and a two-note *we're here*. **BEHAVIORS:** Rarely on the ground. Sallies out for insects from a perch under a tree. **HABITAT:** In West Texas, mixed montane forest. In South Texas, riparian forest. **STATUS:** Rare and local summer resident in the Davis Mountains and very rare and irregular summer resident in the Chisos Mountains. Very rare winter visitor to the Lower Rio Grande Valley. In the Trans-Pecos, breeding birds are present from April through late July. In the Lower Rio Grande Valley, birds have been found between early November and early April.

Ash-throated Flycatcher

(Myiarchus cinerascens)

IUCN RED LIST STATUS (2020): Least Concern
POPULATION TREND: Increasing

Ash-throated Flycatchers are secondary cavity nesters, using cavities excavated by other species. They can nest in smaller cavities and readily use man-made ones like nest boxes, the ends of pipes, crevices in buildings, and so on. They have a tolerance for heat, do not need to drink water, and can use desert scrub as long as there are nest sites and food.

LENGTH: 8.5 inches. **WINGSPAN:** 12 inches.
ADULT: Light brown head and pale, almost whitish throat. Thin dark bill. Pale gray on breast and very pale yellow on belly and underwing coverts. Whitish wingbars and white edges on secondaries. Tail is rufous with dark tip. **JUVENILE:** (June–Sept.) Like adult but wingbars are not as bright or well defined. **VOICE:** Soft *brik-brr-brr* and a *BIP-brrrrrrr*. **BEHAVIORS:** Flight is rapid and direct. Acrobatic when pursuing prey. **HABITAT:** Arid and semiarid scrub and open woodlands, riparian woodlands in arid and semiarid regions. Outside of breeding season, generally similar but a wider variety of woodland types. Avoids denser humid forests. **STATUS:** Common to uncommon summer resident in the western half of the state. Rare to locally uncommon in winter along the Rio Grande in the Trans-Pecos, the South Texas Brush Country, and along the coastal prairies. Breeding birds begin to arrive in mid-March, and most have dispersed by late July.

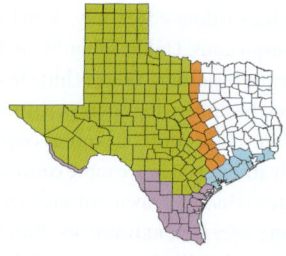

Great Crested Flycatcher

(Myiarchus crinitus)

IUCN RED LIST STATUS (2021): Least Concern
POPULATION TREND: Stable

Great Crested Flycatchers are often inconspicuous because they are typically in the upper canopy, but their clear and relatively loud calls often reveal their presence. They are cavity nesters and the female does most of the nest building. Nests can be built from an amazing variety of materals, including paper, plastic wrap, duct tape, dog and cat fur, and very often shed snake skins.
LENGTH: 8.75 inches. **WINGSPAN:** 13 inches.
ADULT: Brown-headed with dark gray face and throat. Bill is pale at the base. Belly is bright yellow and extends further up the breast than on other *Myiarchus* flycatchers. Extensive rufous on tail and primaries. Bright white wingbars. **JUVENILE:** (June–Sept.) Like adult but wingbars are largely absent. **VOICE:** Clear *wee-eeep* or *quee-EEEP*. **BEHAVIORS:** Spends little time on the ground. Sallies from a perch for prey. When not pursuing prey, often glides effortlessly. **HABITAT:** Open deciduous woodland, mixed woodland, and the edges of clearings. **STATUS:** Common and uncommon summer resident in the eastern half of the state, including the eastern Panhandle and eastern Rolling Plains, south to the Nueces River drainage. Spring migrants arrive as early as mid-March, and fall migrants have departed by mid-October.

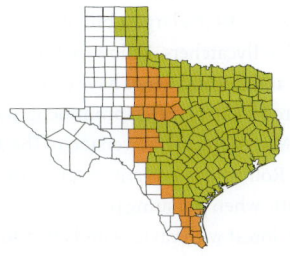

Brown-crested Flycatcher

(Myiarchus tyrannulus)

IUCN RED LIST STATUS (2016): Least Concern
POPULATION TREND: Increasing

Until 1983, the Brown-crested Flycatcher was known as Wied's Crested Flycatcher. Like the Great Crested Flycatcher, it is often in the canopy and its presence is easier to detect by its call than by sight. It's not tied closely to any particular habitat, though; the availability of suitable nest sites, usually abandoned woodpecker holes, is the critical factor.
LENGTH: 8.75 inches. **WINGSPAN:** 13 inches.
ADULT: Similar to Ash-throated Flycatcher but larger, with the largest bill of any of the *Myiarchus* flycatchers. Yellow on the belly is bright, and tail and wings have extensive rufous. Brown crest is often prominent. **JUVENILE:** (May–Sept.) Drab version of the adult. **VOICE:** Rolling *pur-prprprpr*. **BEHAVIORS:** Acrobatic when pursuing prey. **HABITAT:** Subtropical forest with cavities and abandoned woodpecker holes large enough for nesting.

STATUS: Common summer resident in the Lower Rio Valley. Uncommon summer resident through the South Texas Brush Country and the western Edwards Plateau. Birds arrive on the breeding ground in mid-April and depart between mid-August and early October.

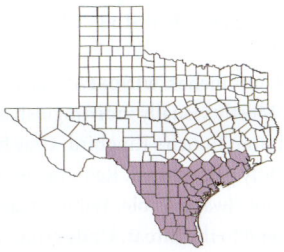

Great Kiskadee

(Pitangus sulphuratus)

IUCN RED LIST STATUS (2018): Least Concern
POPULATION TREND: Increasing

The Great Kiskadee was formerly known as the Derby Flycatcher. It and several similar tropical flycatchers are still, as a group, often referred to by that name. For a flycatcher, it has a diet much more like a jay. Its diet includes perched and flying arthropods, snails, small fish, frogs and toads, small reptiles, and fruits in season.

LENGTH: 9.75 inches. **WINGSPAN:** 15 inches.
ADULT: Bold black-and-white-striped head with white stripe above the eye. Small and not always visible yellow crown patch. Throat is white. Back, wings, and tail are plain rufous. Belly and underwing coverts are bright yellow. **JUVENILE:** (May–Aug.) Like adult but lacks yellow crown patch. **VOICE:** Loud *KREE-ta-peee* or *eat-your-meat*, often repeated. **BEHAVIORS:** Flight is powerful. Able to maneuver easily through foliage. Long-distance flight over foliage. **HABITAT:** Thorn forest and riparian forest, especially along waterways. **STATUS:** Locally common in the Lower Rio Grande Valley. Uncommon on the coast north to Nueces County and on the Rio Grande north to Val Verde County. Its range is expanding, and there are now regular nesting records up the coast as far as Chambers County. There have been several long-term pairs in Harris County.

Tropical Kingbird

(Tyrannus melancholicus)

IUCN RED LIST STATUS (2022): Least Concern
POPULATION TREND: Increasing

Until 1979, Tropical Kingbird and Couch's Kingbird were considered cospecific under the name Tropical Kingbird. At the time of the split of the species, though, there was but a single record for Tropical Kingbird from Texas, in 1909. By 1998 they had become regular enough to be removed from the Texas Bird Records Committee review list. The differences between these two species are subtle, and they are difficult to distinguish visually in the field. The best way to distinguish them is by their calls. **LENGTH:** 9.25 inches. **WINGSPAN:** 14.5 inches. **ADULT:** Mostly visually indistinguishable in the field from Couch's Kingbird. Gray head with slightly paler throat. Red-orange crown patch is not always visible. Yellow breast and belly extend to the throat. Underwings are yellow. Upper wings, back, and tail are grayish green. Tail is slightly more notched than that of a Couch's Kingbird. **JUVENILE:** (May–Oct.) Like adults but red-orange crown patch is absent. **VOICE:** Rapid trill *tzip-tzip-tid-tid-tid-tid-tid*. **BEHAVIORS:** Typical flycatcher, sallying out from perch to capture prey in the air. **HABITAT:** In Texas, prefers human-made areas like golf courses, electric substations, and sports fields. **STATUS:** Uncommon and local resident in the Lower Rio Grande Valley and north on the coastal prairies to Nueces County.

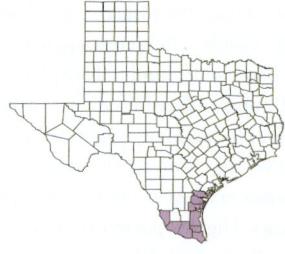

Couch's Kingbird

(Tyrannus couchii)

IUCN RED LIST STATUS (2022): Least Concern
POPULATION TREND: Stable

Until 1979, Couch's Kingbird was considered cospecific with Tropical Kingbird. At the time of the split, Couch's Kingbird was the only one of the two species regular in Texas. Now Tropical Kingbird is common enough in South Texas that one is best to wait for a bird to call before pronoucing its species. **LENGTH:** 9.25 inches. **WINGSPAN:** 15.5 inches. **ADULT:** Mostly visually indistinguishable in the field from Tropical Kingbird. Gray head with slightly paler throat. Red-orange crown patch is not always visible. Yellow breast and belly extend to the throat. Underwings are yellow. Upper wings, back, and tail are grayish-green. Tail is slightly less notched than that of a Tropical Kingbird. **JUVENILE:** (May–Aug.) Like adults but red-orange crown patch is absent. **VOICE:** Accelerating series *PIP PIP PIP pre-cheeerr* or just *pre-cheeerr*. **BEHAVIORS:**

Typical hawking or true fly-catching. Sights prey and sallies out to capture and consume it immediately. Flies back to perch to consume larger items. **HABITAT:** Less urban than the similar Tropical Kingbird and uses denser habitat. **STATUS:** Common to uncommon summer resident in the Lower Rio Grande Valley. Locally uncommon north along the coast and the South Texas Brush Country. The breeding range is slowly expanding north in Texas, with breeding records north to Chambers County. Spring migrants arrive in mid-March, and those that don't overwinter depart by mid-November.

Cassin's Kingbird

(Tyrannus vociferans)

IUCN RED LIST STATUS (2021): Least Concern
POPULATION TREND: Stable

Large and noisy Cassing's Kingbird is conspicuous on the breeding grounds. It is found at mid and high elevations in the sky islands of West Texas. It is named after John Cassin, a Pennylvania orniothologist whose name has been applied to four other bird species. **LENGTH:** 9 inches. **WINGSPAN:** 16 inches. **ADULT:** Dark gray head and breast. White throat patch is often prominent. Yellow belly and underwing coverts. Back is gray and the upper wings are green-gray. Pale edges on upper wing coverts make a scalloped pattern on the wing of perched birds. Tail is dark with pale tip. **JUVENILE:** (June–Aug.) Nearly identical to adults, but slightly browner on the back and upper wing coverts. **VOICE:** Dawn song is a *grrrrrrr WHER-weWEEEE.* **BEHAVIORS:** Typical flycatcher hawking; sallies out to prey and then back to perch or drops to the ground then returns to a perch. **HABITAT:** Open grasslands with trees for perches. **STATUS:** Uncommon to locally common summer resident in the mid and upper elevations of the Guadalupe, Davis, and Chinati mountains. Uncommon to rare migrant in the Trans-Pecos and western portion of the High Plains. Spring migrants pass through from late March to mid-May. Fall migration is from late August to early October.

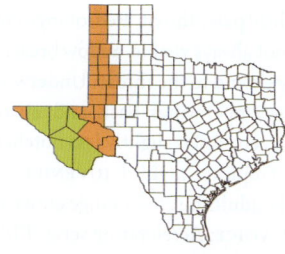

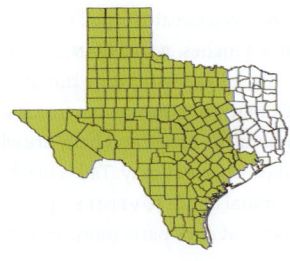

Western Kingbird

(Tyrannus verticalis)

IUCN RED LIST STATUS (2020): Least Concern
POPULATION TREND: Decreasing

Western Kingbird is a conspicuous bird found in most of Texas, except the Pineywoods. It is easily distinguished from the other yellow-bellied kingbirds by its black tail with white outer tail feathers. It overlaps in range with Cassin's Kingbird, but it is much more widespread in Texas and found at lower elevations.

LENGTH: 8.75 inches. **WINGSPAN:** 15.5 inches.
ADULT: Paler gray head than Cassin's Kingbird. Pale gray breast and white throat. Pale yellow on belly and underwing coverts. Back and upper wings are greenish-gray. Tail is black with white outer feathers. **JUVENILE:** (June–Oct.) Essentially the same as adults but paler overall. **VOICE:** An accelerating *pip pip pip pip-WI-WI-WIP*. **BEHAVIORS:** Typical flycatcher hawking; sallies out after prey, captures it in the air, and returns to a perch.

HABITAT: Wide variety of open habitats, almost always associated with tall trees or human-made structures. **STATUS:** Common to uncommon summer resident in the western two-thirds of the state. Common to uncommon migrant in Texas east to the Pineywoods, where they are rare migrants. Spring migration is late March–mid-May, fall late July–October.

Eastern Kingbird

(Tyrannus tyrannus)

IUCN RED LIST STATUS (2021): Least Concern
POPULATION TREND: Decreasing

Eastern Kingbirds are in the genus *Tyrannus*, which means "tyrant, absolute ruler, oppressor, or despot." All the kingbirds in the genus are aggressive, but the Eastern Kingbird is especially aggressive in defending its territory and nest from intruders. They will drive off birds many times their size, sometimes aided by other Eastern Kingbirds from adjacent territories. There are many accounts of Eastern Kingbirds driving off Red-tailed Hawks, Great Egrets, and even Great Horned Owls.
LENGTH: 8.5 inches. **WINGSPAN:** 15 inches.
ADULT: Black head with white throat and cheek. Orange-red crown patch that is normally concealed. Breast and belly are white. Back is dark charcoal-gray. Tail is black with white terminal band. **JUVENILE:** (July–Oct.) Like adults but dark parts more dark gray

than black. Lacks the orange-red crown patch.
VOICE: High-pitched *pzzzzt* and high-pitched, trilling *pikapikapikapikapika*. **BEHAVIORS:** Hawks for insects. Rarely glides. Never walks.
HABITAT: Open habitats, fields with scattered shrubs, trees, and hedgerows. **STATUS:** Common to locally uncommon summer resident in the eastern half of the state. Common to uncommon migrant in the eastern two-thirds of the state. Spring migration is late March–mid-May, fall mid-August–early October.

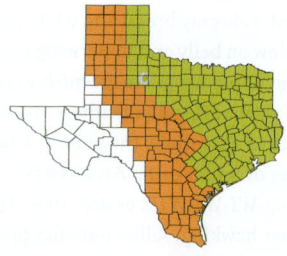

Scissor-tailed Flycatcher

(Tyrannus forficatus)

IUCN RED LIST STATUS (2021): Least Concern
POPULATION TREND: Decreasing

When driving along a country road, the Scisssor-tailed Flycatcher is a familiar sight in Texas. Perched high on a wire with its impossibly long tail, it can be recognized with ease. Scissor-tailed Flycatchers are a great boon to farmers as their diet consists almost exclusively of grasshoppers and beetles, more so than other North American flycatchers.
LENGTH: 10 inches. **WINGSPAN:** 15 inches.
ADULT: Nearly white head and breast. Belly and underwing coverts are pale pink. Back and upper wings are gray. Tail is very long and deeply forked. Outer tail feathers are white. Female tails average 30% shorter than males.
JUVENILE: (May–Dec.) Like adults but belly is more yellow than pink. Tail is shorter and not as deeply forked. **VOICE:** Shrill *pup pup pup pup preeewheeer*. **BEHAVIORS:** Flies with rapid wingbeats and folded tail; able to make abrupt turns in midair. **HABITAT:** Open savannahs with occasional trees and shrubs. Also, towns, agricultural fields, pastures, and golf courses. **STATUS:** Common to locally abundant summer resident throughout the eastern three-quarters of the state. Spring migrants begin arriving in late February and most have departed by the end of November.

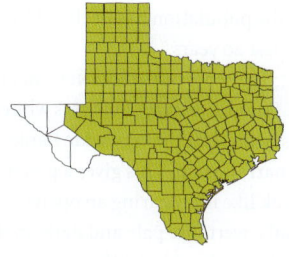

Olive-sided Flycatcher

(Contopus cooperi)

IUCN RED LIST STATUS (2017): Near Threatened

POPULATION TREND: Decreasing

The Olive-sided Flycatcher is so prominent when present—often perched high or sallying back and forth after prey—they are hard to miss. If their visibility isn't enough, their clear, piecing call of *quick, three beers* is an attention-getter. This conspicuous presence makes it seem counterintuitive that this is a species headed for trouble. The International Union for Conservation of Nature (IUCN) estimates the population has declined by 76.6% over the last 20 years.

LENGTH: 7.5 inches. **WINGSPAN:** 13 inches. **ADULT:** Dark olive-gray. Throat is white. Breast is white, but the very wide flanks make it just a narrow strip. This gives a perched bird a look like it is wearing an open vest. Undertail coverts are pale and dark-spotted. White tufts on the sides of the rump are often exposed on perched birds. **JUVENILE:** (June–Nov.) Like adults but average browner overall. **VOICE:** Shrill *wip WEE-WEIR*. **BEHAVIORS:** Fly-catches from a high perch, usually a bare branch high on a tree. **HABITAT:** In migration, uses a wide variety of open habitats with suitable high perches in open areas. **STATUS:** Uncommon to rare spring and fall migrant statewide. Uncommon summer resident at the upper elevations of the Guadalupe Mountains. Rare and irregular summer resident in the Davis Mountains. Spring migration is late April–early June, fall early August–mid-October.

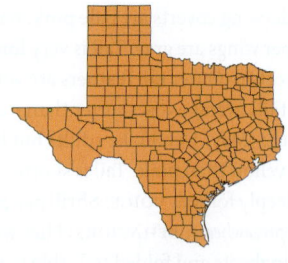

Western Wood-Pewee

(Contopus sordidulus)

IUCN RED LIST STATUS (2021): Least Concern
POPULATION TREND: Decreasing

The Western Wood-Pewee is a common inhabitant of open forests, forest edges, and riparian corridors in the western edge of Texas. Nearly identical to Eastern Wood-Pewee, Western Wood-Pewees average a bit darker, but there is enough overlap in the two species that their calls are the only surefire way to distinguish between them in the small area where they overlap.

LENGTH: 6.25 inches. **WINGSPAN:** 10.5 inches. **ADULT:** Green-gray above with pale belly. Weak eye ring. Pale throat. Narrow grayish wingbars. Dusky smudges on undertail coverts. **JUVENILE:** (June–Nov.) Like adult but wingbars are buffy-grayish and upper wingbar is weaker. **VOICE:** Ringing *breeee-yur*. **BEHAVIORS:** Sallies from an open perch and returns to the same perch. Doesn't flick its tail like phoebes and *Empidonax* flycatchers.

HABITAT: A generalist, it uses woodlands and forests except absent from dense forests. **STATUS:** Rare to locally common summer resident in the Davis, Chinati, and Guadalupe mountains. Common to uncommon migrant through the High Plains, western Rolling Plains, and the Trans-Pecos. Spring migration is late April–early June, fall late July–early October.

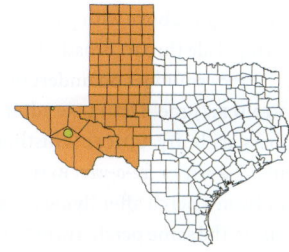

Eastern Wood-Pewee

(Contopus virens)

IUCN RED LIST STATUS (2021): Least Concern
POPULATION TREND: Decreasing

Eastern Wood-Pewees are nearly identical to Western Wood-Pewees and along with Tropical Pewee in Central and South America form a superspecies. The population of Eastern Wood-Pewees has declined by about 45% over the last fifty years. A potential cause is the huge population of white-tailed deer in North America, which overbrowse forests and affect the foraging of Eastern Wood-Pewees. **LENGTH:** 6.25 inches. **WINGSPAN:** 10 inches. **ADULT:** Green-gray above with pale belly. Weak eye ring. Pale throat. Broad whitish wingbars. Dusky smudges on undertail coverts. **JUVENILE:** (June–Oct.) Like adults but wingbars are buffy. **VOICE:** Whistling, drawn-out *pee-a-wee* or *fee-o-wee*. **BEHAVIORS:** Sallies out from a perch after flying insects and returns to the same perch. **HABITAT:** All types of forests, associated with clearings and edges. **STATUS:** Uncommon and declining summer resident in the eastern half of Texas and the Edwards Plateau, west to the Rolling Plains. Uncommon migrant in the eastern two-thirds of the state. Rare migrant on the High Plains. Spring migration is mid-April– mid-May, fall late July–early October.

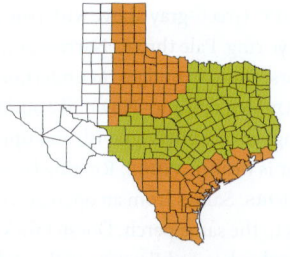

Acadian Flycatcher

Empidonax Flycatchers

Yellow-bellied Flycatcher (*Empidonax flaviventris*)
Acadian Flycatcher (*Empidonax virescens*)
Alder Flycatcher (*Empidonax alnorum*)
Willow Flycatcher (*Empidonax traillii*)
Least Flycatcher (*Empidonax minimus*)
Hammond's Flycatcher (*Empidonax hammondii*)
Gray Flycatcher (*Empidonax wrightii*)
Dusky Flycatcher *(Empidonax oberholseri)*
Western Flycatcher (*Empidonax difficilis*)

The family of Empidonax flycatchers is without a doubt the most challenging genus of flycatchers to identify in the field. Ten species occur in Texas, nine regular and one review species (Buff-breasted Flycatcher may be extripated from Texas now). The field marks that differentiate the individual species are difficult to illustrate in a field guide. Differences are at best subtle and generally have a high degree of subjectivity. Identification by calls is the best method.

LENGTH: 5.25-6 inches. **WINGSPAN:** 7.75-9 inches. **ADULT:** Small, large-headed, greenish to green-gray to grayish. Eye ring that varies from broken to complete to teardrop-shape. Dusky breast, from greenish to grayish. Pale belly, sometime yellowish. Short wings do not extend to tail, or barely onto tail. Two wingbars from white to grayish. **FIRST WINTER:** Similar to adults but usually wingbar-buffy. Eye rings are generally less distinct. Less contrast. **VOICE:** Yellow-bellied: hoarse *cheberk*. Acadian: High-pitched *pee-zit*. Alder: Burry *FRITZ-bew*. Willow: Harsh *rhea-BEWW*. Least: Dry *CHEbek*. Hammond's: High, hoarse *tsi-pik* followed by *grr-vik*. Gray: Emphatic

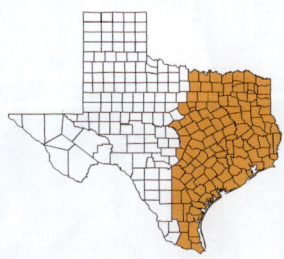

Yellow-bellied Flycatcher

chur-rip followed by *tidoo*. Dusky: three phrases, usually in order, *tisbip querrrp sue-weet*. Western: two phases, *pseep* and *pre-eat*. **BEHAVIORS:** Mostly under or low in canopy. Sallies out and returns to perch. Most species have some tail wagging or bobbing. **HABITAT:** Wide variety of woodlands and scrub. **STATUS:** Yellow-bellied, Acadian, and Alder are common to uncommon migrants in the eastern half of the state. Willow and Least are common to uncommon migrants statewide. Hammond's, Gray, Dusky, and Western are common to uncommon migrants in the western parts of the Trans-Pecos and High Plains. Acadian migrate late March–mid-May and mid-August–late September. Other Empidonax species migrate late April–early June and late July–early October. Acadian is common to uncommon summer resident in the eastern third of the state south to Harris and Bexar counties. Gray is locally common summer resident in the Davis and Guadalupe mountains. Western is locally uncommon summer resident in the Chisos, Davis, and Guadalupe mountains. Least is rare winter resident on the coast.

Acadian Flycatcher

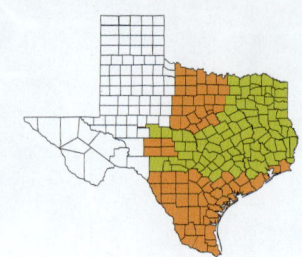

Least Flycatcher

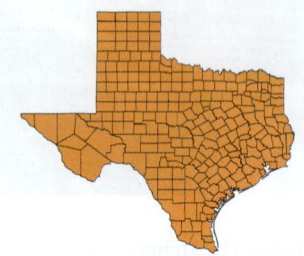

Hammond's Flycatcher

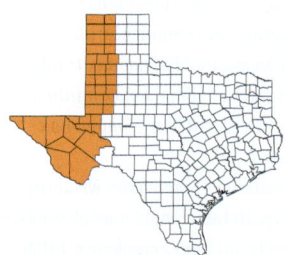

Gray Flycatcher

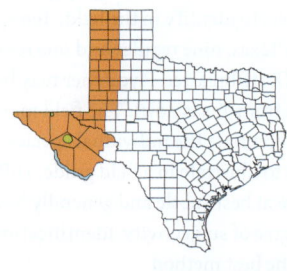

Probable Alder Flycatcher

Dusky Flycatcher

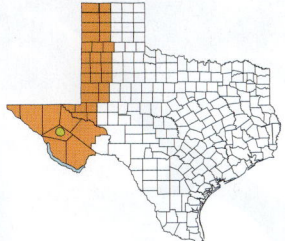

Western Flycatcher

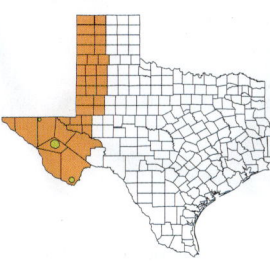

Alder Flycatcher

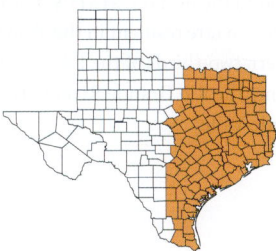

Willow Flycatcher

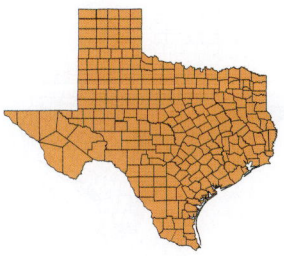

Probable Least Flycatcher

Western Flycatcher

Black Phoebe

(Sayornis nigricans)

IUCN RED LIST STATUS (2021): Least Concern
POPULATION TREND: Increasing

Black Phoebe is the only black-and-white North American flycatcher. Always associated with water, they use nest sites like sheltered rock face, streamside boulders, and even hollow trees to build a cuplike mud nest in. Bridges over small streams and even large culverts over streams make ideal nest structures. In West Texas, if there is water, mud, and a bridge you are likely to find Black Phoebes. **LENGTH:** 7 inches. **WINGSPAN:** 11 inches. **ADULT:** Dark, almost black head and breast. Back, upper wings, and tail are dark blackish-gray. Belly and undertail are white. **JUVENILE:** (Apr.–Sept.) Like adults but wings have two buffy wingbars. Back feathers and especially rump feathers have buffy edges. **VOICE:** High, thin *sitsew* and *sitSU-susew*. **BEHAVIORS:** Flight is direct with head raised. When perched, pumps tail frequently.

HABITAT: Always associated with water: streamside, creeks, streams, lake borders, ephemeral ponds, even fountains. Requires mud for nest building and sheltered "rock face" to build the nest on. **STATUS:** Locally uncommon to rare resident in the Trans-Pecos and western two-thirds of the Edwards Plateau south along the Rio Grande River to the Lower Rio Grande Valley.

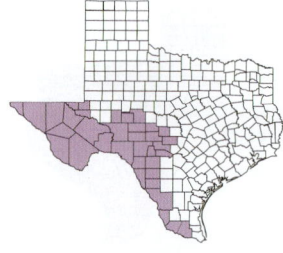

Adult

Fresh fall plumage

Eastern Phoebe

(Sayornis phoebe)

IUCN RED LIST STATUS (2021): Least Concern
POPULATION TREND: Increasing

Although unobstrusive and drab-colored, the Eastern Phoebe is a familiar bird because it regularly nests on human dwellings. Its ability to use man-made structures may be why it has been able to expand onto the Great Plains, where natural nest sites are rare.
LENGTH: 7 inches. **WINGSPAN:** 10.5 inches. **ADULT:** Dark head with pale throat. Smudgy on sides of breast, paler on belly. Freshly molted fall birds can be very yellow below, fading to white in spring. Gray to gray-brown on back with darker tail. **JUVENILE:** (June–Nov.) Like adults but belly and undertail are pale yellow. Two variable reddish-brown wingbars. Less contrast between head and back. **VOICE:** Emphatic *phree-reep* or *phree-de-reep*. **BEHAVIORS:** Perches low but in the open, always conspicuous. Bobs tail almost constantly. Flights are direct and often quite low. **HABITAT:** Requires suitable nest sites with some woody cover and a proximity to mud for best building. The nest site is a niche or overhang, either natural or man-made; sides of buildings under eaves are common sites. In winter, uses diverse woodland habitats but usually in proximity to water. **STATUS:** Uncommon to common summer resident in the northern two-thirds of Texas, except for the Panhandle. Common migrant in the eastern two-thirds of the state, becoming less westward, and rare in the western Trans-Pecos. Common winter resident in the eastern two-thirds of the state.

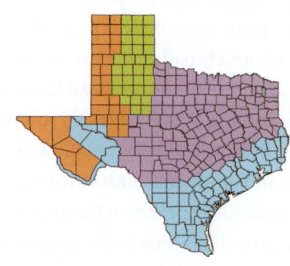

Say's Phoebe

(Sayornis saya)

IUCN RED LIST STATUS (2021): Least Concern
POPULATION TREND: Increasing

Say's Phoebe has one of the widest breeding ranges of any flycatcher. They commonly nest from northern Mexico to the Alaskan tundra. Their range seems only limited by availability of nest sites. Unlike the other two phoebe species in Texas, Say's Phoebe avoids watercourses and does not use mud to build nests. **LENGTH:** 7.5 inches. **WINGSPAN:** 13 inches. **ADULT:** Gray head with lighter gray throat and back. Darker around eye and lores. Upper wings are gray. Belly is pale rufous. Tail is black and broad at the tip. **JUVENILE:** (Apr.–Aug.) Like adults but head lacks the dark face pattern. Wings have two rufous wingbars. **VOICE:** Two whistled phrases: *pidepeeew* and *pre-dreeeep*. **BEHAVIORS:** Often perches quite low, or even on the ground, when foraging. Flies direct with no undulation. Often hovers while foraging. **HABITAT:** Open country; avoids watercourses. Prairies, ranchlands, sagebrush, barren foothills, and desert borders, all with open perches for foraging. **STATUS:** Common to uncommon resident in the Trans-Pecos and western edge of the Edwards Plateau. Rare and local on the western edge of the Panhandle High Plains. Uncommon in winter and migration in South Texas Brush Country. Rare but regular winter resident on the central coast and Blackland Prairies. Fall migrants arrive mid-September to mid-October. Winter birds depart between early March and mid-May.

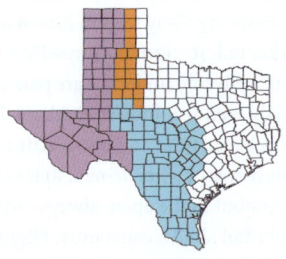

Male

Female

First-year male

Vermilion Flycatcher

(Pyrocephalus rubinus)

IUCN RED LIST STATUS (2021): Least Concern
POPULATION TREND: Increasing

The male Vermilion Flycatcher is easily the most spectacular flycatcher in Texas, and perhaps even one of the most spectacular birds in Texas. The genus name *Pyrocephalus* comes from Ancient Greek and means "flame-headed."

LENGTH: 6 inches. **WINGSPAN:** 10 inches.
ADULT: Male is bright vermilion-red on head and below. Dark face mask and nape. Dark back and wings. Tail is black. Female has gray-brown head with white throat and pale streaked breast. Pink belly and vent. Gray-brown on back and wings. Tail is black with white outer tail feathers. **FIRST YEAR:** Male is like adult females but variable in red below, starting with belly and vent. Female is like adult female but yellow instead of pink below. **JUVENILE:** (May–Sept.) Like adult female but no colored wash below. Breast is spotted instead of streaked. **VOICE:** Trilling *pit pit pit dibly-you.* **BEHAVIORS:** Rarely on the ground. Short, swift foraging flight; usually returns to the same perch. **HABITAT:** Arid scrub, parks, golf courses, savannah, farmland, usually near water. **STATUS:** Uncommon summer resident in South Texas Brush County, Edwards Plateau and southern Trans-Pecos. Uncommon winter resident from Hudspeth County in Trans-Pecos along the Rio Grande through South Texas Brush Country and along the coastal prairies to the upper Texas coast. Spring migration is early March–early April. Summer residents begin to depart in early October.

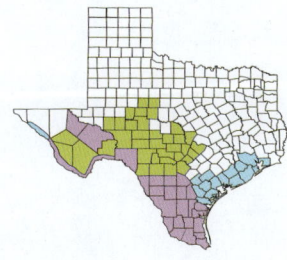

VIREONIDAE (VIREOS, SHRIKE-BABBLERS, AND ERPORNIS)

The vireos were a bit of a taxonomic puzzle. They were thought to be linked to the Old World corvids and shrikes but were endemic to the New World. Recent DNA work has shown that the shrike-babblers and erpornis of Asia are in the same family and provide the missing Old World link.

The name vireo comes from latin word *virere*, which means "to be green." Vireo is an apt name, as most vireos are shades of olive green, gray, and yellow. They are similar in size to the warblers but have heavier bodies and stouter hooked bills. The bills have two toothlike notches near the tip. While warblers flit through the treetops, vireos move more deliberately, gleaning insects from under leaves. Many species rely heavily on fruit during migration and the nonbreeding season.

There are sixty-one species of vireos, shrike-babblers, and erpornis worldwide. Fifteen species of vireos have been recorded in Texas. The Black-whiskered Vireo is considered a rare visitor to Texas, and the Yucatan Vireo has been recorded only once in the state.

Male

Female

Black-capped Vireo

(Vireo atricapilla)

IUCN RED LIST STATUS (2019): Near Threatened
POPULATION TREND: Decreasing

The Black-capped Vireo is a near-Texas-endemic species. It once bred into Kansas but the range has shrunk. It now mostly ranges across a swath of central Texas and small areas in Oklahoma and Mexico.
LENGTH: 4.5 inches. **WINGSPAN:** 7 inches. **ADULT:** Male has black face and crown. White lores and white teardrop-shaped eye ring. Red iris. Back and wings are olive with yellowish wingbars. Flanks are yellowish and white below. Tail is short but extends well beyond the wingtips. Female is similar to male but black cap is more dark gray than black. **FIRST WINTER:** (Aug.–Mar.) Like females but head and crown are olive and do not contrast with the back. **VOICE:** Song is rapid and harsh: *tik-a-purrreer-chik*, in many variations. **BEHAVIORS:** Hops and flutters between branches when foraging. Often hangs upside down or even hovers when foraging. Flight is slightly undulating, with rapid wingbeats. **HABITAT:** Open, scrubby habitat with heights up to about 6 feet and open space between shrubs and thickets. Early successional habitat or rocky slopes to maintain the scrubby structure. Heavy use by deer or cattle changes the structure to make it unsuitable for nesting. **STATUS:** Rare to uncommon summer resident from Brewster County across the Edwards Plateau north to Taylor and Palo Pinto counties. Rarely recorded in migration outside the breeding range.

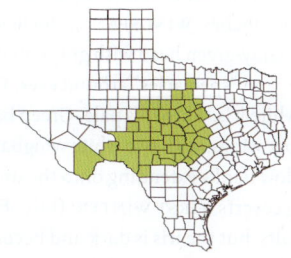

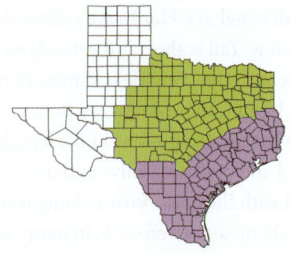

White-eyed Vireo

(Vireo griseus)

IUCN RED LIST STATUS (2020): Least Concern
POPULATION TREND: Increasing

The very vocal White-eyed Vireo is more often heard than seen. It favors dense, scrubby habitat. The careful observer, though, is often rewarded with a close eye-to-eye sighting since the preferred habitat is rarely more than head-high. Describing their call is a challenge. Not only do they have an amazing vocabulary of their own, they often incorporate notes from other birds into their song, including in one instance a near-perfect Carolina Chickadee imitation. **LENGTH:** 5 inches. **WINGSPAN:** 7.5 inches. **ADULT:** Gray-green head with gray collar. Yellow lores and "spectacles" around eyes. Eye has bright white iris. Throat is pale. Green back and upper wings, with two white wingbars. Pale yellow flanks extending onto the underwing coverts. **FIRST WINTER:** (July–Feb.) Like adults, but the iris is dark and becomes pale during winter. **VOICE:** Song is extremely variable; generally starts with a *chik* note, includes long *kerreer* note, and ends with sharp *wit* note. Raspy, scolding *che-che-che*. **BEHAVIORS:** A skulker, staying well inside dense vegetation. **HABITAT:** Generally secondary deciduous scrub. Overground pastures, clearing edges, mature hedgerows. **STATUS:** Common to locally abundant summer resident in the eastern two-thirds of Texas. Locally uncommon to rare winter resident in the southern Pineywoods, coastal prairies, and south of the Edwards Plateau. Common to uncommon migrant in the eastern two-thirds of Texas. Spring migration is mid-March–early May, fall mid-August–mid-October.

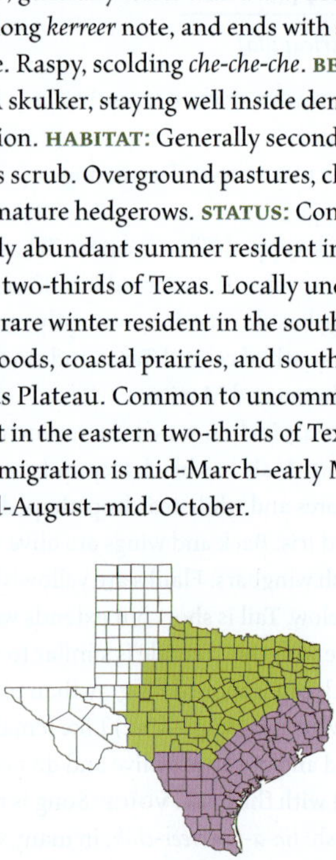

Bell's Vireo

(Vireo bellii)

IUCN RED LIST STATUS (2020): Least Concern
POPULATION TREND: Increasing

Bell's Vireo was discovered by John James Audubon and named for his friend and companion John Bell. Drab, with a preference for dense vegetation, its presence is known by its persistent song. The population of Bell's Vireo has been declining in Texas, particularly in the east. Like other vireos, they are heavily parasitized by Brown-headed Cowbirds.
LENGTH: 4.75 inches. **WINGSPAN:** 7 inches.
ADULT: Small gray head with weak white eye line and broken white eye ring. Greenish back and upper wings with a single bright wingbar. Pale below. Variable with eastern birds being brighter and greener than western birds. **VOICE:** Chatty *chew-dee chew-dee chew-dee cheedle-doo*. **BEHAVIORS:** Hops while foraging between branches. Undulates slightly in flight. Eastern birds gently bob their tails, while western birds flip and wag their slightly longer tail like gnatcatchers. **HABITAT:** Dense low vegetation, often along a drainage or near water. **STATUS:** Locally common to uncommon summer resident in the southern Trans-Pecos, Edwards Plateau, Rolling Plains, and north-central Texas. Rare summer resident in the northern Pineywoods. Common to uncommon migrant in the Trans-Pecos, Edwards Plateau, and Rolling Plains, and north-central Texas. Rare migrant in the South Texas Brush Country and on the coast. Spring migrants arrive late March–early May. Fall migration is mid-August–early October.

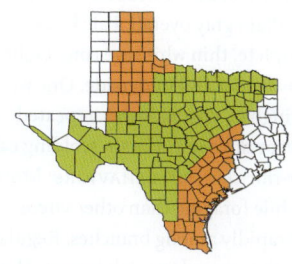

Gray Vireo

(Vireo vicinior)

IUCN RED LIST STATUS (2019): Least Concern
POPULATION TREND: Increasing

Gray Vireo is a cryptic species that hides in plain sight. Its preferred habitat is arid, harsh, and often inaccessible. While having a distinctive call, silent birds are easily confused with other small gray birds like gnatcatchers, bushtits, and titmice that use the same habitat. Its species name, *vicinior*, comes from *vicinis,* meaning "neighboring or related," referring to its resemblance to these other species.
LENGTH: 5.5 inches. **WINGSPAN:** 8 inches.
ADULT: Plain gray overall. Pale lores and complete, thin white eye ring. Light dull-gray throat and underside. One weak white wingbar. **VOICE:** Sweet, repeated *cheer-reet* or *cheer-row*. Harsh, scolding call: *chir-chir-chir-chir-chir.* **BEHAVIORS:** More active while foraging than other vireos, moving rapidly among branches. Regularly drops to the ground to catch insects. Hovers at leaf clusters to glean insects. **HABITAT:** Mixed juniper-piñon and oak scrub in hot, arid mountains. **STATUS:** Locally uncommon to rare summer resident in the southern Trans-Pecos east to the western Edwards Plateau. Rare and local summer resident in the Guadalupe Mountains. Rare winter resident along the Rio Grande River from Big Bend to Presidio. Spring birds begin arriving in late March and most depart by early September.

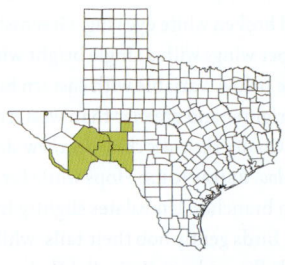

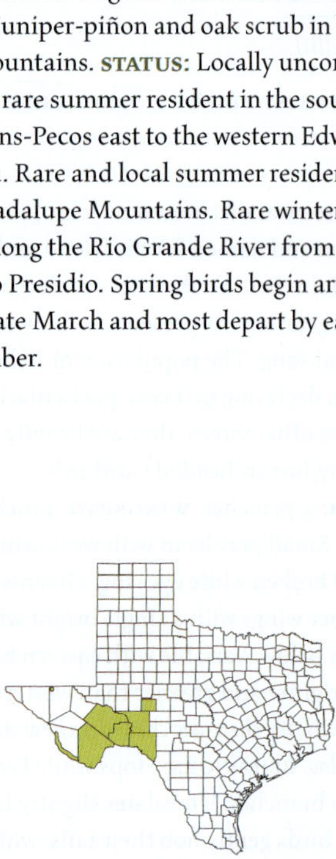

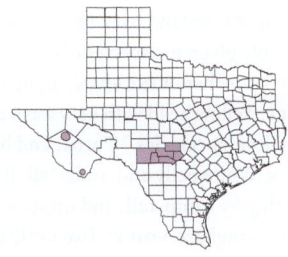

Hutton's Vireo

(Vireo huttoni)

IUCN RED LIST STATUS (2020): Least Concern
POPULATION TREND: Increasing

Hutton's Vireo has two populations in North America, a Pacific coast population that is essentially nonmigratory and an interior population that ranges along the interior mountains of Mexico into the Trans-Pecos and Edwards Plateau of Texas. Easily confused at first glance with a Ruby-crowned Kinglet and nesting earlier than many songbird species, it is potentially overlooked in breeding bird surveys and atlases. Until the 1990s, it was rare outside of the Chisos and Davis mountains, but it has now colonized the southern Edwards Plateau.
LENGTH: 5 inches. **WINGSPAN:** 8 inches.
ADULT: Small and stocky, resembling a kinglet with a larger bill. Drab olive. Pale lores and thin eye ring that flares into a teardrop shape behind the eye. Two distinct wingbars. **VOICE:** Ringing, repeated *CHEERrup*. **BEHAVIORS:**

Wingbeats are rapid in flight, mostly within or between trees. **HABITAT:** Evergreen forests and woodlands with moderate to dense crown closure and understory. **STATUS:** Locally common summer and uncommon winter resident in the Davis and Chisos mountains. Since the 1990s has colonized the southern Edwards Plateau. About half the population leaves Texas for the winter starting in late September and returns in late April.

Yellow-throated Vireo

(Vireo flavifrons)

IUCN RED LIST STATUS (2020): Least Concern
POPULATION TREND: Increasing

Yellow-throated Vireos almost disappeared from the towns and cities of the northeastern United States. The time of the decline correlated with the heavy spraying of insecticides in efforts to control Dutch elm disease in the 1940s and '50s. Biologist Rachel Carson, who later published the landmark conservation book *Silent Spring*, was one of the first to oppose the spraying. Since that time, the population has been steadily increasing. The conversion of farmland back to mature forest has probably given the species a boost. **LENGTH:** 5.5 inches. **WINGSPAN:** 9.5 inches. **ADULT:** Yellow head with gray wash and bright yellow "spectacles." Throat and breast are yellow. White belly and undertail. Back is green with gray rump, tail, and upper wings. Two white wingbars. **VOICE:** Two or three slurred syllables: *rreeeyeooo* or *rreeeye-up*.

BEHAVIORS: Moves slowly and deliberately. Flights are short, from tree to tree and branch to branch. Sings from high in a tree. **HABITAT:** Bottomland and upland mature deciduous and mixed forests. Forest edges along streams, rivers, swamps, roads, parks, and small towns. **STATUS:** Common to uncommon summer resident in the eastern third of Texas north of the coastal prairies and west across the Edwards Plateau. Uncommon to common migrant in the eastern half of Texas. Spring migration is late March–mid-May, fall late August–early October.

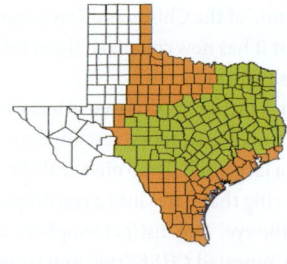

Cassin's Vireo

(Vireo cassinii)

IUCN RED LIST STATUS (2020): Least Concern
POPULATION TREND: Stable

Cassin's Vireo is named for John Cassin, who published the first comprehensive studies of North American birds. Until 1997, Cassin's Vireo—along with Plumbeous Vireo and Blue-headed Vireo—were listed together as Solitary Vireo. The conservation status of this bird is better than many other vireo species, perhaps because it utilizes large tracts of land in the western mountains that are relatively free of Brown-headed Cowbirds that would parasitize nests.
LENGTH: 5.5 inches. **WINGSPAN:** 9.5 inches.
ADULT: Gray head with white "spectacles" around the eyes. Weak contrast between gray face and white throat. Green to gray on the back, depending on feather wear. Two white wingbars. White below with pale yellow flanks. White undertail coverts. **FIRST FALL:** (Aug.–Mar.) Like adults but head is the same color as the back with no contrast between them.
VOICE: Song is a cheery *cheer-reet* or *cheer-wree*. Call is a harsh rattle or piercing *whreeWREET*.
BEHAVIORS: Movement is slow and deliberate through the trees. Flights are direct with strong wingbeats. **HABITAT:** All types of forests. **STATUS:** Uncommon to rare migrant through the western half of the Trans-Pecos, very rare east to the Edwards Plateau. Spring migration is early March–mid-May, fall late August–early November.

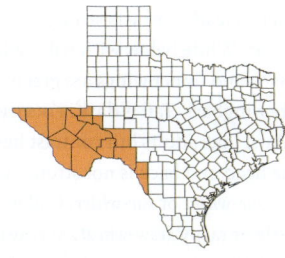

Blue-headed Vireo

(Vireo solitarius)

IUCN RED LIST STATUS (2021): Least Concern
POPULATION TREND: Increasing

Until 1997, Blue-headed Vireo was considered cospecific with Cassin's Vireo and Plumbeous Vireo under the name Solitary Vireo. The split was made after genetic studies showed differences between the distinct plumage forms. Blue-headed Vireo is the only vireo in the Solitary Vireo complex likely to be found east of the Pecos River in Texas.

LENGTH: 5.5 inches. **WINGSPAN:** 9.5 inches.
ADULT: Dark blue-gray head with bold white "spectacles." Head contrasts strongly with white throat. White below with pale yellow flanks. Back and upper wings are green with two white wingbars. Tail is dark. **FIRST FALL:** (Aug.–Mar.) Like adults but contrast between the head and back is not strong. **VOICE:** Emphatic *sue-wheat* or *sue-whier*. Call is a harsh rattle or raspy *keew-wheat*. **BEHAVIORS:** Slow and deliberate movement through the trees and short, direct flights. **HABITAT:** Forests, deciduous or mixed, but oaks are usually present. **STATUS:** Uncommon winter resident in the eastern two-thirds of Texas. Common to uncommon migrant east of the Pecos River. Rare migrant in the Trans-Pecos. Spring migration is late March–mid-May, fall late August–early November.

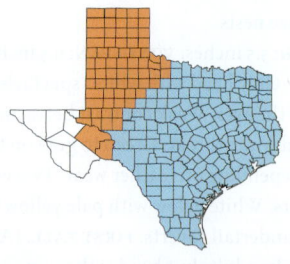

Plumbeous Vireo

(Vireo plumbeus)

IUCN RED LIST STATUS (2021): Least Concern
POPULATION TREND: Decreasing

Plumbeous Vireo is one of the three species that resulted from the split of Solitary Vireo in 1997. Genetics showed that these three forms, usually separately illustrated in field guides, were different enough to be treated as separate species. Distinguishing between the three species—Cassin's, Blue-headed, and Plumbeous Vireo—should be done with caution in the field. Plumbeous is the most distinct of the complex, since it is the one with little or no yellow or green and never shows contrast between the head and the back. **LENGTH:** 5.75 inches. **WINGSPAN:** 10 inches. **ADULT:** Lead-gray overall with white "spectacles" around the eyes. Pale grayish below with gray flanks. Two white wingbars. **FIRST FALL:** (Aug.–Mar.) Like adults but a slight greenish-gray overall with faint yellow wash below. **VOICE:** Raspy *sue-wreet* or *sue-rhee*.

Call is like Cassin's and Blue-headed Vireo. **BEHAVIORS:** Slow and deliberate movements through trees and short flights between trees. **HABITAT:** Montane mixed and coniferous forests. **STATUS:** Common summer resident of the Davis and Guadalupe mountains. Common migrant through the Trans-Pecos. Uncommon to rare migrant through the western High Plains and the western Edwards Plateau. Spring migration is early March–early May, fall late August–early October.

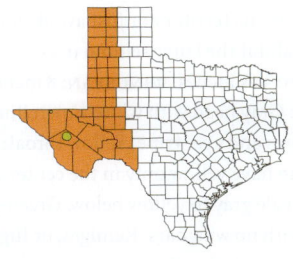

Drab adult

Bright adult

Philadelphia Vireo

(Vireo philadelphicus)

IUCN RED LIST STATUS (2021): Least Concern
POPULATION TREND: Increasing

Philadelphia Vireo was named by John Cassin from a specimen that was collected in Philadelphia. Ironically this vireo turns out to be rare there. Philadelphia Vireo closely resembles Red-eyed Vireo in both appearance and song. They can often occupy the same range and, because the Red-eyed Vireo is the more abundant bird, Philadelphia Vireo can easily be overlooked. The Philadelphia Vireo modifies its behavior to exclude Red-eyed Vireo from its territory or just avoids it by using habitat the latter seldom uses.
LENGTH: 5.25 inches. **WINGSPAN:** 8 inches.
ADULT: Gray head with white supercilium. Dark lores and eye stripe. Yellow throat; drab birds can have yellow only in the center of the throat. Pale gray to yellow below. Green-gray above with no wingbars. Remiges, or flight feathers, are darker than upper wing coverts.

VOICE: Multiphase song like Red-eyed Vireo but higher-pitched with longer pauses: *cheer-rhee cheer-rhe-oo*. Call a raspy *rheet rheet rheep*. **BEHAVIORS:** Moves through foliage in short flights. Hops along small twigs. **HABITAT:** Primarily encountered in coastal migrant traps. Inland, prefers second-growth, small trees, and thickets, often bordering water. **STATUS:** Uncommon to rare spring and rare to very rare fall migrant in the eastern third of Texas. Mostly found on the coast during migration. Spring migration is early April–late May, fall late August–late October.

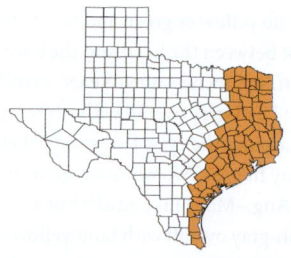

Eastern adult

Warbling Vireo

(Vireo gilvus)

IUCN RED LIST STATUS (2021): Least Concern
POPULATION TREND: Increasing

Warbling Vireo is one of the drabest of vireos, yet its warbling, bunting-like song is conspicuous for its persistence. It will even sing from the nest. Warbling Vireo may actually be two species, a western and an eastern species based on differences in the overall size, size of the bill, voice, and genetics. The American Ornothological Society has considered but not yet accepted proposals to split the species into two. They have deemed that more study is needed on the disjunct population in Mexico and on the closely related Brown-capped Vireo.
LENGTH: 5.5 inches. **WINGSPAN:** 8.5 inches.
ADULT: Gray head with weak supercilium and pale lores. Little or no contrast with the gray or greenish-gray back. Upper wing coverts and remiges, or flight feathers, have little or no contrast. Pale below with pale yellow on flanks. **VOICE:** Song is bunting-like and variable. Call is a thin, raspy *wreeeeet*. **BEHAVIORS:** Foraging is variable; it adapts to what's available. A large percentage of food is gleaned from leaves while hovering. **HABITAT:** Mature mixed deciduous woodlands. In migration, uses a wide variety of forested and shrubby habitats. **STATUS:** Uncommon to rare summer resident in the Davis and Guadalupe mountains and in the eastern Panhandle across the north part of north-central Texas. Uncommon to common migrant across Texas.

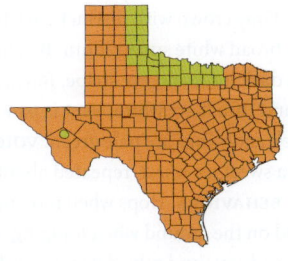

Red-eyed Vireo

(Vireo olivaceus)

IUCN RED LIST STATUS (2019): Least Concern
POPULATION TREND: Increasing

Red-eyed Vireos are one of the most common songbirds in North America. Generally it stays high in the trees during breeding season and is not easily observed, but sings persistently and longer into the season than many other songbirds, revealing its presence. Red-eyed Vireos leave North America after the breeding season to winter in the Amazon Basin in South America.
LENGTH: 6 inches. **WINGSPAN:** 10 inches.
ADULT: Gray crown with distinct dark border and broad white supercilium. Bright red eye and dark lores and eye stripe. Bill is large. Breast and throat are white. Vent is pale yellow. Back and wings are pale green. **VOICE:** Song is a sweet *see-wheat*, repeated about every second. **BEHAVIORS:** Hops when moving in trees and on the ground when foraging. **HABITAT:** Deciduous and mixed deciduous forest. Prefers an abundant understory. In migration, uses more habitats but still prefers deciduous habitat. **STATUS:** Locally abundant to uncommon summer resident in the eastern half of Texas south to the Victoria River. Statewide in migration but very rare in the western half of Texas. Spring migration is early April–late May, fall late July–mid-October.

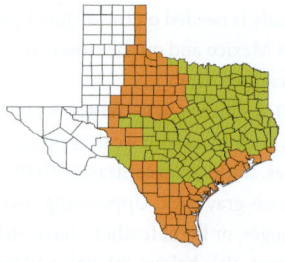

LANIIDAE (SHRIKES)

The shrikes are a family of passerines that have evolved into miniature diurnal raptors. Convergence in evolution has made them similar to many larger raptors. Shrikes can use a wide variety of open and edge habitats as long as there are perches to hunt from. They generally have large, strong heads and beaks and strong feet for capturing and ripping apart their prey. Most species rely heavily on insects, but many can capture small vertebrates, which they kill by severing the spine at the base of the skull with their powerful sharp bills.

The family name Laniidea comes from the Latin word for "butcher." This is a reference to the family's trait of hanging prey like a butcher—on thorns, barbed-wire fences, or even crooks in branches. In many places, shrikes are colloquially referred to as "butcher birds" even today.

Worldwide there are thirty-four species of shrikes. The family is sister to the corvids, or the jays, magpies, and crows. Only two species occur in Texas.

Loggerhead Shrike

(Lanius ludovicianus)

IUCN RED LIST STATUS (2020): Near
Threatened
POPULATION TREND: Decreasing

Loggerhead Shrike is the only shrike species
that occurs exclusively in North America.
"Loggerhead" refers to its large head that supports the muscles for its stubby but powerful
bill. It hunts from perches, taking prey nearly
as large as itself.
LENGTH: 9 inches. **WINGSPAN:** 12 inches.
ADULT: Large gray head with broad black
mask. Bill is stubby. Pale throat sharply contrasts with black mask. Wings are black with
white patch at base of the primaries. Back
and rump are gray. Some subspecies have
paler rumps. Tail is black with white tips on
the tail feathers. **JUVENILE:** (Apr.–Sept.) Like
adults but finely barred on the breast. Washed
in tan overall. A single tan wingbar. **VOICE:**
Cheerful two-syllable *cheer-drup.* **BEHAVIORS:**
Hunts from a high perch and dives on prey.

Hops on the ground to subdue prey. Known
for impaling prey on sharp objects and narrow V-shaped forks. **HABITAT:** Open country
with short vegetation. The nest site is usually
an isolated tree or large shrub. **STATUS:** Rare
to locally common resident statewide, except
the northwestern portion of the South Texas
Brush Country and the southwestern portion
of the Edwards Plateau, where they are winter
residents. Local populations increase in winter
from an influx of migrants from the northeastern United States. Fall migrants begin to
arrive in early August and begin to depart in
mid-March through mid-April.

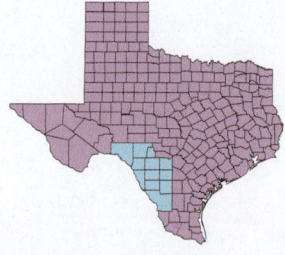

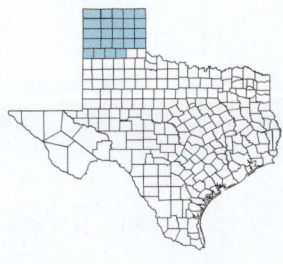

Northern Shrike

(Lanius borealis)

IUCN RED LIST STATUS (2017): Least Concern
POPULATION TREND: Stable

As the name implies, the Northern Shrike is the northern breeding cousin of the Loggerhead Shrike. The shrikes breed in the taiga and taiga-tundra of North America, and their winter range just reaches the top of the Texas Panhandle. Like the Loggerhead Shrike, it impales prey on sharp objects like thorns and barbed wire.
LENGTH: 10 inches. **WINGSPAN:** 14.5 inches. **ADULT:** Similar but larger than Loggerhead Shrike. The face mask is narrower and does not extend above the eye. Breast is faintly scalloped. **FIRST WINTER:** (July–Apr.) Gray form is like adults but lacks white in the wings. Brown wash overall and gray face mask. Distinct scalloped pattern on the breast. Brown form is brown-gray overall, distinctly scalloped in the breast. The face mask is only a shadow of that of an adult. **VOICE:** Call is a harsh, scolding *REET REET REEP*. Song is a repeating *plea-plea-plea-plid*. **BEHAVIORS:** Similar to Loggerhead Shrike. From a perch, drops down and flies low to pounce on prey. Attacks aerial prey by chasing or towering from below. **HABITAT:** Edges of deciduous and mixed forest adjacent to fields and agricultural areas. **STATUS:** Rare to very rare winter resident in the Panhandle. Primary wintering area is the northern Panhandle with little observer coverage; actual status in Texas is uncertain. Wintering birds appear starting in mid-November and most depart by early March.

CORVIDAE
(CROWS, JAYS, AND MAGPIES)

Crows, jays, and magpies are a worldwide family that likely orginated in Australia. They occupy nearly every habitat on earth, from arctic tundra to arid desert to dense jungles. They are the largest of the perching birds. The Common Raven actually has a larger wingspan than a Red-tailed Hawk. They have stout all-purpose bills, and most species can consume almost any food resource available to them.

Crows, jays, and magpies are among the most intelligent of birds. Ravens and magpies have been shown to be able to count nonverbally up to seven; they can recognize groups by size about as well as humans. Some species of jays have been documented not only making tools to gather food but storing the tools for later use.

There are 130 species of crows, jays, and magpies worldwide. Habitat loss is the main threat to the family, and twenty-six species are threatened. The Hawaiian Crow is extinct in the wild and is currently breeding in captivity. Fourteen species have been documented in Texas, of which five are rare visitors: Brown Jay, Pinyon Jay, Clark's Nutcracker, Black-billed Magpie, and Tamaulipas Crow. Brown Jay and Tamaulipas Crow were formerly residents in South Texas but have undergone a range constriction and are currently not resident in the state.

Green Jay

(Cyanocorax yncas)

IUCN RED LIST STATUS (2017): Least Concern
POPULATION TREND: Increasing

Green Jay is one of the specialties of the Lower Rio Grande Valley that birders are seeking when they travel to South Texas. Green Jays have the unusual habit of keeping related but nonbreeding birds in the family flock until the next generation is fledged. Young birds from the previous year will help patrol the family territory until breeding starts, and will continue their defense when the breeding parents are tending the nest. When the birds of the year fledge, the male rejoins the defense of the territory and begins to drive off the one-year-old birds.

LENGTH: 10.5 inches. **WINGSPAN:** 13.5 inches.
ADULT: Blue on crown and cheeks. Wide area of black around eye. Breast is black. Green belly and back. Undertail and sides of upper tail are yellow; yellow tail is especially visible in flight. **JUVENILE:** Similar to adults but colors are duller. **VOICE:** Makes a variety of electric mechanical calls. Often makes a harsh *jeek jeek jeek jeek jeek*. **BEHAVIORS:** Flight is woodpecker-like and undulating. Hops in trees and on the ground when foraging. **HABITAT:** Prefers open woodlands, dense secondary growth, and brushy thickets dominated by mesquite, huisache, and ebony. **STATUS:** Common resident of the Lower Rio Grande Valley. Uncommon and local north to Corpus Christi and the Nueces River, south of US 90 to Del Rio.

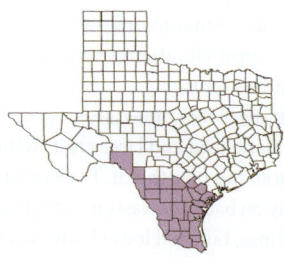

Steller's Jay

(Cyanocitta stelleri)

IUCN RED LIST STATUS (2017): Least Concern
POPULATION TREND: Increasing

Steller's Jay is named for Georg Wilhelm Steller, the German-born naturalist on Vitus Bering's second Kamchatka expedition. In 1741 the expedition reached Kayak Island in Alaska and Steller was able to spend ten hours on the island, where he collected the first specimen and made notes that after his death allowed a new species to be described as Steller's Crow. Steller's Jay is only regularly found in two mountain ranges in Texas—the high Davis and Guadalupe mountains—where one must reach the upper elevations to encounter these bold, raucous birds.
LENGTH: 11.5 inches. **WINGSPAN:** 19 inches.
ADULT: Dark head and large crest. White forehead marks and pale throat. Dark on breast. Dark gray on back contrasting with pale blue rump. Wings, tail, and lower body are blue.
JUVENILE: (May–Aug.) Like adults but head and upper body are drab gray. Crest is smaller than adult crests. **VOICE:** Harsh, raspy *skraaa skraaa skraaa*. Or a rapid, woodpecker-like *chut-chut-chut-chut*. **BEHAVIORS:** Hops from branch to branch in trees. Hops while on the ground. Flights are short but strong and deliberate. **HABITAT:** Coniferous and mixed coniferous-deciduous forest. **STATUS:** Locally common resident of the higher elevations of the Davis and Guadalupe mountains. Irregular winter visitor to El Paso County.

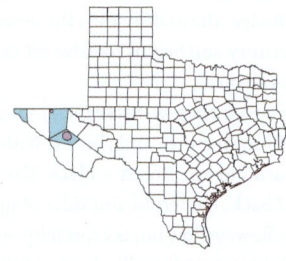

Blue Jay

(Cyanocitta cristata)

IUCN RED LIST STATUS (2016): Least Concern
POPULATION TREND: Stable

The Blue Jay is ubiquious in eastern North America. They are loud, flashy, and seem to prefer urban and residential areas. They have an endless repertoire of sounds and a penchant for imitating Red-shouldered Hawks. While they seem almost clownlike in our backyards, they do have a dark side. They are omnivores and might eat a morsel from a pet's bowl or raid other birds' nests for eggs and nestlings. **LENGTH:** 11 inches. **WINGSPAN:** 16 inches. **ADULT:** Blue crest and whitish face and throat. Throat is surrounded by dark "necklace" that extends to the crest. Dark lores and eye line that extends to back of the necklace. Breast is gray. Upper parts are blue. Upper wings are blue with white secondaries and a single white wingbar. White crescent on underwing. Tail is blue with white corners. **JUVENILE:** (May–Aug.) Like adults with smaller crest and grayer back. **VOICE:** Wide variety of calls, but the most common is a loud *skray skray*. **BEHAVIORS:** Bounces around trees by rapidly hopping from branch to branch. **HABITAT:** Deciduous, coniferous, and mixed forest and woodlands. More numerous in towns and residential areas. **STATUS:** Common to locally uncommon resident in the eastern three-quarters of Texas south to the Nueces River. Absent from the western edge of the High Plains. In the Edwards Plateau and High Plains, primarily found in urban areas.

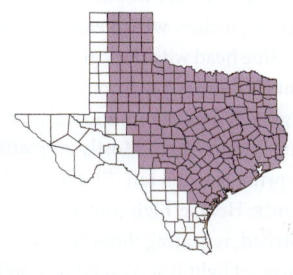

Woodhouse's Scrub-Jay

(Aphelocoma woodhouseii)

IUCN RED LIST STATUS (2016): Least Concern
POPULATION TREND: Decreasing

Woodhouse's Scrub-Jay has quite the taxonomic history. In 1931 three groups of scrub-jays were lumped into one superspecies known only as Scrub-Jay. In 1995 the superspecies was split into three species: Florida Scrub-Jay, Island Scrub-Jay, and Western Scrub-Jay. In 2016, Western Scrub-Jay was further split into Calfornia Scrub-Jay and Woodhouse's Scrub-Jay, named for naturalist and physician Samuel Woodhouse.
LENGTH: 11.5 inches. **WINGSPAN:** 15.5 inches.
ADULT: Blue head with dark gray cheeks. Thin white supercilium. Gray from throat to vent. Back is gray. Wings, rump, and tail are blue.
JUVENILE: (Mar.–Aug.) Mostly gray with blue only on primaries and tail. Cheeks are dark gray. **VOICE:** Hoarse, high-pitched *SKEET*. High pitched, repeating *chep-chep-chep-chep*.
BEHAVIORS: Flight is several rapid wingbeats followed by a short glide. Flight undulates as altitude is lost in glides. When not flying, hops on the ground. Less social than other jays. In winter, forms flocks of five to fifteen individuals. **HABITAT:** Oak and oak-piñon-juniper woodlands. **STATUS:** Common resident of the Edwards Plateau west to the mountains in the Trans-Pecos. Uncommon resident in the canyonlands of the southern Panhandle. In the Rolling Plains, High Plains, and elsewhere in the Trans-Pecos, they can occur in irruptions in fall and winter between mid-September until as late as early April.

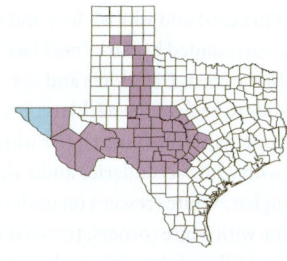

Mexican Jay

(Aphelocoma wollweberi)

IUCN RED LIST STATUS (2016): Least Concern
POPULATION TREND: Decreasing

While Mexican Jay isn't endemic to Texas, it has one of the smallest ranges inside Texas of any resident species. With the exception of two records, it doesn't stray from the Chisos Mountains in Big Bend National Park. There may be no more sedentary species in Texas. A Mexican Jay may live for twenty years and spend that whole time in the territory where it hatched and among the same family group. **LENGTH:** 11.5 inches. **WINGSPAN:** 19.5 inches. **ADULT:** Blue head with dark lores, light gray breast and belly. Rump is whitish. Upper wings and tail are blue. Back is gray but does not contrast as much as in Woodhouse's Scrub-Jay. **VOICE:** Raspy *skeenk skreenk skreenk*. **BEHAVIORS:** Hops on the ground. In flight, usually several strong wingbeats are followed by gliding in an undulating pattern. **HABITAT:** Pine-oak-juniper woodlands. **STATUS:** Common resident of the Chisos Mountains in Big Bend National Park.

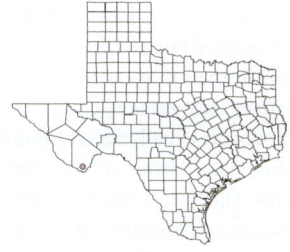

American Crow

(Corvus brachyrhynchos)

IUCN RED LIST STATUS (2018): Least Concern
POPULATION TREND: Increasing

American Crow is familiar to most Texans, except in the South Texas Brush Country, Trans-Pecos, and western Edwards Plateau. They are known for their cleverness and intelligence. They adapted to being widely persecuted as pests in the 19th and early 20th centuries by moving into towns and villages, where there was protection from guns and a ready supply of food.
LENGTH: 17.5 inches. **WINGSPAN:** 39 inches.
ADULT: All glossy black. Heavy bill. Relatively short and rounded tail. **JUVENILE:** (June–Aug.) Body feathers thin and threadlike. Other feathers have a sooty or brownish look, not black. **VOICE:** Usually the classic *carr* or *caaw*; sometimes a drawn-out *caaaaaawwwww*.
BEHAVIORS: Flies with a regular wingbeat, rowing like doing a breaststroke through the air. **HABITAT:** Variety of habitats with an open quality and scattered trees. **STATUS:** Common to abundant resident in the eastern half of the state to the central Panhandle and eastern edge of the Edwards Plateau, and south on the coast to Guadalupe River. Rare to locally uncommon migrant and winter visitor to the High Plains, Rolling Plains, and eastern Edwards Plateau. Common to uncommon winter visitor to El Paso County. In fall, migrants begin arriving in mid-October. Winter birds depart by late March.

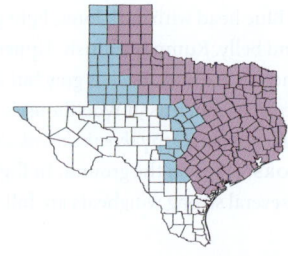

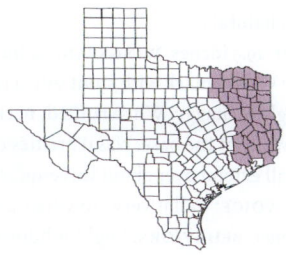

Fish Crow

(Corvus ossifragus)

IUCN RED LIST STATUS (2016): Least Concern
POPULATION TREND: Increasing

Fish Crow reaches the western edge of its range in Texas. Over most of its range in the southeastern part of the United States it is distinctly coastal. In Texas, it is rarely found on the coast in Jefferson County. Instead, it is mostly found along the rivers of East Texas and in the towns east of the San Jacinto and Trinity rivers. In the field, Fish Crow and American Crow are difficult to distinguish except by call. Even in the call there are similarities in many of the common vocalizations, except the distinct nasal *Cha-Uh* call.
LENGTH: 15 inches. **WINGSPAN:** 36 inches.
ADULT: All glossy black with bluish sheen. Heavy bill. Shaggy throat feathers. Relatively short and rounded tail, proportionally longer than an American Crow's. **VOICE:** Only reliably separated from American Crow by voice. More nasal-sounding than American Crow.

Its most distinct call is the two-note *Cha-Uh*.
BEHAVIORS: Very similar to American Crow in habits but perches often on wire, which American Crows rarely do. **HABITAT:** Usually found near water. **STATUS:** Uncommon and local resident east of the San Jacinto and Trinity rivers in Texas.

Showing neck's white under-feathers

Chihuahuan Raven

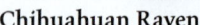

(Corvus cryptoleucus)

IUCN RED LIST STATUS (2016): Least Concern
POPULATION TREND: Stable

The Chihuahuan Raven is one of two raven species found in Texas. It was formerly known as the White-necked Raven because the base of the neck feathers are white, though they can only be seen when the feathers are ruffled; the species name *cryptoleucus* means "hidden white." Common Raven and Chihuahuan Raven overlap in range but typically use very different habitats. Chihuahuan Raven prefers open treeless habitat, grasslands, and open desert scrub. Common Raven is found in wooded habitats.
LENGTH: 19.5 inches. **WINGSPAN:** 44 inches.
ADULT: Glossy black with stout bill. Throat is only slightly shaggy. When ruffled, neck feathers are white at the base. Nasal bristles on top of the bill extend over about three-quarters of the bill. **VOICE:** Call is very crow-like, a hoarse *craaa craaa*. **BEHAVIORS:** Flight is buoyant with measured wingbeats. **HABITAT:** Dry open grasslands, unbroken desert scrub. **STATUS:** Uncommon to common resident in the western half of Texas south to Eagle Pass and the Lower Rio Grande Valley.

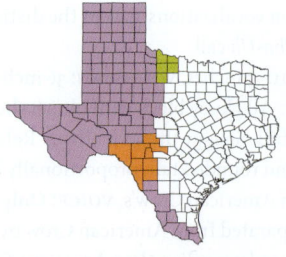

Common Raven

(Corvus corax)

IUCN RED LIST STATUS (2017): Least Concern
POPULATION TREND: Increasing

Common Raven is one of the most widely distributed birds in the world, found over much of the northern hemisphere. It is also the largest of the passerines, or perching birds. Common Ravens are slightly larger than your average Red-tailed Hawk. A habitat generalist, Common Ravens use wilderness areas and are comfortable in urban and suburban areas. They do so well in some urban areas that they are considered pests.
LENGTH: 24 inches. **WINGSPAN:** 53 inches. **ADULT:** Glossy black. Feathers on neck are very shaggy. Tail is longer and more wedge-shaped than Chihuahuan Raven. Bill is long and heavy with nasal bristles that extend just past the halfway point of the bill. **JUVENILE:** (June–Aug.) Like adults but can be distinctly brownish. **VOICE:** Deep, croaking *kraah*. **BEHAVIORS:** Even wingbeats; often soars high. Frequently makes half rolls and full rolls in flight. Known to fly upside down. **HABITAT:** Habitat generalist but prefers wooded areas and areas with structures for nesting. **STATUS:** Uncommon to common resident in the Trans-Pecos mountain sky islands and on Edwards Plateau. Rare to uncommon in the western Rolling Plains.

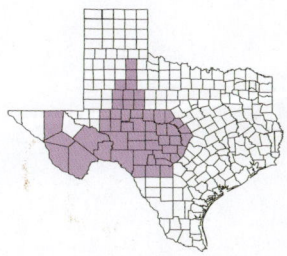

REMIZIDAE (PENDULINE-TITS)

The penduline-tits are a small family named for the hanging nests most members build. The family of eleven species is mostly found in Eurasia and Africa, with a sole species found in North America: the Verdin, which appears in Texas. Recent DNA studies have confirmed that the family is closely related to the tits, chickadees, and titmice.

Verdin

(Auriparus flaviceps)

IUCN RED LIST STATUS (2022): Least Concern
POPULATION TREND: Decreasing

Verdin is one of the smallest songbirds in North America and the only member of a family of Eurasian and African birds that occurs in North America. They are conspicuous because of their nearly continuous foraging and the nests they build. The large, ball-like nests are usually on the edge of a thorny scub and can be up to 10 inches in diameter. Outside of breeding season, they build roost nests that they continuously add to, sometimes by recycling older nests or stealing from another Verdin's nest.
LENGTH: 4.5 inches. **WINGSPAN:** 6.5 inches.
ADULT: Yellow head with dark lores and small pointed bill. Light gray overall with small reddish shoulder patch. **JUVENILE:** (Apr.–Aug.) Plain gray overall. Lacks the yellow head, dark lores, and wing patch of adults. **VOICE:** Song is a piercing *sweet sweet sweet*. Call is a rapid, ticking *tsip-tsip-tsip-tsip-tsip*. **BEHAVIORS:** Moves like a chickadee, rarely on the ground; gleans from live foliage. **HABITAT:** Thorn forest, scrub oak, oak-juniper, and chaparral. **STATUS:** Uncommon to common resident in the western two-thirds of the state, south to the Rio Grande Valley, west through the Trans-Pecos, and north to the Panhandle.

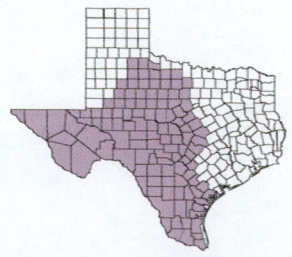

PARIDAE (TITS, CHICKADEES, AND TITMICE)

The tits, chickadees, and titmice species are familiar to humans. As a group, most species adapt well to humans and often visit bird feeders. They are vocal and often lead mixed-species flocks foraging through forests. In North America, species with crests are called titmice and those without are called chickadees. In Europe and Asia, they are called tits.

Most species are found in forests but a few are found in desert scrub. All are cavity nesters and most use cavities excavated by other species. A few species will excavate their own if they can find a decaying tree with softened wood. All are insectivores but switch readily to seeds and berries when insects become scarce. No species is migratory.

There are sixty-three species worldwide. Four are of conservation concern. Habitat destruction and alteration is the main concern for these very-limited-distribution species. Six species have been recorded in Texas. One, the Black-capped Chickadee, is represented in Texas by a single hundred-year-old specimen record.

Carolina Chickadee

(Poecile carolinensis)

IUCN RED LIST STATUS (2016): Least Concern
POPULATION TREND: Stable

Carolina Chickadees are a familiar backyard bird in the eastern half of Texas. They are able to use a wide variety of woodland habitats, and as long as there are dead snags with woodpeckers' holes for nesting, Carolina Chickadees will usually be present. They will use nest boxes and are a frequent visitor to feeders, dashing in to grab a sunflower seed and retreating to eat it.

LENGTH: 4.75 inches. **WINGSPAN:** 7.5 inches. **ADULT:** Black head and throat with white cheek extending below the eye to the lore. Whitish breast and wide, light-buffy flanks. The line between the dark throat and the light breast is clean and crisp. Pale gray above with short tail. **VOICE:** Song is a high-pitched, whistling *see-dee-see-dah*. There are several variations of calls similar to *see-che-che-che* or *see-see-see che-che-che*. **BEHAVIORS:** Short undulating flights. Moves rapidly through trees. Responds aggressively to "pishing." **HABITAT:** Wooded habitats, often swamps and riparian, parks, and well-wooded residential areas. **STATUS:** Common resident of the eastern half of Texas south to Live Oak and Aransas counties and north to the southeastern part of the Panhandle.

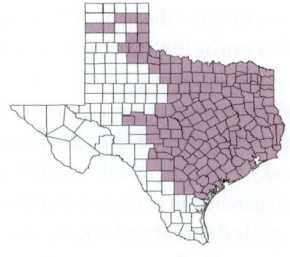

Mountain Chickadee

(Poecile gambeli)

IUCN RED LIST STATUS (2017): Least Concern
POPULATION TREND: Stable

Mountain Chickadees in Texas are restricted to the upper elevations of the Davis and
Guadalupe mountains. They store pine seeds
for winter and form social groups that defend
a winter territory and their seed caches from
other social groups. When seeds are scarce, the
younger birds may be forced out of the group
and descend to lower altitudes.
LENGTH: 5.25 inches. **WINGSPAN:** 8.5 inches.
ADULT: Black head with large white cheek
extending almost to the nape. White supercilium that can be indistinct on worn birds in
the late summer. Light gray below and slightly
dark gray above. Tail is short. **VOICE:** Song is a
clear descending whistle: *dzeee dzeee dzeee*. Call
is a squeaky *dzeee cheee chee chee*. **BEHAVIORS:**
Hops and perches on twigs and on the ground.
Hangs upside down when foraging. A weak
but maneuverable flier. **HABITAT:** Montane

coniferous forests. **STATUS:** Locally common
to uncommon resident in the higher elevations of the Davis and Guadalupe mountains. Some move to lower elevations in winter, including the desert scrub around the
Guadalupe Mountains.

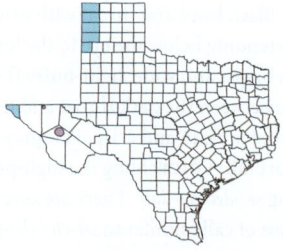

Juniper Titmouse

(Baeolophus ridgwayi)

IUCN RED LIST STATUS (2017): Least Concern
POPULATION TREND: Stable

Until recently, Juniper Titmouse and Oak Titmouse were lumped together as one species under the name Plain Titmouse. Juniper Titmouse is not very common in its range and only has a small footprint in Texas, in the foothills of Guadalupe Mountains National Park. Pairs defend a territory year-round. Unlike other titmouse species, they do not form winter flocks.

LENGTH: 5.75 inches. **WINGSPAN:** 9 inches. **ADULT:** Gray overall. Small stout bill. Gray crest. Proportionally long tail. **VOICE:** Song is a rapid, sweet *weep-weep-weep-weep-weep-weep-weet.* **BEHAVIORS:** Actively moves from tree to tree and branch to branch. Flights between trees are undulating and shallow. **HABITAT:** Juniper and juniper-piñon woodlands. **STATUS:** Locally uncommon to rare resident in the foothills of the Guadalupe Mountains.

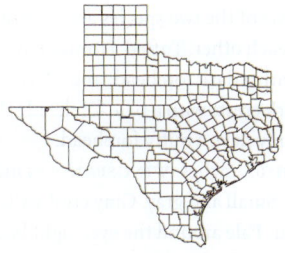

Juvenile

Adult

Tufted Titmouse

(Baeolophus bicolor)

IUCN RED LIST STATUS (2016): Least Concern
POPULATION TREND: Increasing

From the 1990s until the early 2000s, Black-crested and Tufted Titmouse were considered to be one species, based on the narrow zone of hybrids found where the range of these two species overlaps. That was reconsidered and the species were once again split. This zone of hybrids is only about 12 miles wide, mostly along the Balcones Escarpment and south along the Guadalupe River drainage, and most of the titmice present are hybrids. The crests of the two species are reverse images of each other. Tufted Titmice have gray crests with a dark forehead, and Black-crested have dark crests with a pale forehead. Hybrids commonly have a buffy forehead.
LENGTH: 6.5 inches. **WINGSPAN:** 9.75 inches.
ADULT: Small and gray. Gray crest with black forehead. Pale around the eye. Light below with orange flanks. **JUVENILE:** (May–Aug.)

Like adults but paler orange or no orange on the flanks. The dark forehead is reduced, brownish, or absent. **VOICE:** Song is a clear, whistled *peter peter preter*. Sometimes a raspy *wheir wheir*. Call is a raspy *rheeep rheeep rheeep*. **BEHAVIORS:** Flies from branch to branch and tree to tree. Long flights, direct and non-undulating. **HABITAT:** Mostly deciduous woods. **STATUS:** Common resident in the eastern third of the state south to Refugio County and west to the Balcones Escarpment and the edge of the Rolling Plains.

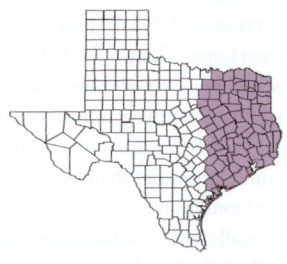

Hybrid Tufted x
Black-crested Titmouse

Black-crested Titmouse

(Baeolophus atricristatus)

IUCN RED LIST STATUS (2017): Least Concern
POPULATION TREND: Stable

Black-crested Titmouse is one of the Texas specialties that birders and birdwatchers travel here to see. It was considered cospecific with Tufted Titmouse until the early 2000s. In the narrow zone where the two species meet, hybrids of the two are more common than pure birds. There is much variation in these hybrids, but most have a gray to dark gray crest with a chestnut or brown forehead. **LENGTH:** 6.5 inches. **WINGSPAN:** 9.75 inches. **ADULT:** Small and gray. Crest is black with pale forehead. Orange on flanks. **JUVENILE:** (May–Aug.) Like adults. Forehead is pale but crest is pale gray. **VOICE:** Sometimes a loud *peew peew peew peew*. Also a rapid *wep-wep-wep-wep-wep-wep*. Call is nearly identical to Tufted Titmouse, a raspy *wheir wheir wheir*. **BEHAVIORS:** Flies direct without undulating. Hops along branches. Often hangs upside down when foraging. **HABITAT:** Most types of deciduous and semi-deciduous forests, particularly oaks and mesquite. **STATUS:** Common resident in the western two-thirds of the state. South of Refugio County to the Lower Rio Grande Valley, north to Clay County and the southeast corner of the Panhandle. In the Trans-Pecos, local to Chisos, Davis, and Del Norte mountains and the Stockton Plateau.

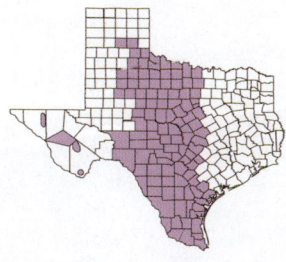

ALAUDIDAE (LARKS)

The lark family is a mostly Old World family of birds with the largest concentration in Africa. They are birds of the open country. Larks mostly walk or run on the gound in short grass or bare dirt. In most species the males fly high in the air and sing when doing breeding displays, known as skylarking.

Larks have a uniquely structured syrinx (the vocal organ of birds) and scales on both the front and back of the tarsus. This characteristic is considered primitive, and because of it larks were formerly considered closely related to swallows, but recent studies have shown they are not. The lark family is sister to the monotypic family of bearded reedling.

There are 93 species of larks worldwide, most found in Africa and Asia. Thirteen are of conservation concern, as habitat loss poses a threat. The Horned Lark is the only North American species in the family.

Horned Lark

(Eremophila alpestris)

IUCN RED LIST STATUS (2019): Least Concern
POPULATION TREND: Decreasing

Horned Lark is a Holarctic species and the only member of the lark family native to North America.
LENGTH: 7.25 inches. **WINGSPAN:** 12 inches.
ADULT: Male has brown crown and nape. Face is white with black lores and cheeks. Black-and-white tufts, or "horns," above the eyes. Throat is yellow. Black breast band. White below with brown flanks. Tail is dark with light brown central tail feathers and narrow white edge on the outside. Female is brown above with yellow throat and brown breast band. White below. Dark tail with pale brown center and white outer tail feathers. **JUVENILE:** (Apr.–Aug.) Crown and cheeks are brown with white spots. Back is dark with white spots. Brown breast and white throat. Belly is white with brown flanks. Tail is like adults. **VOICE:** Song is a couple of *chips* followed by a tinkling, rising warble. Call is a quiet *tsip-tsip*. **BEHAVIORS:** Walks the ground. Flight is undulating. **HABITAT:** Open, barren areas. Prefers bare ground to grass higher than an inch. Frequents recently burned sites. **STATUS:** Common to uncommon resident locally in the Trans-Pecos, from the Panhandle south to the Concho Valley, and along the coast south to the Lower Rio Grande Valley. Locally common migrant and winter resident in central Texas south to Guadalupe County. Fall migration is from early October, and spring migrants leave before early April.

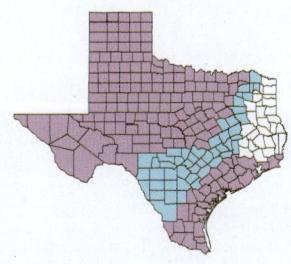

HIRUNDINIDAE (SWALLOWS)

Swallows, like swifts, treeswifts, and woodswallows, are specialized aerial insectivores. They are highly migratory because they need aerial insects to feed on. Like other aerial predators they have long pointed wings, wide short bills, and short inconspicuous legs and feet. They forage on the wing in long swooping flights. Within the family they are often divided into groups by how they nest. There are mud nesters, cavity nesters, and burrow nesters.

Taxonomically, swallows have been hard to place. They are likely closely related to Bush Warblers, Long-tailed Tits, and Leaf Warblers. Much work still needs to be done to identify a sister family.

Worldwide there are 88 species of swallows. Habitat destruction is the primary threat to the seven species of conservation concern. White-eyed River Martin, known only from its wintering locations in Thailand, has not been seen since the 1980s and is likely extinct. Ten species of swallows have been recorded in Texas. The Gray-breasted Martin is a very rare visitor, and the South American Blue-and-white Swallow is known from a single record.

Bank Swallow

(Riparia riparia)

IUCN RED LIST STATUS (2019): Least Concern
POPULATION TREND: Decreasing

The Bank Swallow is almost completely cosmopolitian in distribution, one of the most widely distributed of the passerines in the world. In the Old World, it is known as the Sand Martin.
LENGTH: 5.25 inches. **WINGSPAN:** 13 inches. **ADULT:** Brown above, with the rump paler than the back. White throat and brown breast band. The white of the throat wraps around the back of the auriculars, or the cheek. Breast band has a streak that extends down slightly from the band. Belly and undertail are white. Tail is long and slender. **JUVENILE:** (June–Dec.) Like adults but feathers of the tertials, wing coverts, rump, and upper tail coverts edged in buffy or pale brown. **VOICE:** Scratchy, dry-sounding, repeated *wit wit drrrrr drr drr*. Call is a dry *zreert zreert zreert*. **BEHAVIORS:** Flight is fast but fluttery; the tail is spread when gliding and looks square. **HABITAT:** Breeds in lowland areas with vertical banks, cliffs, and bluffs with soils that are easily crumbled for making burrows. Also uses artificial sites like sand and gravel quarries and road cuts. In migration, seen in a variety of open and water-associated habitats. **STATUS:** Locally uncommon summer resident along the Rio Grande north to Del Rio and Val Verde County. Common to uncommon migrant statewide. Spring migration is mid-March–end of May, fall early July–late October.

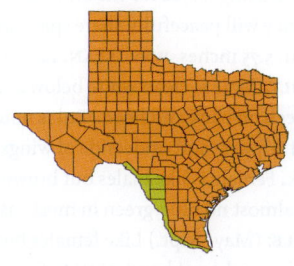

Tree Swallows resting during migration

Tree Swallow

(Tachycineta bicolor)

IUCN RED LIST STATUS (2016): Least Concern
POPULATION TREND: Stable

The Tree Swallow is a bird of open places: fields, meadows, and marshes. It doesn't construct a nest like other swallows but uses abandoned cavities in dead trees. It also readily uses nest boxes. Instead of fighting the swallow, many bluebird enthusiasts pair boxes close together. Since both bluebirds and Tree Swallows are territorial, they won't let another pair of the same species occupy the neighboring box, leaving it free for the other species, which they will peacefully share space with. **LENGTH:** 5.75 inches. **WINGSPAN:** 14.5 inches. **ADULT:** Male is clean white below and blue-green on back and head. Tail is notched in flight. Upper- and underside of wings are dark. Female is like males but browner with almost no blue-green in most cases. **JUVENILE:** (May–Sept.) Like females but with pale gray breast band. **VOICE:** Song is a clear *twit-twit-twit-twedle-twedle-twit.* Call is a rapid *chit-it chit-it chit-it.* **BEHAVIORS:** Tends to glide more than other swallows. **HABITAT:** Usually breeds near bodies of water. Standing dead trees with nest cavities needed for nesting. Winter habitat is similar, often large bodies of water and marshes. **STATUS:** Rare to locally uncommon summer resident in northeast Texas south to Houston and Harris counties. Rare to uncommon winter resident in the southern third of Texas, and common migrant throughout. Spring migration is late February–mid-May, fall early August–mid-November.

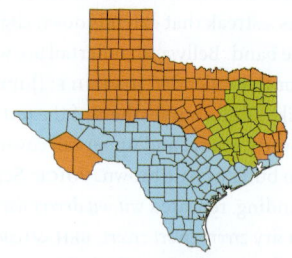

Violet-green Swallow

(Tachycineta thalassina)

IUCN RED LIST STATUS (2017): Least Concern
POPULATION TREND: Increasing

The western counterpart of the Tree Swallow, the Violet-Green Swallow occurs only rarely outside the central and western Trans-Pecos in Texas. Its name refers to its violet rump and sea-green back. While a common migrant in its range, due to its preference to nest in inaccessible habitat it is probably the least studied swallow in North America.
LENGTH: 5.25 inches. **WINGSPAN:** 13.5 inches. **ADULT:** Male has green crown and back. Rump is violet. All clean-white below. White undertail coverts extend up onto rump. The white in the throat extends onto the auriculars (cheeks) and above the eye. Some females are similar to males, and some dull, mostly green-gray with grayish cheeks. **JUVENILE:** (May–Oct.) Like a gray female. Dusky and not clean-white below. **VOICE:** Song is a sweet-sounding *chirt chirt chirt dreee dchirt*. Call is a chirping *chit-chit-chit-chit-chit*. **BEHAVIORS:** Perches often on wires and tree branches. Can fly at many heights but circles at heights greater than other swallows. **HABITAT:** Any kind of woodlands but prefers coniferous forests. Most common above 6500 feet. **STATUS:** Uncommon to common summer resident in the mountains of the Trans-Pecos. Common migrant in the central and western Trans-Pecos. Spring migration is early March–mid-May, fall early July–early October.

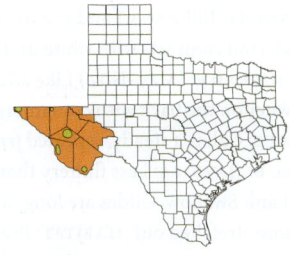

Northern Rough-winged Swallow

(Stelgidopteryx serripennis)

IUCN RED LIST STATUS (2016): Least Concern
POPULATION TREND: Decreasing

The Northern Rough-winged Swallow is distinguished mostly by its lack of distinguishing marks. The most common adjective describing it in most guides is "dingy." Its name describes something an observer can't detect in the field: its primary feathers have a rough texture that can only be felt.
LENGTH: 5.5 inches. **WINGSPAN:** 14 inches.
ADULT: Drab gray-brown above. Plain, dingy, pale gray-brown below. Dark lores. White undertail coverts. Tail is short and square, and dark undertail contrasts with white undertail coverts. **JUVENILE:** (May–Nov.) Like adults but browner on the back with cinnamon wingbars. **VOICE:** Call is a steady, repeated *frrip frrip frrip*. **BEHAVIORS:** Less fluttery than the similar Bank Swallow. Glides are long, with wings more stretched out. **HABITAT:** Prefers open areas, including open woodland, that have crevices or burrows for nesting. Open areas in migration. **STATUS:** Locally rare to uncommon summer resident in the eastern half of the state and west along the Rio Grande from Val Verde County. Rare winter resident in the South Texas Brush Country and west along the Rio Grande. Common to uncommon migrant statewide. Spring migration is mid-February–mid-May, fall late July–late October.

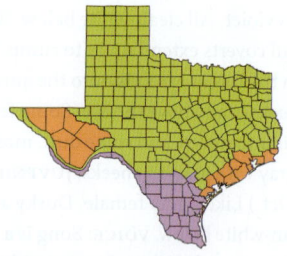

Purple Martins going to roost

Male (below) and female

Purple Martin

(Progne subis)

IUCN RED LIST STATUS (2016): Least Concern
POPULATION TREND: Stable

Purple Martins are cavity nesters that use abandoned woodpeckers' nests and depend on dead snags with holes for nesting. The primary subspecies in Texas is now enterly dependent on nest boxes for nesting. Because of this, they are almost exclusively found around humans during the breeding season. In the late summer they form enormous roosts until fall migration starts. Roosts of more than half a million birds have been recorded.

LENGTH: 8 inches. **WINGSPAN:** 18 inches.
ADULT: Male is dark purple overall. Wings are almost black. Tail is moderately long and slightly notched. Female is dingy gray below, with gray collar and forehead. Smudgy marking on belly and undertail. Dark purple on crown and back. **JUVENILE:** (May–Feb.) Like females but lacking purple. Belly is white with fine streaks. **VOICE:** Song is a low-pitched, liquid gurgle. Call is a loud *cheert cheert cheert*.
BEHAVIORS: Flies at various heights, often in big looping circles. **HABITAT:** Almost entirely dependent on birdhouses for nesting and now mostly found in areas of human habitation and open areas for foraging. **STATUS:** Common to uncommon summer resident in most of Texas east of the Pecos River. Only absent from the Trans-Pecos and the most western part of the High Plains. Common to abundant migrant in the eastern two-thirds of the state. Spring migration is early January–mid-April, fall early July–mid-September.

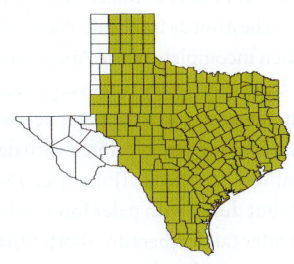

Taking a break after completing a flight across the Gulf of Mexico on the Texas coast

Barn Swallow

(Hirundo rustica)

IUCN RED LIST STATUS (2019): Least Concern
POPULATION TREND: Stable

Barn Swallows are one of the true cosmopolitain species found in Texas. They breed in North America, Europe, Asia, and North Africa, and populations winter in South America, Africa, and Australia.
LENGTH: 6.75 inches. **WINGSPAN:** 15 inches.
ADULT: Sexes are similar. Long deeply forked tail, longer in males. Iridescent blue-black upper side. Black face mask with dark eye. Rich chestnut forehead. Bill is dark. Throat is same rich chestnut as forehead. Narrow breast band, often incomplete in the American subspecies. Underparts vary from very pale to rich chestnut; females tend to be paler below than males. Underwings are bicolored with dark flight feathers. **JUVENILE:** (June–Dec.) Similar to adults but duller with paler forehead and throat. Outer tail feathers are shorter than adults, with white band across tail. **VOICE:**

Song is a rapid twittering, often ending in a dry rattle. Main contact call is a *witt-witt* or *vit-vit.* **BEHAVIORS:** Can fly from very low to very high. Flight is easy, and they fly a straight path longer than other swallow species. Can dramatically alter their direction when needed because of the long, deeply forked tail. **HABITAT:** Needs an open area to forage for insects. Nests almost exclusively on buildings under overhangs. **STATUS:** Rare to common summer resident in all regions of Texas. Migrants begin to appear mid-March, and most birds are gone by mid-October.

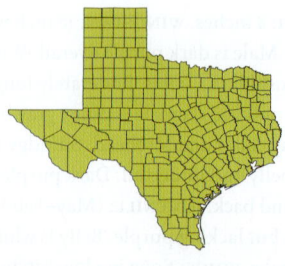

White forehead Cliff Swallow downed
in a spring migration event

Darker forehead Cliff Swallow

Cliff Swallow

(Petrochelidon pyrrhonota)

IUCN RED LIST STATUS (2016): Least Concern
POPULATION TREND: Increasing

Cliff Swallows are one of the most social land-birds in North America, forming colonies sometimes in the thousands of nests. This is the bird known for its mythical return to Mission San Juan Capistrano in California on March 19 each year. They are brood parasites, laying eggs and even moving eggs from their own nest to neighboring nests.
LENGTH: 5.5 inches. **WINGSPAN:** 13.5 inches. **ADULT:** Sexes are the same. Bright white forehead, dark crown, pale nape and collar. Cheeks are a dark tawny color. Throat is dark with darker spots. Pale underneath with spotted undertail. Dark above with pale tawny rump and dark square tail. **JUVENILE:** (June–Dec.) Variable, but all have dark ear patches. Often has some degree of white speckling on throat. Forehead is dark. Dark back with pale rump and dark square tail. **VOICE:** Song is thin and drawn out, shorter and simpler than a Barn Swallow's. Purr call is a husky *veer.*
BEHAVIORS: Flies at various heights, from just above the ground to almost 200 feet. Agile flier with short, gliding flights. **HABITAT:** Historically inhabited open areas with cliffs for nesting. Able to use artificial structures and is now found in a variety of habitats with suitable resources for nests (mud) and areas for foraging. **STATUS:** Common and locally abundant statewide. Migrants arrive mid-March and depart at the end of October.

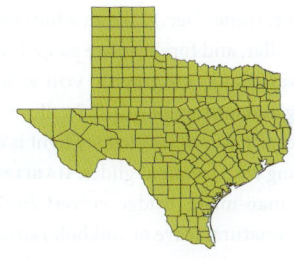

Cave Swallow

(Petrochelidon fulva)

IUCN RED LIST STATUS (2019): Least Concern
POPULATION TREND: Increasing

Cave Swallows have been expanding in Texas since the 1990s. It has been suggested that their range expansion follows the large overpass structures of interstate highways, which provide nesting sites. **LENGTH:** 5.5 inches. **WINGSPAN:** 13 inches. **ADULT:** Dark on crown and back. Throat and collar are pale tawny. Crown is a darker tawny. Flanks are light rufous or tawny. Rump is rufous. Belly is pale. Tail is short and square. **JUVENILE:** (June–Dec.) Like an adult, but throat, collar, and forehead are paler. Rump is pale tawny. Flanks are grayish. **VOICE:** Song is creaking and rattling like Cliff Swallows. Call is a sweet *suit suit*. **BEHAVIORS:** Flight is deep and strong with frequent glides. **HABITAT:** Nests in man-made (bridge, culvert, building, or silo) or natural (cave or sinkhole) structures. When using a man-made structure, they choose areas with twilight-like light levels that match natural caverns. **STATUS:** Common to locally abundant migrant and summer resident in the southern half of Texas, north to the southern Rolling Plains and west through the Trans-Pecos and most of the South Plains. Uncommon migrant and summer resident through the southern Post Oak Savannah. Rare migrant and summer resident on the upper Texas coast. Rare to locally common winter resident in the southern half of the state. Spring migration is early February–mid-April, fall early September–November.

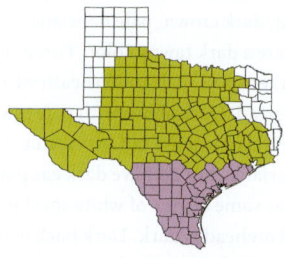

AEGITHALIDAE
(LONG-TAILED TITS)

Long-tailed tits were for a long time placed in the tit family. Differences in nest structure and ge-netics led to the group being placed in its own family. Tits and long-tailed tits are only distantly related, the latter likely being a sister family to the Old World Bush Warblers. Generally, they are small chickadee-like birds with proportionally long tails, large heads, and short conical bills. They are mostly insectivores, gleaning insects from branches and leaves, often while hanging upside down.

Long-tailed tits build large globular nests. Some species are cooperative breeders at times. Most are highly social, and failed breeders may become helpers of other breeding pairs.

Worldwide there are eleven species of long-tailed tits, none of conservation concern. The Bushtit is the only member of the family found in North America and Texas.

Bushtit

(Psaltriparus minimus)

IUCN RED LIST STATUS (2021): Least Concern
POPULATION TREND: Decreasing

Bushtits are the only New World member
of the family of long-tailed tits. They are
polymorphic, meaning they come in multiple forms in the same population. From
the Texas Trans-Pecos and south through
Central America, many have dark auriculars,
the feathers of the cheek and over the ear.
These dark-eared birds were once considered
a separate species, the Black-eared Bushtit,
Psaltriparus melanotis.

LENGTH: 4.5 inches. **WINGSPAN:** 6 inches.
ADULT: Small, plump, and gray-brown with a
large head, stubby bill, and long tail. Usually,
a cheek patch varies from very pale brown to
completely black. Males have dark eyes; females have light eyes. **JUVENILE:** (Apr.–Aug.)
Like adults but always with dark eyes. Many
juvenile males in Texas have a dark cheek
or ear patch. **VOICE:** Sharp, dry *tsit tsit tsit.*

BEHAVIORS: Movements are chickadee-like.
Flies short distances within a tree and from
tree to tree. **HABITAT:** Variety of habitats.
Prefers open woodlands with some evergreen
shrubby understory. **STATUS:** Uncommon
to common resident from the southern
Panhandle through the Rolling Plains and
Edwards Plateau. Locally abundant in
oak-juniper habitat in the Trans-Pecos.

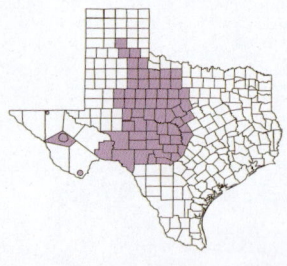

PYCNONOTIDAE
(BULBULS)

Bubuls are a large family found in Africa and Southeast Asia. The name is derived from the Persian word for "nightingale." Though known for their song, not all species are great songsters. They are known for consuming fruit, but insects often make up a large part of their diet. Some have rather plain plumage; others have bold head patterns.

Bulbuls may be most closely related to the cisticolas of Africa and Southeast Asia. Worldwide there are 156 species of bulbuls. Habitat destruction is the primary threat to the twenty-seven species of conservation concern. There is no native North American species, but two—the Red-whiskered Bulbul and Red-vented Bulbul—have become established in North America. In Texas, the Red-vented Bubul can be found in the Houston area.

Red-vented Bulbul

(Pycnonotus cafer)

IUCN RED LIST STATUS (2018): Least Concern
POPULATION TREND: Increasing

A popular cage bird, Red-vented Bulbul has become established in the Inner Loop area of Houston. They primarily occupy older neighborhoods with mature live oak trees. The process of becoming established was relatively quick; from the first birds being noticed to a recognized established population only took about twenty years.

LENGTH: 8.5 inches. **WINGSPAN:** 12.5 inches.
ADULT: Black or dark-charcoal-colored with crest. Bright red undertail. Back and breast feathers often edged pale. Brown cheek patches. **JUVENILE:** (June–Jan.) Like adults but brownish on upper wings and back. **VOICE:** Song is a musical *PEEK-two-rue*. **BEHAVIORS:** Hops on branches in trees and shrubs or on the ground to forage. **HABITAT:** Primarily urban and suburban areas with many mature live oak trees. **STATUS:** Uncommon introduced species inside the Inner Loop of Houston in Harris County. Rare in the Houston suburbs.

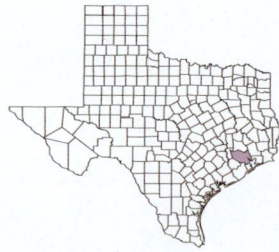

REGULIDAE (KINGLETS)

Kinglets are some of our smallest birds. In some species it would take as many as five individuals to make an ounce, and you could mail them for one US postal stamp! Only some species of hummingbird are smaller. All species of kinglets are similar in plumage and habits. They are mostly insectivores. All seem to share the same fidgety wing flicking, constantly on the move as they forage, often hovering to snatch prey from under leaves.

Kinglets for a very long time were placed among the Old World warblers but have recently been elevated to their own family. Some studies now indicate they might be most closely related to the waxwings, although distantly.

Worldwide there are six species of kinglets. John James Audubon in his *Birds of North America* included a Cuvier's Kinglet, drawn from a specimen he collected in Pennsylvania, that cannot be matched to any known bird. Two species occur in North America and Texas: Ruby-crowned Kinglet and Golden-crowned Kinglet.

Caught in a spiderweb

Male fiercly raising crest

Ruby-crowned Kinglet

(Corthylio calendula)

IUCN RED LIST STATUS (2021): Least Concern
POPULATION TREND: Increasing

Ruby-crowned Kinglets are common and easily recognized winter residents in all parts of Texas. Their small size, frenetic movements, and constant wing flicking make them familiar to many. A generalist when it comes to winter habitat, it's more specialized when nesting—high in trees in mature spruce-fir forests. It is one of the smallest of North America's kingbirds but lays one of the largest clutches of any songbird. Clutches of up to 12 eggs have been recorded.
LENGTH: 4 inches. **WINGSPAN:** 7.5 inches.
ADULT: Small and drab grayish-green. Face is plain with thin eye ring. Two white wingbars. Dark bar on secondaries. Males have a bright ruby crest that is usually concealed but is raised and visible when agitated. **JUVENILE:** (June–Aug.). Like adults but light parts washed with buffy. **VOICE:** Song is a clear, very high-pitched *sii sii sii sii beer beer beer beer beeri*. Call is a rapid, dry *chit-chit chit-chit chit-chit*.
BEHAVIORS: Moves with short hops amid foliage. Often flicks wings. Gleans for insects by hovering at branch tips. **HABITAT:** Various woodlands, hedgerows, and thickets. **STATUS:** Common to abundant migrant and winter resident statewide. Fall migrants begin to arrive in mid-September and spring birds depart between mid-March and mid-May.

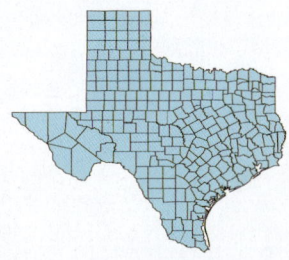

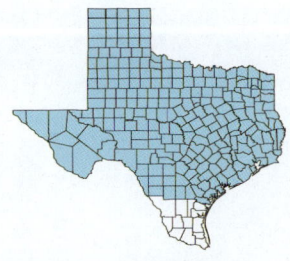

Golden-crowned Kinglet

(Regulus satrapa)

IUCN RED LIST STATUS (2022): Least Concern
POPULATION TREND: Stable

The Golden-crowned Kinglet formerly bred almost exclusively in remote boreal spruce-fir forests in North America. Sensative to logging and other habitat disturbances, it has declined in the western portion of its range but has benefited from reforestation of spruce in the east. In winter, it joins mixed species flocks in Texas in a variety of habitat types, but almost always with mature trees.

LENGTH: 4 inches. **WINGSPAN:** 7 inches. **ADULT:** Pale grayish-green. Face is boldly striped with golden crown patch with black borders, white supercilium, black eye line, and pale lores. Throat is pale. Single pale wingbar. Undertail is whitish. **JUVENILE:** (June–Aug.) Like adults but lacks golden crown patch. **VOICE:** Song is a thin, very high-pitched *see see see-tsi-tsi-tsi chedi chedi chedi*. Call is a ringing, high-pitched *tchi-tschi-tschi*. **BEHAVIORS:** When feeding, often hangs upside down. Movements are more deliberate than Ruby-crowned Kinglets. Hovers while gleaning from twigs. **HABITAT:** Winters in a wide variety of forested habitats. **STATUS:** Uncommon to locally common migrant and winter resident in the eastern two-thirds of Texas. Uncommon to rare west to El Paso. Rare winter visitor to the South Texas Brush County and the Lower Rio Grande Valley. Sometimes irruptive, it can be scarce some years and common in others. Fall migrants arrive in early October and most leave in mid-March, while a few linger until mid-May.

BOMBYCILLIDAE (WAXWINGS)

Waxwings get their name from the secondary feathers of their wings, where the tips are fused with keratin and have the texture of sealing wax. They feed almost entirely on fruit and are the only birds outside of the tropics that feed fruit to their young. They nest later than other birds, waiting until fruit is most abundant. In winter they form large nomadic flocks and can be abundant where fruit is found.

Waxwings are most closely related to silky-flycatchers, which they stongly resemble. Worldwide there are three species, two of which have been recorded in Texas. One, the Bohemian Waxwing, is a rare visitor.

Feeding on berries

Cedar Waxwing

(Bombycilla cedrorum)

IUCN RED LIST STATUS (2021): Least Concern
POPULATION TREND: Increasing

Cedar Waxwings do glean and hawk insects, but the majority of their food is small, sugary fruits. Depending on the amount of available fruit, they express little territoriality. In winter, they form large flocks that descend on trees and consume the fruit, assimilating nutrients from it and passing the seeds intact. This makes them one of the least favorite birds for car owners who happen to be parked under a fruiting tree when they arrive. **LENGTH:** 7.25 inches. **WINGSPAN:** 14.5 inches. **ADULT:** Plain gray-brown upper parts. Long crest. Black mask from bill to base of the crest. Gray on rump and tail. Tail is yellow-tipped; some birds have orange tips. Undertail coverts are whitish. **JUVENILE:** (July–Jan.) Plain gray-brown overall, paler below. Wide streaks below with yellow tail tip. **VOICE:** High-pitched series of *sreee* notes. **BEHAVIORS:** Flocks from less than a dozen to hundreds. Wanders in search of food; moves on when an area is out of fruit. **HABITAT:** Found where there are fruit-bearing shrubs and trees. **STATUS:** Common to abundant migrant and winter resident east of the Pecos River. Irregular and uncommon winter resident of the Lower Rio Grande Valley. Year to year their number fluctuates. Fall migrants arrive as early as mid-August. Spring migration is late March–early May.

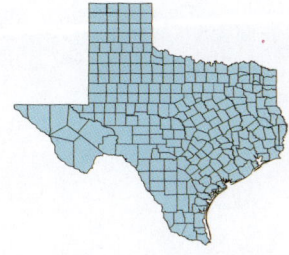

PTILIOGONATIDAE (SILKY-FLYCATCHERS)

Silky-flycatchers are a small group of medium-size songbirds of Central America and the southwestern United States. They have an upright posture and most species have crests, giving them a distinct profile. They have broad, short bills and soft, silky plumage in black, gray, or brown. Most species eat more fruit than insects, except in the breeding season. Insects are caught on the wing, hawking from perches.

There are four species of silky-flycatchers. The most recent research seems to indicate their closest living family is the waxwings. Two species of silky-flycatchers have been recorded in Texas, with only a single record of Gray Silky-flycatcher.

Male

Female

Phainopepla

(Phainopepla nitens)

IUCN RED LIST STATUS (2021): Least Concern
POPULATION TREND: Stable

Phainopeplas have an unusual breeding biology among Texas birds. They breed twice a year, first in the early spring in the deserts of the Trans-Pecos that have fruiting trees and shrubs. In the desert they are territorial and aggressively defensive. In late spring and early summer, they move into the foothills of the mountains to breed in loose colonies.
LENGTH: 7.75 inches. **WINGSPAN:** 11 inches.
ADULT: Male is like a glossy-black Northern Cardinal. Red eye is prominent. Crest is ragged. Tail is long and square. Wings have white patches in the primaries. Female is like a dark-gray version of a male. No white primary patches in the wings. All wing feathers have narrow white edges, giving the wings a frosted look. **VOICE:** Most vocalizations have a thrasher-like quality. Song is an irregular *cheer-reet* and descending *CHEERLL*. Call is a sweet-sounding *CHUREET*. **BEHAVIORS:** Territorial in desert habitats, where individuals spend the greater part of the day perched high in trees and shrubs, calling and defending their territory. In montane habitats, forms loose breeding colonies. **HABITAT:** In late spring and summer, semiarid woodlands in the mountain foothills. The rest of the year is spent in deserts with fruiting shrubs. **STATUS:** Rare to uncommon resident throughout the Trans-Pecos. Can be common in the foothills of the Chinati, Davis, Del Norte, and Guadalupe mountains.

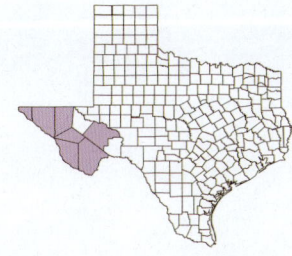

SITTIDAE (NUTHATCHES)

Nuthatches are small songbirds that at first glance resemble small, stout woodpeckers. The two species are not closely related, though both forage on tree bark. Woodpeckers always forage up the trunk of a tree, while nuthatches have little to no respect for gravity and look perfectly at home foraging head-down on a tree trunk or walking upside down under a large branch. The etymology of the name nuthatch isn't clear, but it may derive from "nuthack," applied to the European species when it would wedge a nut into a crevice to break it open.

Nuthatches forage mostly on insects and other invertebrates in the cracks and crevices of trees. In winter when these become scarce, they switch to seeds and nuts and will visit bird feeders. In pairs and small groups, or even in mixed flocks of other species, nuthatches call constantly, keeping in touch with each other.

There are twenty-nine species of nuthatches worldwide. Their closest relatives are not well resolved. They may be most closely related to treecreepers, wallcreepers, or perhaps even wrens. Human pressure on limited habitat is the major concern for the eight species of conservation concern. Four species occur in Texas.

Red-breasted Nuthatch

(Sitta canadensis)

IUCN RED LIST STATUS (2018): Least Concern
POPULATION TREND: Increasing

Formerly known as the Canada Nuthatch or the Red-bellied Nuthatch, Red-breasted Nuthatch breeds in fir or spruce forests. Of North America's four nuthatch species, it's the only one that is irruptive when food is scarce, and it is present in winter as far south as the Gulf Coast.
LENGTH: 4.5 inches. **WINGSPAN:** 8.5 inches. **ADULT:** Male is blue-gray above and reddish-orange below. Bold white supercilium and white cheek. Dark eye line from the bill across the ear patch (auriculars). Tail is short and square; center of the tail is blue-gray like the back, with black-and-white diagonals. Female are like males but the crown is close to the back in color, not black. **VOICE:** Song is a nasal, drawn-out *REEK REEK REEK*. Call is a nasal *chep-chep-chep-chep-chep*. **BEHAVIORS:** Flights are short and undulating, mostly between trees. Walks up and down tree trunks, often headfirst and upside down while foraging. **HABITAT:** Mature coniferous forests. **STATUS:** Irruptive throughout the state in fall and winter. Rarely present in the South Texas Brush Country. Abundance varies from year to year. Fall migrants are first seen in late September and wintering birds depart during March, with some lingering into May.

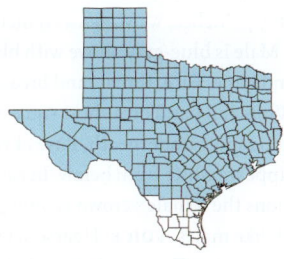

White-breasted Nuthatch

(Sitta carolinensis)

IUCN RED LIST STATUS (2016): Least Concern
POPULATION TREND: Increasing

White-breasted Nuthatch is the largest of North America's nuthatches and, unlike the Red-breasted Nuthatch, doesn't excavate a burrow but uses old woodpecker burrows. Breeding birds form year-round pairs and defend their territory year-round. Some part of the population does migrate or irrupt in winter, but little is known about these birds, including what their survivorship is and whether they return to their natal areas. **LENGTH:** 5.75 inches. **WINGSPAN:** 11 inches. **ADULT:** Male is blue-gray above with black crown and nape. Face, throat, and breast are white. White tail band on a short square tail, with rufous spots underneath. Base of the primaries appears white from below. In eastern populations the female's crown is blue-gray, not black like males. **VOICE:** Hoarse *YAK YAK YAK*. **BEHAVIORS:** Typical of all nuthatches, especially in walking upside down on tree trunks. **HABITAT:** Mature deciduous woodlands. **STATUS:** Common to uncommon resident in the Chisos, Davis, and Guadalupe mountains and northeast Texas. Rare to locally uncommon in the central and southern part of the Pineywoods and eastern Panhandle. Rare to uncommon winter visitor to the northern half of Texas. Fall migration is from late August to early October. Wintering birds leave during March.

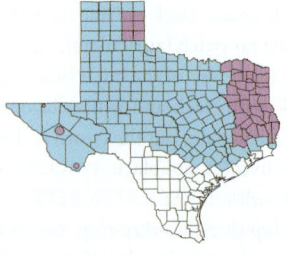

Pygmy Nuthatch

(Sitta pygmaea)

IUCN RED LIST STATUS (2016): Least Concern
POPULATION TREND: Increasing

Pygmy Nuthatch was formerly a resident of the upper elevations of the Davis Mountains. It is thought that these birds arrived in winter invasions and became temporary residents. A significant invasion has not occurred since the 1990s, and there has been no reliable record of Pygmy Nuthatch there since 2005. Today, it can only be regularly found in the upper elevations of the Guadalupe Mountains. **LENGTH:** 4.25 inches. **WINGSPAN:** 7.75 inches. **ADULT:** Gray-brown head and blue-gray back. Dark eye line from the lores through the eye and to the auriculars. White cheek and throat. Buffy below. Dark tail with white tail band. Flanks are blue-gray. **VOICE:** Rapid, high-pitched *PIP-PEE PIP-PEE*. **BEHAVIORS:** Movements typical of a nuthatch, especially head-down movement on tree trunks. **HABITAT:** Almost exclusively long-leaf pine forests. **STATUS:** Uncommon resident at the upper elevations in the Guadalupe Mountains in the Trans-Pecos.

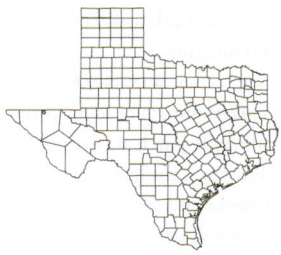

Brown-headed Nuthatch

(Sitta pusilla)

IUCN RED LIST STATUS (2016): Least Concern
POPULATION TREND: Decreasing

Brown-headed Nuthatch is very similar to Pygmy Nuthatch, and closely related, but there is no overlap in range. Brown-headed Nuthatch reaches the very western edge of its range in the East Texas Pineywoods. It is one of the few cooperative breeding birds in North America. Between 25% and 50% of nesting pairs have a helper in the territory. They are also one of the few birds with documented tool use. Brown-headed Nuthatches are known to use flakes of pine bark to pry off other bark flakes when foraging.
LENGTH: 4.5 inches. **WINGSPAN:** 7.75 inches. **ADULT:** Dark brown head and blue-gray back. Small white patch at base of neck. Dark eye line from the lores across the auriculars. Cheeks and throat are white. Pale buffy wash below, with blue-gray flanks and belly. Little white in the tail compared to other nuthatches. **VOICE:** Squeaky *CHI-WIT CHI-WIT*. Very much like a dog's squeak toy. **BEHAVIORS:** Movement typical of other nuthatches. Flight is weak and slow, mostly within a tree or between trees. **HABITAT:** Mature pine and pine-hardwood forests. **STATUS:** Locally common to rare resident in the Pineywoods.

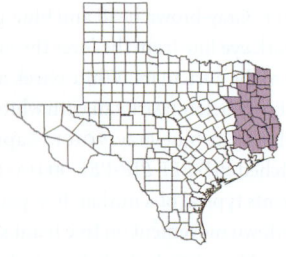

CERTHIIDAE (TREECREEPERS)

Treecreepers are small birds of temperate forests in the northern hemisphere. Most are speckled with brown and buff for excellent camouflage on tree trunks. Like most other tree-creeping families, they have stiff tails for support as they inch up a tree trunk. They probe in the bark with thin decurved bills for small arthropods.

Worldwide there eleven species of treecreepers. Only the Sichuan Treecreeper is a conservation concern, due to a limited range of small forest patches threatened by development. Only the Brown Creeper occurs in North America and Texas.

Brown Creeper

(Certhia americana)

IUCN RED LIST STATUS (2021): Least Concern
POPULATION TREND: Increasing

Brown Creepers are widespread and not rare in their range, but their amazing camouflaged appearance makes them very inconspicuous. Their call is loud but so high-pitched that many observers don't hear it well. It's almost identical to the Eurasian Treecreeper in appearance, and the two were long considered the same species. Recent studies have shown otherwise, and they are no longer considered to be cospecific.

LENGTH: 5.25 inches. **WINGSPAN:** 7.75 inches.
ADULT: Variable from gray to brown. Streaky above with long stiff tail. Thin decurved bill. Pale supercilium. White below with buffy undertail. White underwing coverts. Bold buffy band is visible on upper and underwings.
JUVENILE: (May–Aug.) Nearly identical to adults. May have a light scaly appearance below. **VOICE:** Song is a high-pitched, piercing *see-see dee-dee-dee see*. **BEHAVIORS:** Almost always hops up a tree trunk facing up. Flies to the base of a tree and starts over. **HABITAT:** Variety of mature forest types with large trees. **STATUS:** Uncommon to rare migrant and winter resident throughout the northern three-quarters of the state. Uncommon summer resident in the Guadalupe Mountains. Sometimes locally common in the Pineywoods as a winter resident. Fall migrants arrive late October to early December. Winter birds depart between late February and late March.

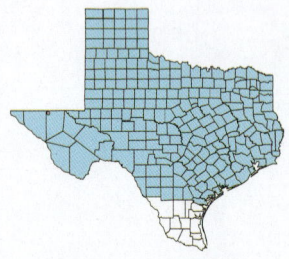

POLIOPTILIDAE (GNATCATCHERS)

The gnatcatchers are small jaunty birds, generally in blues, blue-grays, and browns. Most have long tails that flick from side to side as they move energetically through the forest. They are insectivores with light bills and legs, and they glean insects and other arthropods from leaves and branches, often by hovering. The name gnatcatcher is misleading, as they rarely take flying insects on the wing.

Orginally gnatcatchers were thought to be antbirds, then later they were placed in the Old World warblers. They have since been elevated to their own family, and molecular DNA analysis indicates they are likely most closely related to the wrens. They are restricted to the Americas and there are twenty-one species. Two are endangered: Iquitos Gnatcatcher was first described in 2005 and occupies a tiny patch of Amazonian Peru. The Creamy-bellied Gnatcatcher of southeastern Brazil is near-threatened and rapidly declining due to forest conversion. Two species are resident in Texas.

Blue-gray Gnatcatcher

(Polioptila caerulea)

IUCN RED LIST STATUS (2016): Least Concern
POPULATION TREND: Increasing

Blue-gray Gnatcatchers is widespread in
North America and the only gnatcatcher
that migrates.
LENGTH: 4.5 inches. **WINGSPAN:** 6 inches. **ADULT BREEDING:** (Mar.–July) Bright
blue-gray above, paler blue-gray below. Throat
is whitish. White eye ring. Black line from the
bill to above the eye. Tail is dark with white
outer tail feathers. Undertail is mostly white.
Female is like males but pale blue-gray and
lacks the black eye stripe. **NONBREEDING:**
(Aug.–Feb.) Male and female are like breeding
females. **VOICE:** Song is a soft series of *see, seep,*
and *seet* notes. Call is a raspy, drawn-out *spee
spee spee.* **BEHAVIORS:** Flight is fluttery, almost
mothlike. Constantly in motion in foliage.
HABITAT: Variety of wooded habitats, though
generally not present in conifer-exclusive
habitats. **STATUS:** Rare to locally common

summer resident in the eastern half of Texas,
Edwards Plateau, and northern portion of
the Rolling Plains. Summer resident in the
oak motts in Kleberg and Kenedy counties
and along the Rio Grande River in the Lower
Rio Grande Valley. Common to uncommon
summer resident in the Guadalupe and
Chisos mountains. Common migrant in the
eastern half of the state, uncommon in the
western half. Uncommon winter resident in
the coastal prairies, southern Pineywoods,
and South Texas Brush Country. Spring migration is mid-March–late May, fall early
August–mid-October.

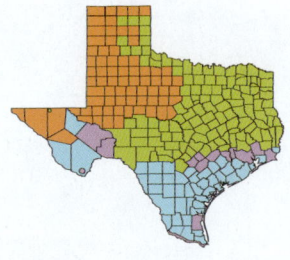

Black-tailed Gnatcatcher

(Polioptila melanura)

IUCN RED LIST STATUS (2018): Least Concern
POPULATION TREND: Decreasing

Black-tailed Gnatcatcher is one of the small foliage-gleaning songbirds of the arid desert scrub. With a raspy *SPEEP* call and a constant wagging of its tail, it is one of the smallest of our songbirds. Despite this, it is an aggressive defender of its territory against much larger birds.
LENGTH: 4.5 inches. **WINGSPAN:** 5.5 inches. **ADULT BREEDING:** (Feb.–July) Male is blue-gray with dark black cap and white eye ring. Upper tail is mostly black with white tips to the outer tail feathers. Undertail is mostly black. Female is like males without the black cap. Brownish wash overall. **NONBREEDING:** (Aug.–Jan.) Male is like breeding male without a black cap. Dark stripe over eye. Female is like breeding female. **VOICE:** Song is a raspy *sip-sip-sip-sip SPREE*. Call is a raspy *SPEEP*. **BEHAVIORS:** A foliage gleaner; very active.

Hops from branch to branch inspecting surfaces of leaves, flicking its tail from side to side. **HABITAT:** Semiarid or desert scrub. **STATUS:** Common to uncommon resident in the western part of the South Texas Brush Country and the southwestern Edwards Plateau, and along the Rio Grande River in the Trans-Pecos.

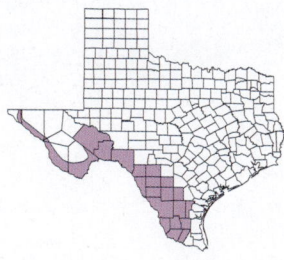

TROGLODYTIDAE (WRENS)

Wrens occupy every habitat in the New World, but only one species occurs in the Old World. They are small to medium-size and usually brownish birds. Most are some shade of brown above and lighter below; some species are spotted below. Most species cock their tail over their back. Most make musical sounds that seem impossibly loud for birds of their size.

The family name Troglodytidae is derived from the Greek word *troglodytes*, for "cave dweller," and indeed some species seek shelter in crevices and nest in hidden nooks and crannies, usually making a cup-shaped nest. Others nest on open scrub and shrubs, and most of these make a domed or ball-shaped nest with a side opening. Some species place the nest near a wasp nest for protection. Most species are monogamous, but there are cooperative-breeding and polygynous species.

Wrens are most closely related to the gnatcatchers. There are eighty-six species worldwide. Eleven are of conservation concern; all of these species have very low populations in very small habitat patches. Ten species have been recorded in Texas. One, the Pacific Wren, was recently split from Winter Wren and is a rare visitor to Texas.

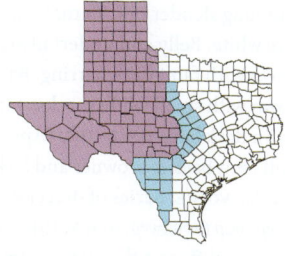

Rock Wren

(Salpinctes obsoletus)

IUCN RED LIST STATUS (2021): Least Concern
POPULATION TREND: Decreasing

Rock Wrens are more often heard than seen, their strong ringing call echoing off the rocky hillsides and cliffs they favor. Well camouflaged among the rocks, they often perch in the open, calling and singing as the masters of their domain.

LENGTH: 6 inches. **WINGSPAN:** 9 inches.
ADULT: Gray-brown above with fine speckles. Indistinct pale supercilium. Pale buffy below with fine brown streaking on breast. Undertail is barred dark. Tail feathers have pale buffy tips. **JUVENILE:** (Apr.–Aug.) Like adults but lack streaking in the breast. **VOICE:** Call is a ringing *treee treee treee*. Song is a series of variations on *chee-chee-chee, skree skree skree, tweee tweee twee*, repeated three to six times.
BEHAVIORS: Bobs up and down with deep knee bends. Runs and hops when foraging. Glides downhill and lands with a spread tail.

HABITAT: Rocky arid and semiarid habitats. Will often occupy a small island of rocks, even artificial rock piles and embankments with rocks. **STATUS:** Common resident in the western half of the state. Most birds retreat from the northern half of the Texas range in winter. Rare and local in winter along the Rio Grande to Hidalgo County. Fall migrants begin movements in mid-October and remain sometimes into late April.

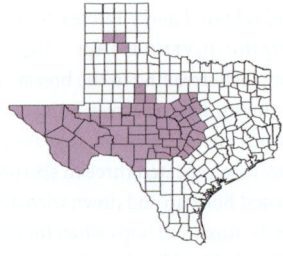

Canyon Wren

(Catherpes mexicanus)

IUCN RED LIST STATUS (2021): Least Concern
POPULATION TREND: Stable

The call of a Canyon Wren echoes across canyons and cliffs, but the bird is often hard to observe high up on near-vertical rock faces. Aided by strong toes and long claws, they nimbly walk up, down, and across rock walls. Due to the inaccessibility of its preferred habitat, there is much research yet to be done on this species.
LENGTH: 5.75 inches. **WINGSPAN:** 7.5 inches.
ADULT: Gray head with pale charcoal-gray mask and long slender bill. Throat and upper breast are white. Belly and undertail are dark brown with faint dark barring. Back is gray-brown with small white specks. Tail is brown with dark bars. **JUVENILE:** (Apr.–Aug.) Like adults but back is browner and lacks small specks. **VOICE:** Series of descending notes: *twep twep twep twep.* **BEHAVIORS:** Moves rapidly across cliffs and flies across canyons, nimble and able to cling to vertical rock faces.
HABITAT: Steep rocky hillsides and cliffs. More vertical than the habitat of Rock Wrens.
STATUS: Uncommon to locally common resident in the Trans-Pecos through the Edwards Plateau. Common resident in the southeastern Panhandle canyonlands.

Cactus Wren

(Campylorhynchus brunneicapillus)

IUCN RED LIST STATUS (2021): Least Concern
POPULATION TREND: Decreasing

The nonmigratory Cactus Wren is a true desert species. It is able to get enough water from its food alone: insects, small reptiles, fruit pulp and seeds, and cactus juice from wounds on the plant. It is so associated with deserts that movie sound engineers often add the call of Cactus Wrens to the soundtrack to let viewers know they are now in the desert. **LENGTH:** 8.5 inches. **WINGSPAN:** 11 inches. **ADULT:** Brown crown with bold white supercilium. Gray-brown on back with dense white streaks. Pale on breast with dense dark spots. Buffy belly and undertail with lines of small spots. Tail is gray with dark bars. **JUVENILE:** (Apr.–Aug.) Like adults but spotting on the breast is less dense. **VOICE:** Rapid, rhythmic *krrr krrr krrr krrr krrr krrr*. **BEHAVIORS:** Flights are short and direct. Hops on the ground to forage. Calls during the heat of the day. **HABITAT:** Lowland habitats with cholla, prickly pear, acacia, and mesquite. **STATUS:** Uncommon to locally common from the Lower Rio Grande Valley north through the Edwards Plateau and the southeastern Panhandle, and in the Trans-Pecos.

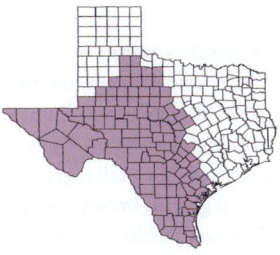

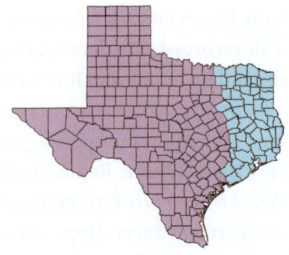

Bewick's Wren

(Thryomanes bewickii)

IUCN RED LIST STATUS (2021): Least Concern
POPULATION TREND: Decreasing

Bewick's Wren was first described by John James Audubon from a specimen from Louisiana and named for the British engraver Thomas Bewick. In the last century, the Bewick's Wren was common and beloved in the Appalachians and the Midwest. It has declined throughout that range and East Texas by as much as 80%. Competition with Northern House Wrens is the most likely cause of this decline of the distinctive eastern population.
LENGTH: 5.25 inches. **WINGSPAN:** 7 inches.
ADULT: Brown head with gray on the side of the neck. Bold white supercilium. White throat and pale grayish below. Undertail coverts are barred. Long tail is grayish and barred. Back is grayish-brown in western birds and brown in eastern birds. **JUVENILE:** (Apr.–Aug.) Like adults. Breast feathers narrowly tipped in dark, giving a faint scaly look. Undertail coverts not barred. **VOICE:** Songs are varied, always a rising buzz and a slow trill. Call is a harsh *chit chit chit, chet chet*, or *cheeet cheet*. **BEHAVIORS:** Highly active, quick hops between perches. Flights are short. Tail is wagged side to side constantly and often up. **HABITAT:** Brushy areas, scrub, thickets, open woodlands, and chaparral. **STATUS:** Uncommon to common resident in the western two-thirds of Texas. Rare east of a line from the mouth of the Colorado River north to Grayson County.

Carolina Wren

(Thryothorus ludovicianus)

IUCN RED LIST STATUS (2021): Least Concern
POPULATION TREND: Increasing

Carolina Wren is a generalist species that can be found around humans and in wild places. They are energetic and always on the move. Pairs are monogomous and defend territories year-round. Its ringing call makes it easier to hear than to see. Perhaps no bird has a greater ratio of volume to size.
LENGTH: 5.5 inches. **WINGSPAN:** 7.5 inches.
ADULT: Reddish-brown above with long white supercilium. Buffy-orange below. Throat is white. Tail is short and brown with dark bars. Undertail coverts are white with dark bars. **JUVENILE:** (Apr.–Aug.) Like adults but undertail coverts not barred. Not as bright below. **VOICE:** Rolling series of phrases like *tea kettle tea kettle* or *tweedo tweedo tweedo*. Call is a musical, rolling trill and a raspy *cheat cheat cheat*. **BEHAVIORS:** Hops and flits on or near the ground. Moves rapidly in and out of dark spaces. Flights are short. **HABITAT:** Wide range of habitats with dense shrubs and brushy cover. **STATUS:** Common resident in the eastern two-thirds of Texas from Val Verde County to the eastern Panhandle. In arid portions of the range found in the riparian corridors.

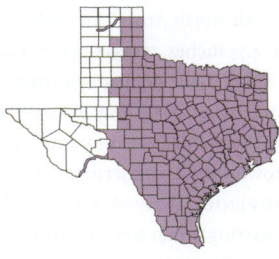

Northern House Wren

(Troglodytes aedon)

IUCN RED LIST STATUS (2017): Least Concern
POPULATION TREND: Increasing

Northern House Wrens breed fom Canada down to the tip of South America. Over such a wide latitudinal range there is a huge number of subspecies (more than forty are recognized by some texts). Northern House Wrens don't occupy large contiguous forests but instead prefer the edges. As an edge specialist, they coexist well with the supreme maker of edge habitat: man. Widespread, common, and compatible with humans, it is one of the best studied of all North American birds. **LENGTH:** 4.75 inches. **WINGSPAN:** 6 inches. **ADULT:** Small plain wren, gray-brown above with faint barring, plain gray below. Weak supercilium. Some birds are more rufous below. Tail is brown with dark barring. Undertail is barred. **JUVENILE:** (May–Aug.) Like adults but lack barring under the tail. **VOICE:** Song is varied, a rolling series of rattle and trills, usually with descending trills at the end. Call is a rapid, raspy *chi-chi-chi-chi* or a very rapid *tri-tri-tri-tri-tri*. **BEHAVIORS:** Flight is low above the ground, direct and steady when crossing openings. Hops off the ground. **HABITAT:** In or near forest edges. **STATUS:** Uncommon summer resident above 7000 feet in the Davis Mountains; rare in the Guadalupe Mountains. Summer resident in the Canadian River drainage in the Panhandle. Common to uncommon migrant statewide. Common to rare winter resident throughout the state south of the Panhandle outside of the Canadian River drainage.

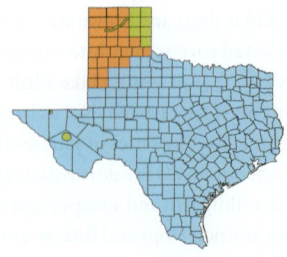

Winter Wren

(Troglodytes hiemalis)

IUCN RED LIST STATUS (2021): Least Concern
POPULATION TREND: Increasing

Winter Wren seems more mouse than bird as it creeps around the forest floor, scrambling through tangles and over fallen logs. Its cryptic color makes it difficult to spot, but it calls attention to itself with a superb song. Formerly considered a Holarctic species, it was split into three species: the Old World population became the Eurasian Wren and the western North American population became Pacific Wren, leaving Winter Wren in eastern North America.
LENGTH: 4 inches. **WINGSPAN:** 5.5 inches.
ADULT: Small and dark brown. Tail is very short and barred. Distinct but low-contrast supercilium. Dark barring on the flanks. **JUVENILE:** (June–Aug.) Like adults but dark; barring and supercilium less distinct. **VOICE:** Song is a long series of very high, tinkling trills and warbles. Call is a short, hard note, usually doubled: *jip-jip.* **BEHAVIORS:** Moves mouselike on the ground and through tangles. The short tail is usually cocked up. **HABITAT:** Uses varied habitats, usually near water and with fallen trees and large mature trees. **STATUS:** Uncommon to common migrant and winter resident in the eastern two-thirds of Texas south to the Nueces River drainage. Fall migrants arrive in early October and are present until mid-March.

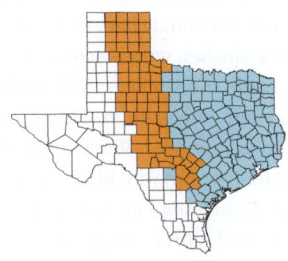

Sedge Wren

(Cistothorus stellaris)

IUCN RED LIST STATUS (2016): Least Concern
POPULATION TREND: Increasing

Formerly known as the Short-billed Marsh Wren, this species was renamed to distinguish it from the closely related Marsh Wren (then known as the Long-billed Marsh Wren). The Sedge Wren is a Goldilocks species when it comes to habitat. The grass can't be too short, it can't be too high. It can't have too many woody shrubs. It can't be too dry. Everything has to be just right. The ephemeral nature of this habitat makes for very low site fidelity for Sedge Wren, as habitat changes over time. **LENGTH:** 4.5 inches. **WINGSPAN:** 5.5 inches. **ADULT:** Small, short-billed, and light brown. Indistinct buffy supercilium. Throat is pale. Crown is finely streaked. Wings and tail are barred. Back is brown with black streaks. **JUVENILE:** (June–Aug.) Like adults but patterns are muted. **VOICE:** Song is an accelerating series of *chit chit chit chi-chi-chi-chi*. Call is a monotone, repeating *chit chit chit chit*. **BEHAVIORS:** Runs on the ground to escape disturbances. When flushed, flies only a short distance before crashing into the grass. **HABITAT:** Wet or damp grassy fields. Grassy edges of marshes, ponds, and lakes. **STATUS:** Common to uncommon migrant through the eastern half of the state. Uncommon to common winter resident on the coastal prairies. Fall migrants arrive in late September. Spring migration is between late March and late May.

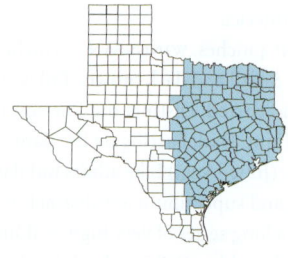

Marsh Wren

(Cistothorus palustris)

IUCN RED LIST STATUS (2021): Least Concern
POPULATION TREND: Increasing

Marsh Wren is an impressive songster, with some males having a repertoire of up to 200 songs. Early ornithologists somehow weren't impressed. Alexander Wilson called it "deficient and contemptible in singing." John James Audubon compared the song to "the grating of a rusty hinge." Arthur A. Allen compared the song to an old-fashioned sewing machine. The Marsh Wren mates with several females and sings cheerfully and lustily day and night to attract them.
LENGTH: 5 inches. **WINGSPAN:** 6 inches.
ADULT: Dark brown crown and white supercilium. Cheek and throat are pale. Back is dark with white stripes. Upper wings and tail are brown. Tail is barred dark. Pale brown below. **JUVENILE:** (May–Aug.) Like adult but supercilium is less distinct. Back is brown and lacks streaking. **VOICE:** Call is a dry *check check.*

Song is a single note followed by a gurgling trill: *chit gle-gle-gle-gle-gle.* **BEHAVIORS:** Clings to cattails and reeds. Moves rapidly and easily over the mud and roots at the base. **HABITAT:** Uses a variety of wetlands; prefers to nest in cattails. **STATUS:** Uncommon to locally common resident east of Galveston Bay and from Matagorda County to Aransas County. Rare to locally common winter resident statewide. Fall migrants arrive in early September. Spring migration is from late March to mid-May.

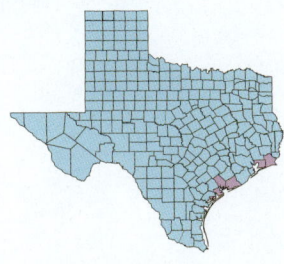

MIMIDAE (MOCKINGBIRDS AND THRASHERS)

The mockingbirds and thrashers, also known as mimids, are medium-size songbirds known for their rich vocal repertoires and mimickery. The Texas state bird, the Northern Mockingbird, is one of the best. They have been recorded mimicking more than a dozen other bird species in a row, along with man-made sounds like car alarms and ringing telephones.

The mimids include the mockingbirds, thrashers, catbirds, and tremblers. Most are gray and brown, but the Blue Mockingbird and the Blue-and-white Mockingbird are bright blue. Generally, they have short, rounded wings, long legs, and a long and often decurved bill. Most species sing from an exposed perch but disapper into thick cover when not singing.

The most recent work has placed the mimid family as a sister family to the starlings. Mimids are found exclusively in the New World. There are thirty-four species, eight of conservation concern, all due to habitat loss. The population of the Cozumel Mockingbird was reduced to just a few individuals by Hurrican Gilbert in 1988. The last credible sighting was in 2006. Nine species of mimids have been recorded in Texas. The Blue Mockingbird is a rare visitor, and there is a single specimen record of Black Catbird for the state.

Gray Catbird

(Dumetella carolinensis)

IUCN RED LIST STATUS (2016): Least Concern
POPULATION TREND: Stable

Gray Catbirds are named for their catlike meowing call but have an extremely varied and versatile song. Their singing prowess stems from having a syrinx (the bird equvalent of a larynx) with two sides that can operate independently, giving a Gray Catbird the ability to sing with two voices at the same time. They are also one of the few species that learns to recognize Brown-headed Cowbird eggs and ejects them from their nest.
LENGTH: 8.5 inches. **WINGSPAN:** 11 inches.
ADULT: Dark gray overall. Cap is black. Undertail coverts are rufous. **JUVENILE:** Similar to adults. **VOICE:** Call is a distinct mewing *MAeeee*. Song is a series of musical whistles and chirps. **BEHAVIORS:** Flights are short and just above the tops of shrubs and through small spaces. Avoids flying across large open spaces. Hops through shrubs.

HABITAT: Dense shrubby habitats. **STATUS:** Uncommon to rare summer resident in the northern part of the state west to the Rolling Plains. Rare and local breeder in the Pineywoods urban settings. Uncommon to rare winter resident on the coastal prairies south to the Lower Rio Grande Valley. Uncommon to common migrant west of the Trans-Pecos and western High Plains. Spring migration is late March–late May, fall late August–late October.

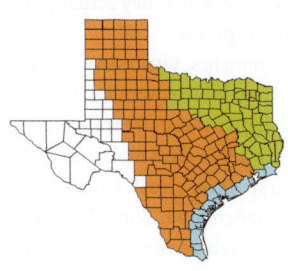

Curve-billed Thrasher

(Toxostoma curvirostre)

IUCN RED LIST STATUS (2018): Least Concern
POPULATION TREND: Decreasing

Curve-billed Thrasher suffers from a bit of an identity crisis. Its name suggests it possesses a standout curved bill, but at least three other species of thrasher, including the Texas resident Crissal Thrasher, have more curved bills. There are two distinct groups of Curve-billed Thrashers, the Sonoran Group of southern Arizona and the Chihuahuan Group of Texas. These two groups are visually separable, and it has been suggested they actually constitute two distinct species.
LENGTH: 11 inches. **WINGSPAN:** 13.5 inches.
ADULT: Gray-brown overall. Long, all-dark, curved bill. Eye is usually orange. Throat is pale with a thin dark malar stripe. Large brownish spots on breast. Undertail is buffy-orange colored. Wings have two narrow white wingbars. Small white spots at tips of the tail feathers. **JUVENILE:** (Mar.–Aug.)

Like adults but browner and lacks wingbars. Spotting is reduced or absent. **VOICE:** Song is hurried with short notes. Gives a distinct *whip-it* call. **BEHAVIORS:** Walks rapidly while foraging. Flight is jerky; flies from bush to bush. **HABITAT:** Found in more open habitat than Brown and Long-billed Thrasher. Forages in recently cleared areas. Found in thorn scrub and thickets at woodland edges. **STATUS:** Common to uncommon resident in the western half of Texas, west of a line from Calhoun County to the eastern Panhandle. Absent from the northeastern Panhandle.

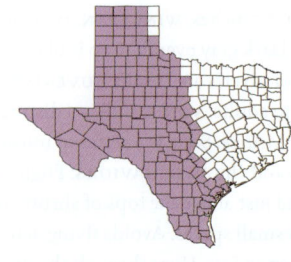

Brown Thrasher

(Toxostoma rufum)

IUCN RED LIST STATUS (2018): Least Concern
POPULATION TREND: Decreasing

Brown Thrasher is a denizen of dense thickets and is more often heard than seen. Its loud, clear songs demonstrate one of the largest repertoires of North American songbirds. It spends much of its time on the ground foraging, sweeping and turning over leaf litter with its bill.

LENGTH: 11.5 inches. **WINGSPAN:** 13 inches. **ADULT:** Bright brown-rufous above. Light buffy below with thin dark-brown streaks. Face is light gray, throat is clear white. Undertail is white with no spots. Lower mandible is mostly pale. Wings have two white wingbars. **JUVENILE:** (June–Aug.) Like adults. **VOICE:** Song is musical and mockingbird-like but with little or no mimicry. Three distinct call notes: a raspy *churrrr*, a sharp *stuck*, and a whistled *weir-we-oooo*. **BEHAVIORS:** Skulks in dense vegetation, spending a lot of time on the ground. Outside of migration, flights are low and heavy. **HABITAT:** Thickets, hedgerows, and shelter belts. **STATUS:** Uncommon to locally common resident in the northeastern two-thirds of Texas, including the eastern Panhandle. Common migrant in the eastern half of the state south to Nueces Bay. Spring migration is early March–late April, fall migration mid-September–early November.

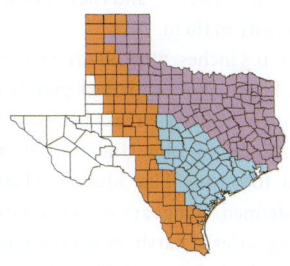

Long-billed Thrasher

(Toxostoma longirostre)

IUCN RED LIST STATUS (2016): Least Concern
POPULATION TREND: Increasing

Long-billed Thrasher is one of the Texas specialties that birders come here to see. They are most common in the dense brush of the lower Rio Grande Valley, but more than 95% of this brush habitat has been lost to mechanical clearing. Its range has been expanding north, though, as former grasslands are being overtaken by mesquite and other brush. These new habitats are less productive for Long-billed Thrashers, and they nest in much lower density in them.
LENGTH: 11.5 inches. **WINGSPAN:** 12 inches.
ADULT: Brown above with dark gray face. Two white wingbars. Black streaks below with a white background color. Undertail coverts are streaked. **JUVENILE:** Like adults but features are less defined. Wingbars are buffy. **VOICE:** Rambling series of harsh, musical phrases, with few paired phrases. Call is a sharp *stick*, hoarse *charrr*, or sweet *eee-yuk*. **BEHAVIORS:** Walks or runs on the ground when foraging. More arboreal than Brown Thrasher and climbs through bushes for berries. Flight is short and jerky, generally near the ground, but may fly above other Long-billed Thrasher territories. **HABITAT:** Riparian woodlands and dense thickets. **STATUS:** Common to uncommon resident of the South Texas Brush Country, Lower Rio Grande Valley, and the southern coastal prairies north to Matagorda County and west to Val Verde County.

Darting from cover

Crissal Thrasher

(Toxostoma crissale)

IUCN RED LIST STATUS (2016): Least Concern
POPULATION TREND: Increasing

The Crissal Thrasher's long scythe-shaped
bill is unique among Texas thrashers. Its
biology is poorly known because of its preference for dense vegetation, where detailed
observations of its life are difficult. Due to a
printing error in 1858, the orginal printing of
the first description of Crissal Thrasher was
published as *Toxostoma dorsalis*, even though
the intended name was *T. crissale*. In the 1920s,
Harry Oberholser insisted that the precedent
of the first published name be used, and it was
known as *T. dorsalis* until 1983, when the name
T. crissale was adopted.
LENGTH: 11.5 inches. **WINGSPAN:** 12.5 inches.
ADULT: Overall gray with light throat and
dark malar stripe. Iris is yellow. Bill is very
long and strongly decurved. Undertail is dark
rufous. Tail is dark gray with light-gray corners. **JUVENILE:** (Feb.–Aug.) Recently fledged

juveniles have shorter, almost straight, bills.
Iris is dark and face pattern is weak. **VOICE:**
Song is a series of random musical notes. The
call is a sweet *pur-reet*. **BEHAVIORS:** Mostly
terrestrial; walks, runs, or hops on the ground
whenever possible. Flight is mostly to and
from singing perches. **HABITAT:** Dense, low,
scrubby desert vegetation and chaparral.
STATUS: Uncommon to locally common resident in the Trans-Pecos. Rare and local in the
northwestern part of the Edwards Plateau to
Midland County.

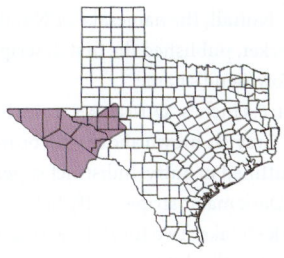

Sage Thrasher

(Oreoscoptes montanus)

IUCN RED LIST STATUS (2016): Least Concern
POPULATION TREND: Decreasing

Originally known as the Mountain Mockingbird, the Sage Thrasher was the subject of naturalist intrigue and competition. The first specimen was collected by John Kirk Townsend, the namesake of Townsend's Warbler and Townsend's Solitaire. While Townsend was still in the field, John James Audubon obtained the specimen and introduced it to the scientific community. Trying to restore credit for his friend Townsend, Thomas Nuttall, the namesake of Nutall's Woodpecker, published the first description under Townsend's name.
LENGTH: 8.5 inches. **WINGSPAN:** 12 inches.
ADULT: Grayish-brown above with brown cheek outlined in buffy. Indistinct supercilium. Dark malar stripe. Buffy below with dark streaks. Two thin white wingbars. Tail is dark with white corners. **JUVENILE:**

(June–Sept.) Like adults but cheek pattern less defined. Streaky above. **VOICE:** Rapid series of warbles, rolling quickly between repeated phrases. Call is a sharp *chep* and a trilling *whirrrrr*. **BEHAVIORS:** Flight is a rapid low dash into cover. Generally, runs rather than flying. **HABITAT:** Arid and semiarid open and semi-open country. Sagebrush, scattered brush, and scrub species dominate. **STATUS:** Uncommon to rare migrant and winter resident in the Trans-Pecos and the western High Plains. Irruptive and variable year to year. Fall migrants arrive in late September. Spring migration is from late February to early April.

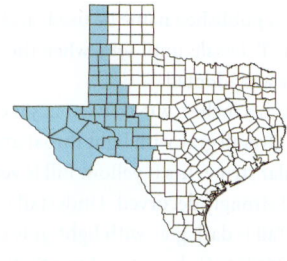

Northern Mockingbird

(Mimus polyglottos)

IUCN RED LIST STATUS (2017): Least Concern
POPULATION TREND: Stable

Northern Mockingbird is the state bird of Texas. It is also the state bird of Arkansas, Florida, Mississippi, and Tennessee. It was adopted as Texas's state bird in 1927 by the legislature at the request of the Texas Federation of Women's Clubs. The resolution declared it to be "a fighter for the protection of his home, falling, if need be, in its defense, like any true Texan." Anyone who has had one nest in their garden can attest to the fierceness of its defense of its nest from all intruders.
LENGTH: 10 inches. **WINGSPAN:** 14 inches.
ADULT: Gray-brownish above and white below. Thin dark eye line. Two white wingbars. Wings have large white patches in the primaries. Tail is dark with white outer tail feathers. **JUVENILE:** (June–Sept.) Like adults but browner overall. Breast is faintly spotted. **VOICE:** Song is a long string of phrases, repeated two to six times. Imitates many species of birds in these strings, along with man-made sounds at times. **BEHAVIORS:** Walks or runs on the ground. Frequently raises wings and flashes its wing patches. Hops easily though vegetation. **HABITAT:** Open habitats, including urban and suburban habitats. **STATUS:** Abundant and common resident statewide. Northern migrants may account for an apparent influx of Northern Mockingbirds in winter.

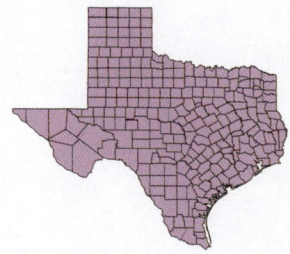

STURNIDAE (STARLINGS)

The Starlings are medium-size songbirds that include the mynas. Most are dark-colored but some have accents of blues, yellow, and even green. They are well-known for their vocal prowess, and several species are skilled mimics, able to spoof human speech. They possess strong legs and are often seen foraging on the ground. Starlings are gregarious, and many species form large flocks outside the breeding season.

Starlings are an Old World species with none native to the Americas. Four species have been introduced to North America, though. The European Starling is the most common species of the family in North America; also introduced are the Hill Myna, Crested Myna, and Common Myna.

Worldwide there are 125 species of starlings and mynas. Seventeen are of conservation concern; most of these are restricted to small islands. The Bali Myna only occurred on Bali Island, and illegal trapping for cage birds made a wild population impossible. Today the population is sustained only by captive breeding. The Black-winged Myna is in a similar situation. The Pohnpei Starling has not been seen regularly on surveys and is now feared extinct. Three species—the Norfolk Island Starling, Mysterious Starling, and Reunion Starling—are extinct. One species of starling occurs in Texas.

Nonbreeding

Juvenile

Breeding

European Starling

(Sturnus vulgaris)

IUCN RED LIST STATUS (2019): Least Concern
POPULATION TREND: Decreasing

European Starlings are an Old World species that was introduced to North America intentionally. All North American European Starlings are descendants of 80 to 100 birds released in New York's Central Park in 1890 and 1891. The release was done by an acclimatization society dedicated to introducing all the birds mentioned in William Shakespeare's writings. By the 1940s, they had spread all the way across Texas.

LENGTH: 8.5 inches. **WINGSPAN:** 16 inches. **ADULT BREEDING:** (Dec.–Aug.) Black with purple iridescence on head, greenish on body. Flight feathers edged buffy. Bill is long and narrow. Tail is short and square. **NONBREEDING:** (Sept.–Feb.) Like breeding but feathers tipped with light buffy or white, giving a spotted appearance. As the bird ages, the tips wear off and give the non-spotted appearance of breeding birds. **JUVENILE:** (May–Aug.) Drab gray-brown overall. Bill is dark. **VOICE:** Call is a harsh rattle. Often gives a slurred whistle. Song is a hissing, clicking chatter. **BEHAVIORS:** Walks or runs quickly; gait is a distinctive waddle. Flight is swift. Forms large flocks. When singing, holds wings out from the body. **HABITAT:** Closely associated with humans; urban, suburban, and agriculture habitats with nesting sites and short mowed or grazed areas for foraging. **STATUS:** Common to abundant resident throughout most of the state. Rare to uncommon in rural areas at times.

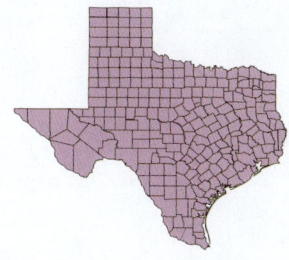

TURDIDAE (THRUSHES AND ALLIES)

The thrushes are generally small to medium-size songbirds. They have alert, upright postures and are frequently seen hopping on the ground. Their bills are usually straight and relatively thin, but not especially pointed. Most are brownish, but several species are blue-colored. Many species are spotted, and almost all are spotted as juveniles. Members of the family are known for their melodius and sometimes ethereal calls.

Thrushes use habitats from dense forests to open grasslands but are not found in arid deserts. Their diets consist mostly of invertebrates, and for species that winter in temperate areas, fruit is a major part of their winter diet. Some species, like bluebirds, will actively hawk for insects.

Thrushes are most closely related to the Old World flycatchers. There are 175 species of thrushes worldwide. Thirty-nine are of conservation concern, and three in Hawaii are already extinct or likely extinct: the Kamao, the Amaui, and the Olomao. Seventeen species have been recorded in Texas. Six—Orange-billed Nightingale-Thrush, Black-headed Nightingale-Thrush, White-throated Thrush, Rufous-backed Robin, Varied Thrush, and Aztec Thrush—are rare visitors to Texas.

Male

Male

Eastern Bluebird

(Sialia sialis)

IUCN RED LIST STATUS (2021): Least Concern
POPULATION TREND: Increasing

Eastern Bluebirds readily use nest boxes and open areas, and will frequently nest in close proximity to humans. They have inspired a loyal and energized organization of volunteers in the North American Bluebird Society (NABS). NABS constructs bluebird trails, long transects of nest boxes to foster bluebird conservation and the conservation of other cavity-nesting species. **LENGTH:** 7 inches. **WINGSPAN:** 13 inches. **ADULT:** Male is bright blue above with rufous throat and breast. Belly and undertail coverts are white. Tail is blue with dark blue tip. Female is like male but blue-gray instead of blue. **JUVENILE:** (June–Aug.) Brownish-gray with white spots. Blue wings; the flight feathers have rufous edges. Blue tail. **VOICE:** Song is a soft *chi-hit WEEW-we-wit.* Call is a clear *CHUR-reet.* **BEHAVIORS:** Hops sideways; turns 180 degrees while hopping. Flights are generally low in open areas. **HABITAT:** Open habitat with little or no understory. A secondary cavity nester that needs cavities or artificial nest boxes. **STATUS:** Uncommon to locally common summer resident in the eastern half of the state to the eastern Edwards Plateau and south along the coastal prairies to Kenedy County. Common to uncommon migrant and winter resident east of the Trans-Pecos. Rare to uncommon winter resident in the Trans-Pecos and Lower Rio Grande Valley. Fall migrants arrive late September through October. Spring migration is from mid-February through March.

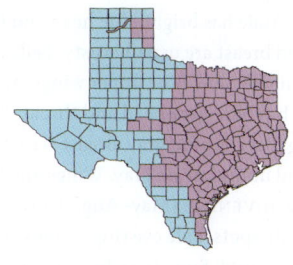

Male

Mixed flock

Western Bluebird

(Sialia mexicana)

IUCN RED LIST STATUS (2021): Least Concern
POPULATION TREND: Increasing

Western Bluebirds are rare in Texas away from the Guadalupe and Davis mountains. Less likely to breed close to man, they aren't as well studied as the Eastern Bluebird. While building bluebird trails for Eastern Bluebirds goes back a century, this idea only became commonplace for Western Bluebirds in response to declining populations in the Pacific Northwest of North America.

LENGTH: 7 inches. **WINGSPAN:** 13.5 inches.
ADULT: Male has bright blue head and throat. Back and breast are usually rusty. Belly and undertail are blue. Blue upper wings and tail. Female has blue-gray head and throat. Pale eye ring. Back is blue-gray. Breast is rusty, belly and undertail are gray. Wings and tail are blue. **JUVENILE:** (May–Aug.) Gray overall with white spots. Pale eye ring. Wings and tail are blue. **VOICE:** Song is a whistled *pew pew pew* or *pepew pepew pepew*. **BEHAVIORS:** Flight is usually low with slow wingbeats. Hops rather than walks when on the ground. **HABITAT:** Open coniferous and deciduous woodlands. On edges with scattered trees. **STATUS:** Uncommon and local winter resident in the Guadalupe and Davis mountains. Above 6000 feet in breeding season, lower elevations in other seasons. An irruptive winter resident in the rest of the Trans-Pecos. Rare winter resident in the western Edwards Plateau and High Plains. Fall migrants begin arriving in Mid-September and depart by early April.

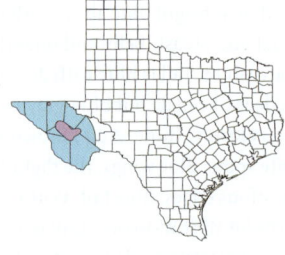

Male (above) and female

Mountain Bluebird

(Sialia currucoides)

IUCN RED LIST STATUS (2021): Least Concern
POPULATION TREND: Increasing

Mountain Bluebird, in all its amazing azure-blue color, is one of the hardiest songbird species. It breeds from above the treeline in the Rocky Mountains through the Yukon and Alaska. It is one of the earliest-arriving species, before the snow and ice is gone in most cases. As a secondary cavity nester there is always fierce competition for scarce nesting resources with other cavity nesters. This may be one of the reasons they arrive early on the breeding grounds. **LENGTH:** 7.25 inches. **WINGSPAN:** 14 inches. **ADULT:** Male is unmarked sky blue or azure all over. Paler below. Female is plain gray or gray with a rufous wash. Wings and tail are pale blue. Pale eye ring. **JUVENILE:** (May–Aug.) Like females but grayer, with white spots below and often small white spots on back. **VOICE:** Song is a hoarse *purr-di-dit*. Call

is a mellow *tu-tu* or a chip note of *chep chep*. **BEHAVIORS:** Usually walks on the ground, rarely running or hopping. Flight is fluttering and usually high. **HABITAT:** Wintering habitat is poorly understood. Usually relatively open with some available perches, piñon-juniper and oak-juniper woodlands. **STATUS:** Common to uncommon migrant and winter resident in the Trans-Pecos and the Panhandle south to the northwestern Edwards Plateau. Irruptive, numbers can vary greatly from year to year. Fall migrants arrive from mid-September to mid-November. Most birds depart by mid-March.

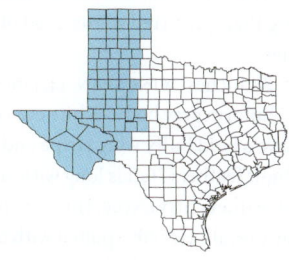

Townsend's Solitaire

(Myadestes townsendi)

IUCN RED LIST STATUS (2021): Least Concern
POPULATION TREND: Stable

Townsend's Solitaire was first collected by John Kirk Townsend in Oregon in 1835. From this single specimen, John James Audubon published the first description of the species in 1838, naming it for Townsend. While drab-colored, they make no attempt to be inconspicuous. They are often perched high on exposed branches to advertise their ownership of the territory and to scan for intruders. They are especially territorial in fall and winter, protecting their preferred winter food of juniper berries.
LENGTH: 8.5 inches. **WINGSPAN:** 14.5 inches.
ADULT: Uniform gray body. White eye ring. Wings have a bold buffy wing stripe and darker flight feathers. Tail is long with white outer tail feathers. **JUVENILE:** (June–Sept.) Dark gray overall. Heavily spotted with buff and white, giving a scaly look. **VOICE:** Song is a long, continuous finchlike warble. Call is a whistled *heet* or raspy *scraaa.* **BEHAVIORS:** Rarely on the ground. Makes short flight from tree to tree in the understory. Long flights are above the treetops. **HABITAT:** Lowlands to 6500 feet in areas with juniper. **STATUS:** Uncommon to rare migrant and winter resident in the Panhandle and Trans-Pecos. Fall migrants arrive as early as late September. Most depart late March through April.

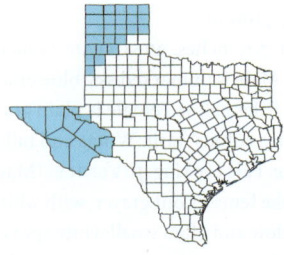

Veery

(Catharus fuscescens)

IUCN RED LIST STATUS (2018): Least Concern
POPULATION TREND: Decreasing

Veery is a tawny thrush with a unique brown color that, with experience, can be recognized as a distinctive "Veery brown." In Texas, most Veeries are encountered in spring in the coastal migrant traps. They stay low to the gound, foraging and sometimes singing in the broadleaf understory.

LENGTH: 7 inches. **WINGSPAN:** 12 inches.
ADULT: Unique reddish brown above, pale gray below. Pale inconspicuous eye ring. Pale gray throat. Weakly spotted brown with a buffy background on the breast. Underwing has bold buffy stripe. **VOICE:** Call is a loud *tsue*. Song is a descending and trilling *veer veer veer veer*. **BEHAVIORS:** Hops like a frog on the ground. Most foraging is in short flights. **HABITAT:** Most often encountered on the coast in hackberry woodlands with a broadleaf understory. **STATUS:** Uncommon to rare spring and a very rare fall migrant in the eastern half of Texas. Spring migration is mid-April–mid-May, fall early September–late October.

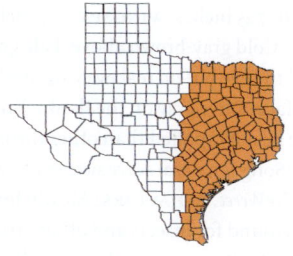

Gray-cheeked Thrush

(Catharus minimus)

IUCN RED LIST STATUS (2021): Least Concern
POPULATION TREND: Decreasing

Gray-cheeked Thrush biology is not well known. It was considered cospecific with Bicknell's Thrush until the 1990s, and most of the studies prior to the split was on the Bicknell's Thrush. It breeds in inaccessible taiga and tundra regions. Little work has been done on its habits during migration. There is some urgency to learn more, as some indications suggest that the population is declining rapidly.
LENGTH: 7.25 inches. **WINGSPAN:** 13 inches.
ADULT: Cold gray-brown above. Pale gray around the eye but no real eye ring. No buffy on the face. Olive gray on the flanks. Breast is more heavily spotted than other thrushes.
VOICE: Song is a nasal *sweee swee seeeee swee-tuu*. Call is *TeeWeeee*. **BEHAVIORS:** Mostly forages on the ground for insects and other arthropods and fruits. **HABITAT:** Favors well-wooded sites with thick understory. **STATUS:** Uncommon to rare spring and very rare fall migrant in the eastern half of Texas. Spring migration is early April–mid-May, fall early September–late October.

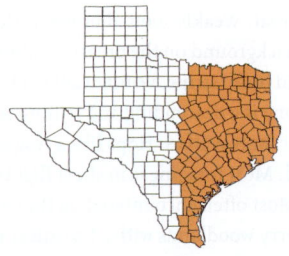

Swainson's Thrush

(Catharus ustulatus)

IUCN RED LIST STATUS (2021): Least Concern
POPULATION TREND: Decreasing

Swainson's Thrush is perhaps the most common of the *Catharus* thrushes in Texas. At times during migration it can be abundant in the coastal migrant traps. While still considered a common bird, its population is declining even where it is considered abundant. Reasons aren't clear for this decline but the trend is clear.

LENGTH: 7 inches. **WINGSPAN:** 12 inches.
ADULT: Gray-olive upper parts. Pale below with olive flanks. Face is buffy with bold buffy spectacles. Throat and breast are buffy with dark brown spots. **JUVENILE:** (June–Aug.) Like adults but breast spots are darker and grayer. Back has pale buffy spots. **VOICE:** Song is a rising, flute-like *pro-ree-reee-reeeee*. **BEHAVIORS:** A near-ground forager but does fly-catch for insects. More likely to forage high in a tree than other *Catharus* thrushes. Call is a sharp *wheat wheat*. **HABITAT:** Most are found during migration in the coastal prairie migrant traps, but appears throughout Texas in habitats with dense understory. **STATUS:** Common to uncommon spring migrant and uncommon to rare fall migrant in the eastern half of Texas. Uncommon to rare spring migrant and rare to very rare fall migrant elsewhere. Spring migration is early April–late May, fall late August–late October.

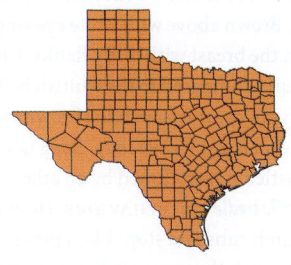

Western

Eastern

Hermit Thrush

(Catharus guttatus)

IUCN RED LIST STATUS (2021): Least Concern
POPULATION TREND: Stable

Hermit Thrush is the only *Catharus* thrush
that winters in North America. They have
a low tolerance for snow, though, and don't
winter in places where snow persists on the
ground for long periods. They sex-segregate in
winter, with females wintering an average of
1.5 degrees latitude farther south than males;
the farther south you go, the greater the ratio
of females to males.

LENGTH: 6.75 inches. **WINGSPAN:** 11.5 inches.
ADULT: Brown above with white eye ring. Bold
spots on the breast with buffy flanks. Undertail is buffy compared to the whitish belly.
Reddish tail contrasts with the brown back.
VOICE: Call is a soft *chup chup*. Song is a fluting, whistled note followed by an ethereal trill:
seeeee tredle tredle tree. **BEHAVIORS:** Hops on
the ground; runs and stops like a plover. Flits
easily through the trees. Flicks wings and calls
when agitated. **HABITAT:** Uses a broad range
of forest type during breeding and wintering.
STATUS: Uncommon summer resident in the
upper Davis and Guadalupe mountains. Common migrant and winter resident statewide.
Uncommon to rare in the western third of the
state. Fall migrants arrive in late September
through mid-November. Spring migration is
from mid-March through mid-May.

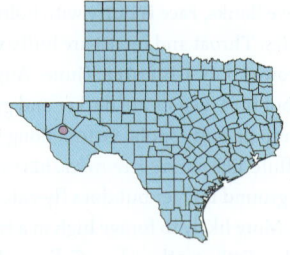

Wood Thrush

(Hylocichla mustelina)

IUCN RED LIST STATUS (2020): Least Concern
POPULATION TREND: Decreasing

Wood Thrushes have a beautiful, ethereal song that can often be heard at dawn and dusk from the forest edges. Unfortunately, the song is also the soundtrack to the decline of neotropical migrants; forest destruction and fragmentation are likely the major causes. Wood Thrushes are especially vulnerable to Brown-headed Cowbird parasitism because the fragmentation allows the edge-loving cowbirds to reach more nests. Loss of primary neotropical forests may also be forcing wintering birds to less-desirable habitats with higher mortality rates.
LENGTH: 7.75 inches. **WINGSPAN:** 13 inches.
ADULT: Brown back and orange-brown head and nape. Bold white eye ring and whitish cheek with dark stripes. White below with bold black spots. Legs are bright fleshy-pink.
VOICE: Song is an ethereal, fluting *wee-o-WEE*

tdreeee. Call is a rapid *wip-wip-wip* or *chup-chup-chup.* **BEHAVIORS:** Forages on the ground by a series of low hops. **HABITAT:** The interior and edges of deciduous and mixed forests, shady with open forest floor. **STATUS:** Uncommon to locally common summer resident in the Pineywoods. Locally uncommon to rare west to Navarro County and in the Lost Pines area of Bastrop County. Uncommon to common spring migrant and uncommon to rare fall migrant. Spring migration is late March–early June, fall mid-September–early November.

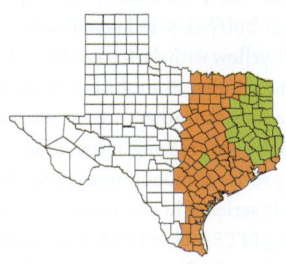

Clay-colored Thrush

(Turdus grayi)

IUCN RED LIST STATUS (2021): Least Concern
POPULATION TREND: Stable

Clay-colored Thrush is a tropical species that
just reaches South Texas. Formerly rare in the
state, it was removed from the Texas review
list in 1998. It is now an uncommon resident
in the Lower Rio Grande Valley and a rare to
uncommon resident along the Rio Grande
to Del Rio and throughout the South Texas
Brush Country.
LENGTH: 9 inches. **WINGSPAN:** 15.5 inches. **ADULT:** Buffy-olive above. Underparts
are lighter buffy-clay. Bill is yellow to
greenish-yellow with dark base. Iris is brown.
JUVENILE: Similar to adults but upper wing
coverts and back feathers are tipped with
buffy spots. **VOICE:** Song a slow, lower-pitched
warbling series of monotonous phrases. Call
is a steady series of hoarse *chu-chu-chu-chu* or
a piercing *EEEE-o-ok* or *EEEE-oooo*. **BEHAV-
IORS:** Habitually flicks wings on landing;
raises and lowers tail. **HABITAT:** Open forest
and edges, gardens and suburban lawns, open
orchards. **STATUS:** Uncommon resident in the
Lower Rio Grande Valley. Rare to uncommon
north to Calhoun County and west to Val
Verde County.

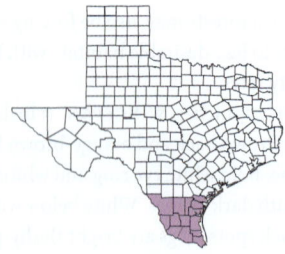

American Robin

(Turdus migratorius)

IUCN RED LIST STATUS (2021): Least Concern
POPULATION TREND: Stable

American Robin is perhaps the most familiar of birds in North America. **LENGTH:** 10 inches. **WINGSPAN:** 17 inches. **ADULT:** Dark charcoal-gray to black head with bold white eye crescent above and below the eye. Bill is yellow-green. Throat is variable from dark to pale gray. Paler females may show a white malar stripe. Breast and belly are brick red. Undertail coverts are pale. Back is brownish-gray. Tail is dark with white corners; western birds may lack white corners in the tail. **JUVENILE:** (May–Sept.) Gray above with pale spotting in wings and back. Pale supercilium. Breast is rufous washed with dark spots. Undertail coverts are pale. **VOICE:** Series of warbled phrases repeated. Call is a high-pitched, rapid *seep-seep-seep-seep* or a slower-paced, lower-pitched *chup chup chup*. **BEHAVIORS:** Runs well on the ground; plover-like, runs and stops and looks around. Short flights through trees when foraging. **HABITAT:** Forests, woodlands, and gardens where grass is short. In winter, flocks roam forests. **STATUS:** Common summer resident in the northern half of Texas south to Nueces County and west to the Pecos River. Uncommon summer resident in El Paso and the Guadalupe Mountains. Common to abundant migrant and winter resident in the northern two-thirds of the state; uncommon and local winter resident in the southern third. Fall migration is late September–mid-October, spring mid-February–late March.

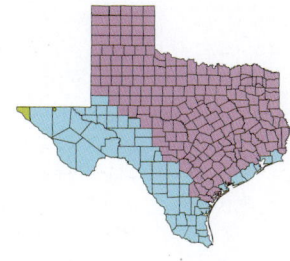

ESTRILDIDAE (WAXBILLS)

The waxbills are finches of Africa and southern Asia. They are a contact species known for huddling together when roosting. They are lively and colorful birds that can live on seeds alone, and can feed their young seeds. This makes them easy to breed in captivity. They are well represented in the pet trade and there are several introduced populations around the world.

They are most closely related to the indigo birds and whydahs. There are 138 species worldwide; 14 are of conservation concern. One, the Scaly-breasted Munia, has become established in Texas.

Scaly-breasted Munia

(Lonchura punctulata)

IUCN RED LIST STATUS (2016): Least Concern
POPULATION TREND: Stable

Scaly-breasted Munia is known by a couple of different common names—Nutmeg Mannikin, Spotted Munia, and Scaly-breasted Mannikan are in common usage. A native to Southeast Asia, Scaly-breasted Munia is a popular cage bird. Either deliberately or accidentally, it has been introduced to many parts of the world. There are occasional sightings in Austin, San Antonio, and Corpus Christi. There is an established and apparently growing population in Houston, primarily in the southwest corner of Harris County and adjacent counties.

LENGTH: 4 inches. **WINGSPAN:** 7 inches.
ADULT: Dark brown above with black face and throat. Thick conical bill is dark brown. Breast and belly are pale with dark-brown scales. Undertail is pale gray. **JUVENILE:** Like adults. No black on the face. Upper breast, throat, and underparts are pale brown. **VOICE:** Call is a high-pitched *pee-IT*. **HABITAT:** In Texas, mostly found in suburban parks near neighborhoods with weedy fields and abundant bird feeders. **STATUS:** There is an established and expanding population in the southwest portion of Harris County and adjacent areas of Fort Bend and Brazoria counties.

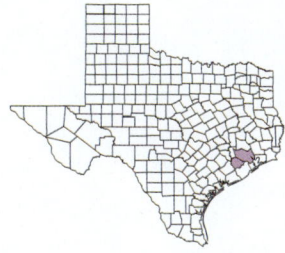

PASSERIDAE (OLD WORLD SPARROWS)

The Old World sparrows resemble the New World sparrows but are not closely related. Old World sparrows are bulkier. They have short legs and short bills, and the culmen, or top of the bill, is decurved. Most are drab-colored. Their diet consists mostly of grain, which has allowed the family to use almost every habitat in the world, though in the breeding season they feed insects and invertebrates to their young.

There are forty-three species of Old World sparrows. Formerly they were considered part of the weavers and finches but have been elevated to their own family. While at first glance the New World Sparrows seem to be closely related, they are not. Old World Sparrows are likely most closely related to the sister families of pipits and finches. Several species coexist very well with humans. One introduced species, the House Sparrow, is found in Texas. It was introduced to North America in the 1800s and has been become ubiquitous in urban habitats.

Male

Female

House Sparrow

(Passer domesticus)

IUCN RED LIST STATUS (2019): Least Concern
POPULATION TREND: Decreasing

The House Sparrow was introduced to North America in the 1850s with releases in Brooklyn, New York. These were followed by other introductions in the United States, all east of the Mississippi River save one site. Beginning in 1867, they were released in Galveston, Texas, and were reported statewide by 1905. **LENGTH:** 6.25 inches. **WINGSPAN:** 9.5 inches. **ADULT BREEDING:** (Mar.–Sept.) Male has gray crown with brown nape. White cheek with black face and bill. Throat and breast are black. Upper wings are brown with bold white wingbar. Back is brown with white and dark stripes. Rump and tail are gray. Female has plain drab-brown crown. Face is dingy with pale supercilium. Dingy gray below. Back is dingy brown with buffy and dark streaks. Rump and tail are brownish-gray. **NONBREED-ING:** (Sept.–Mar.) Male is like breeding males but face pattern less defined. Dark breast feathers are concealed by pale tips. Female is like breeding females. **JUVENILE:** (May–Sept.) Like females but browner. Face pattern is more sharply defined. **VOICE:** Song is a long series of *cheap-chert cheap-chert cheap-chert*. Call is an emphatic *cheat*. **BEHAVIORS:** Hops on the ground, rarely walks. Flight is direct, wingbeats are rapid. **HABITAT:** Residential and urban areas, farms, and agricultural areas. **STATUS:** Common to locally abundant resident in urban areas statewide. Common to locally rare in rural areas near human populations.

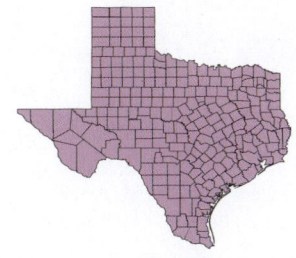

MOTACILLIDAE (WAGTAILS AND PIPITS)

Small to medium-size songbirds, the wagtails and pipits are birds of open country. They walk on the ground constantly pumping their tail. All are primarily insectivores. Wagtails have bright pied or contrasting plumage, or yellow plumage. Pipits are cryptically colored in streaked brown plumages. All members of the family have a large claw on the hind toe.

The name pipit comes from the French name for the bird, derived from the Latin *pipio*, meaning "to peep or chirp." Since they are birds of open country with few singing perches, many species make display flights during which they call continuously.

They are most closely related to finches. There are sixty-nine species of wagtails and pipits worldwide. Ten are of conservation concern, primarily because of habitat loss. Three species have been recorded in Texas. One, the White Wagtail, is a rare visitor.

American Pipit

(Anthus rubescens)

IUCN RED LIST STATUS (2019): Least Concern
POPULATION TREND: Decreasing

The American Pipit was long known as the Water Pipit, a complex of seven subspecies that occurred over much of the northern hemisphere. Studies found that the three North American subspecies and the Siberian species should be regarded as separate. In North America, this new species is called American Pipit.
LENGTH: 6.5 inches. **WINGSPAN:** 10.5 inches.
ADULT BREEDING: (Mar.–June) Gray on back and crown. Gray cheeks with pale buffy supercilium. Buffy eye ring. Tail is dark with gray center and white outer feathers. Legs are dark. Light buffy underparts with bold streaking. Paler individuals vary in the amount of streaking; some birds are completely unstreaked. Cheeks can vary from gray to buffy.
NONBREEDING: (Aug.–Mar.) Like breeding. Darker individuals are whitish below with bold streaking. Paler individuals are buffy below with few streaks. **VOICE:** High-pitched *sip sip sip* or double-note *sip-it sip-it*. **BEHAVIORS:** Walks while foraging. Flights are strong and undulating. Forms dispersed flocks in winter. Pumps tail up and down. **HABITAT:** Coastal beaches and marshes. Stubble, plowed and recently burned fields, and livestock areas. Mowed lawns and golf courses. **STATUS:** Common to uncommon migrant and winter resident in most of Texas. Rare to locally uncommon migrant and winter resident in northern third of the Panhandle. Uncommon winter resident in far southern Texas. Fall migration is late September–mid-November, spring late February–late May.

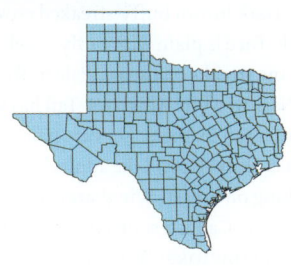

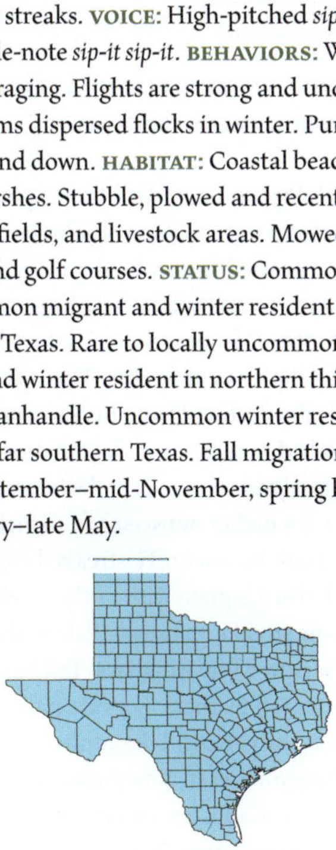

Sprague's Pipit

(Anthus spragueii)

IUCN RED LIST STATUS (2021): Vulnerable
POPULATION TREND: Decreasing

Sprague's Pipit is named for Isaac Sprague, an assistant painter on one of John James Audubon's expeditions who discovered the first nest in North Dakota. The International Union for the Conservation of Nature lists Sprague's Pipit as vulnerable. This species has dramatically declined from historic numbers, most likely due to loss and degradation of the native grasslands it breeds and winters in. **LENGTH:** 6.5 inches. **WINGSPAN:** 10 inches. **ADULT:** Dark brown buffy-streaked crown and back. Face is plain with buffy cheek patch. Fine streaks on breast. Pale buffy below. Narrow white wingbars. Tail has wide white outer tail feathers. Legs are pale pink. **JUVENILE:** (July–Nov.) Like adults but heavier streaking on breast. Wingbars are bold white. **VOICE:** Call is a squeaky *skreet*, often repeated. **BEHAVIORS:** Walks and runs on the ground. Avoids dense litter. Rises from the ground with a stair-step undulating flight. **HABITAT:** A grassland specialist, with grass height less than 18 inches, often overgrazed. **STATUS:** Uncommon to locally common winter resident on the coastal prairies south to the Lower Rio Grande Valley. Rare to locally uncommon winter resident in the Oaks and Prairies. Uncommon migrant in the center of Texas. Fall migration is mid-September–early November, spring late March–late April.

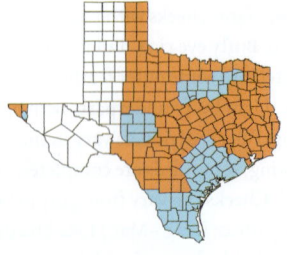

FRINGILLIDAE (FINCHES, EUPHONIAS, AND ALLIES)

Finches are small to medium-size songbirds. Most have conical bills and all are seed-eaters. Plumages vary from drab streaky brown to the gaudy bright euphonias. Finches exhibit sexual dimorphism, meaning the males and females look different. Many species are migratory and nomadic, chasing food where they find it.

There are 235 species of finches worldwide. Thirty-six of them are critically endangered and perhaps already extinct. The Red Siskin was once widespread in the foothills of Venezuela and Guyana but has almost been eliminated by trapping for the cage bird trade. The Warsangli Linnet of Ethiopia and the Yellow-throated Seedeater survive in a few patches of desert hill scrub that is under constant threat from fire and agricultural encroachment.

Thirteen species of finches have been recorded in Texas. Six are rare visitors: Evening Grosbeak, Pine Grosbeak, Gray-crowned Rosy-Finch, Redpoll, White-winged Crossbill, and Lawrence's Goldfinch.

Male

Female

House Finch

(Haemorhous mexicanus)

IUCN RED LIST STATUS (2018): Least Concern
POPULATION TREND: Increasing

House Finch was orginally a bird of the hot deserts and dry open habitats of the Southwest. In 1939 a few birds from California were released from a New York pet store. By the 1990s, that introduced eastern population had reached East Texas. During that time the western population underwent an explosive range expansion. Somewhere during the 1990s, the east and west populations met in central Texas, probably just east of the Balcones Escarpment. **LENGTH:** 5.7 inches. **WINGSPAN:** 10 inches. **ADULT:** Male has red head and breast. Grayish cheeks. Bill is short with curved culmen (upper bill). Pale below with gray-brown streaks. Brownish with faint streaks on back. Rump is red and the tail is brownish-gray. Female is brownish-gray above and pale below with brownish-gray streaks. **VOICE:** Song is a varied warble of short notes, often ending

with a burry *veeerrrr*. Call is a raspy *cheerup*. **BEHAVIORS:** Hops on the ground. Flight is undulating and high, above trees and obstacles. **HABITAT:** In the eastern half of Texas, almost exclusively around human settlement. In the western half, can be found in undisturbed habitats that have structures for nesting and perching. **STATUS:** Uncommon to locally uncommon resident throughout most of Texas. Uncommon to rare in the Pineywoods.

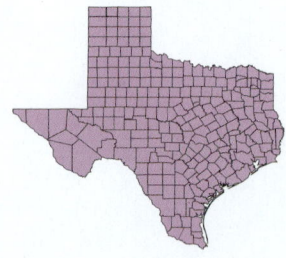

Female

Male

Purple Finch

(Haemorhous purpureus)

IUCN RED LIST STATUS (2016): Least Concern
POPULATION TREND: Decreasing

Purple Finch was formerly more common in Texas than it is now. Prior to the 1980s, winter visitors regularly occurred west to the Rolling Plains and south to the upper Texas coast. Now they only occur in those areas in rare winter irruptions.
LENGTH: 5.7 inches. **WINGSPAN:** 10 inches.
ADULT: Male has extensive purple on head and back. Auriculars (cheeks) are dark. Bill is short and pointed. Breast and throat are red. Belly and undertail are pale whitish. Tail is dark with a distinct notch. Female has high-contrast, brown-and-white head pattern with dark brown auriculars. Heavy dark streaks on breast. Belly and undertail are plain whitish. Back is brown with distinct streaks. **VOICE:** Song is a series of rapid, warbled phrases. Call is a sweet *cheer-re-it* or *cheer-re-reee*. Flight call is a sharp *PIK*.

BEHAVIORS: Hops on the ground. Flight is undulating. **HABITAT:** Wide variety of winter habitats, all kinds of forests, urban and suburban areas, hedgerows. Food resources rather than structure are more important in habitat choice. **STATUS:** Uncommon to rare but irregular migrant and winter visitor to northeast Texas, south through the Pineywoods. Rare but regular migrant and winter visitor in the eastern Panhandle and South Plains. Fall migration is early November–mid-December, spring mid-February–mid-March.

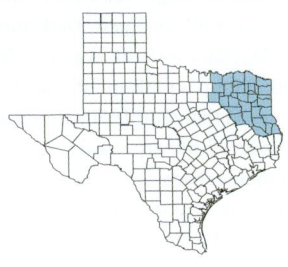

Male

Female

Cassin's Finch

(Haemorhous cassinii)

IUCN RED LIST STATUS (2019): Least Concern
POPULATION TREND: Decreasing

Cassin's Finch was named by Spencer Baird after John Cassin, apparently at Cassin's request. They are rare at best in Texas but occur annually in small numbers in the Davis and Guadalupe mountains. They are irruptive, and in some years they occur in greater numbers in other locations in the Trans-Pecos, but these irruptions are increasingly rare. **LENGTH:** 6 inches. **WINGSPAN:** 11.5 inches. **ADULT:** Male has bright red crown. Pale eye ring. Rosy-pink on breast. Dark auriculars (cheeks). Back is streaked and washed light pink. Rump is pink and streaked. Tail is less notched than Purple Finch. Female is gray-brown with sharp streaks. Whitish below with fine sharp streaks. Pale eye ring. Rump is streaked. **VOICE:** Rapid warble, higher than Purple Finch and faster-paced than House Finch. Call is a squeaky *cheer-reep*.

BEHAVIORS: Hops on the ground. Flight is undulating, usually above the trees. **HABITAT:** Montane coniferous forests. **STATUS:** Rare to very rare winter visitor to the Trans-Pecos. They are irruptive, and numbers vary greatly each year. They occur annually in small numbers in the Davis and Guadalupe mountains. Wintering birds arrive in early November and are present until mid-March.

Male

Female

Red Crossbill

(Loxia curvirostra)

IUCN RED LIST STATUS (2017): Least Concern
POPULATION TREND Stable

Red Crossbill is a Holarctic species with a complicated taxonomy. Modern research has divided Red Crossbill into ten "call types" based on different contact calls and bill structures, which may be evolved to feed on different types of conifer cones. These call types are not subspecies in the classic sense because they overlap in breeding range, but they may represent 10 species. One of these, Cassia Crossbill, has already been elevated to a full species but does not occur in Texas. The conundrum of identifying these call types and the potential to record new species has been vexing birdwatchers for several years.
LENGTH: 6.25 inches. **WINGSPAN:** 11 inches.
ADULT: Male is dull red overall. Grayish auriculars. Wings are darker. Distinct bills with long crossed mandibles. Female is like males but dull yellow. **FIRST YEAR:** Male

is like female but bright yellow. **JUVENILE:** (Jan.–Sept.) Gray-brown with heavy streaking. Indistinct yellowish wingbars. **VOICE:** Flight call is a variable *chipt chipt.* **BEHAVIORS:** Hops on the ground. Uses bill to grab branches like a parrot to climb trees. Flights are rapid and powerful. **HABITAT:** Coniferous forests with large cone crops. **STATUS:** Rare resident in the Davis and Guadalupe mountains. Irregular visitors in irruptions to the northern two-thirds of Texas. Irruptions are unpredictable; fall birds have arrived as early as August and wintering birds have lingered into May.

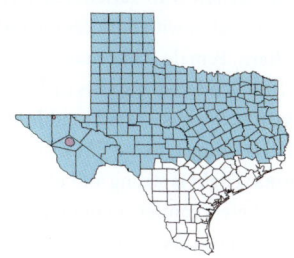

Pine Siskin

(Spinus pinus)

IUCN RED LIST STATUS (2016): Least Concern
POPULATION TREND: Stable

Pine Siskins are highly irruptive. They can be abundant one year and virtually absent the next. Rarely found as single birds, they form flocks; a hundred or more can descend on a bird feeder in winter.

LENGTH: 5 inches. **WINGSPAN:** 9 inches.
ADULT: Male has thin pointed bill. Streaked brown overall. Yellow wingbar. In flight, bold yellow wing stripe on upper and lower wing. Undertail is yellow with dark notched tip. Upper tail is yellow with dark center and tip in a T pattern. Female is like males but wingbars are thin. Underwing shows little or no yellow. **VOICE:** Song is a rapid jumble of husky notes. Call is a buzzy, rising *ZEEEEEEIIIIIIII.*
BEHAVIORS: Flight is strong and undulating.
HABITAT: Prefers conifers in migration and winter but will use deciduous woodlands.
STATUS: Common to abundant migrant and winter visitor in the northern two-thirds of Texas. Uncommon to rare the farther south you go in the state. Annual abundance varies year to year from abundant to virtually absent. Fall migrants arrive as early as October and depart between late March and mid-May.

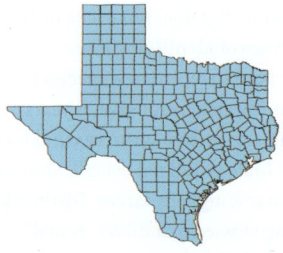

Female (left) and two males

Male

Lesser Goldfinch

(Spinus psaltria)

IUCN RED LIST STATUS (2018): Least Concern
POPULATION TREND: Decreasing

Lesser Goldfinch is generally found in small groups in open scrubby country. Agriculture and residential development seem to favor this species. Western males have greenish instead of dark backs. These green-backed birds have been reported from El Paso County, but it's not clear if they are correctly identified adult males from the western subspecies. **LENGTH:** 4.5 inches. **WINGSPAN:** 8 inches. **ADULT:** Male is bright yellow below, with dark crown and nape. Back and wings are dark. White patches in wings. Tail is white with dark center and tip in a T pattern. Undertail is yellow. Female is like male but yellow-green above. White wing patch is reduced; tail is all dark. **VOICE:** Song is an unmusical, repeated jumble of notes. Call is a high-pitched *tee-sewwww*. **BEHAVIORS:** Undulating flight; often calls and sings in flight. **HABITAT:** Shrublands, agricultural and suburban. **STATUS:** Uncommon to abundant migrant and summer resident in the Trans-Pecos and Edwards Plateau south to the Lower Rio Grande Valley. Uncommon to rare migrant and summer resident north to the southeastern Panhandle. The majority retreat from the northern half of this range. Uncommon to rare winter resident in the southern Trans-Pecos, South Texas Brush Country, and southern coastal prairies. Spring migration is mid-March–mid-May, fall mid-September–mid-November.

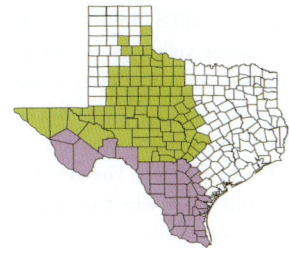

American Goldfinch

(Spinus tristis)

IUCN RED LIST STATUS (2016): Least Concern
POPULATION TREND: Increasing

The bright yellow-and-black male American Goldfinch is a frequently depicted North American songbird. Unfortunately, this cheerful plumage is rare in Texas. Texans mostly see these birds in their yellow-green and gray winter plumage.

LENGTH: 5 inches. **WINGSPAN:** 9 inches.

ADULT BREEDING: (Mar.–Oct.) Male is pale yellow with black forehead. Wings are dark with a narrow white wingbar. Rump and undertail are white. Bill is pinkish. Female is pale yellow below, yellow-green above. White upper wingbar and yellow lower wingbar. Undertail is pink and rump is gray. Bill is pinkish. **NONBREEDING:** (Oct.–Mar.) Male is gray below with yellowish face. Green-yellow on back. Two bold yellow wingbars. Dark wings. Gray bill. Female is yellow-gray overall with buffy wingbars. Pale yellow on face. **VOICE:** Song is a musical, rapid, repeated *toWEE toWEE toWEE toWEE*. Call is a high-pitched *sooo-WEEEEE*. **BEHAVIORS:** Typical finch undulating flight. **HABITAT:** Natural weedy grasslands. Thistle feeders often attract large flocks. **STATUS:** Very rare and local summer resident in northeast Texas. Uncommon to abundant migrant and winter resident statewide. Fall migration is early September–early November, spring early March–mid-May.

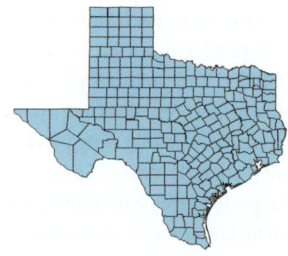

PASSERELLIDAE (NEW WORLD SPARROWS)

The New World sparrows are restricted to the Americas. They are generally more brownish and streaked in the north and become larger and more boldy patterned and colorful toward the tropics. They are all-encompassing in the habitats they occupy and can be found on the arctic tundra and in montane habitats, pine forests, deserts, grasslands, and rainforests. Many species are very similar to each other, posing an identification challenge in some cases. In the vernacular of birders they are often referred to as LBJs, short for "little brown jobs."

The New World sparrows have traditionally belonged to the family Emberizidae, now known as the New World buntings. But DNA studies have shown that they are actually distinct from the Old World buntings. They now appear to be a sister family to a clade or group that includes the wood warblers, blackbirds, and orioles. The genus *Chlorospingus* from South America were long thought to be members of the tanager family, but studies have shown that they are embedded in the New World sparrows.

There are 132 species in the family. Eighteen are of conservation concern, all threatened by habitat loss. Thirty-six species have been recorded in Texas. Yellow-eyed Junco and Golden-crowned Sparrow are considered rare visitors. There is a record of Striped Sparrow that the Texas Bird Records Committee considers a likely escaped cage bird.

Botteri's Sparrow

(Peucaea botterii)

IUCN RED LIST STATUS (2020): Least Concern
POPULATION TREND: Stable

Botteri's Sparrow is named for the naturalist that collected the first specimen, Matteo Botteri. A tallgrass specialist, Botteri's Sparrow is a Mexican species that just reaches the United States in southeast Arizona and Texas. In the United States it is locally abundant, rebounds well from moderate disturbance, and can use a variety of grasslands. Despite its restricted range, the population seems stable in Texas.

LENGTH: 6 inches. **WINGSPAN:** 7.75 inches. **ADULT:** Light grayish-brown overall. Brown crown strip and brown auricular patch. Unmarked below. Strong dark streaks on back. Throat is light gray. Bill is gray. **JUVENILE:** (May–Sept.) Buffy background color. Streaked breast and flanks. Finely streaked crown. Boldly streaked back. Bill is pinkish. **VOICE:** Song has a bouncing-ball cadence, accelerating: *trek trek trek sweet-sweet-sweet-sweet*. **BEHAVIORS:** Walks and runs on the ground. Flight is undulating just above the grass. **HABITAT:** Coastal prairies with scattered shrubs. **STATUS:** Uncommon to locally common summer resident on the coastal prairies from Kleberg County south to Cameron County. Birds return to breed in mid-April and are present until mid-September.

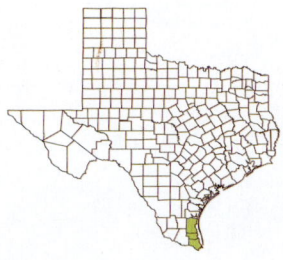

Cassin's Sparrow

(Peucaea cassinii)

IUCN RED LIST STATUS (2020): Least Concern
POPULATION TREND: Decreasing

Cassin's Sparrow was named by Samuel Washington Woodhouse (the namesake of Woodhouse's Scrub-Jay) for his friend John Cassin, then curator of birds for the Academy of Natural Sciences in Philadelphia. Cassin's Sparrows are most conspicuous during the breeding season, when males skylark, flying up high in the air and fluttering down with their musical call.

LENGTH: 6 inches. **WINGSPAN:** 7.75 inches.
ADULT: Grayish-brown to rufous-brown. Finely streaked on crown. Indistinct supercilium. Speckled on back. Breast has fine spots and streaks. Rear flanks are streaked. Throat is white. Pale eye ring. Long, rounded tail has distinct white corners. **JUVENILE:** (May–Sept.) Browner overall than adults. Streaking on breast is more distinct than adults. **VOICE:** Song is a high-pitched *see-seeddddddd*

DE-da-DEE. **BEHAVIORS:** Prefers to run on the ground unless pursued. Males skylark, flying up and singing while fluttering down. **HABITAT:** Arid grasslands with short, scattered shrubs. **STATUS:** Common to abundant summer resident in the Panhandle and western South Plains. Locally common to rare in the remainder of the western two-thirds of Texas and south of Matagorda Bay. Uncommon to rare in winter in the breeding range. Spring migration is mid-March–mid-April. Fall migration starts in early September.

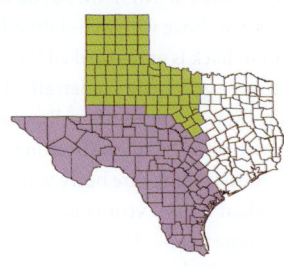

Bachman's Sparrow

(Peucaea aestivalis)

IUCN RED LIST STATUS (2020): Near Threatened
POPULATION TREND: Decreasing

Bachman's Sparrow was named by John James Audubon after his friend John Bachman, who collected the first specimens. Historically the bird was common in mature pine woods that have now mostly been logged. Today it often occurs in clear-cuts and right-of-ways for utilities and pipelines, where the necessary grassy conditions still occur.

LENGTH: 6 inches. **WINGSPAN:** 7.25 inches.
ADULT: Brown above with well-defined gray supercilium. Back is gray-streaked. Breast is buffy with pale gray belly. Undertail is buffy. **JUVENILE:** (May–Sept.) Buffy-yellow background color with bold dark streaking on back and breast. Flanks are buffy with bold streaks. Throat is pale. **VOICE:** Song is a long, drawn-out note followed by a trill *zreeeeeeee whi-whi-whi-whi-whip.* **BEHAVIORS:** Reluctant to fly; will run if possible when pursued. When flushed, drops quickly to the ground and runs. **HABITAT:** A habitat specialist, they require open pine or oak forests with a tall grass understory maintained by prescribed burns or natural fires. **STATUS:** Very rare to locally uncommon resident in the Pineywoods.

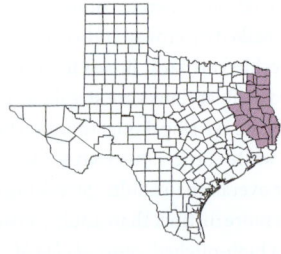

Grasshopper Sparrow

(Ammodramus savannarum)

IUCN RED LIST STATUS (2021): Least Concern
POPULATION TREND: Decreasing

The Grasshopper Sparrow is a cryptic, inconspicuous specialist of grasslands, sticking to the ground when possible. Even though it's a prolific singer during breeding season, its very high-pitched song is hard to hear, with an insect-like quality that might not be recognized as a bird by many. Like many grassland specialists, their decline in numbers matches the loss and degradation of our native grasslands.

LENGTH: 5 inches. **WINGSPAN:** 7.75 inches. **ADULT:** Bright unmarked buffy below. Rufous-spotted with gray streaks on back. Grayish supercilium. Complete eye ring. **JUVENILE:** (May–Sept.) Like adults but breast is streaked with light brown. Brown and gray with no rufous on back. **VOICE:** Song is a very high, insect-like buzz with a sharp *tik* introductory note. **BEHAVIORS:** When flushed, makes a short flutter flight before dropping. Males flutter from singing perch to singing perch. Runs long distances on the ground to flee. **HABITAT:** Semi-open grasslands with patchy bare ground. Avoids shrubby grasslands. **STATUS:** Rare to locally common summer resident east of the Trans-Pecos. Absent from the southern and western South Texas Brushlands. In the Pineywoods, rare breeders in the northern part, absent from the southern part. In Trans-Pecos, rare to uncommon resident in the mid-level grasslands. Uncommon to common migrant and winter resident west of the Pineywoods. Spring migrants are found early April–mid-May, fall migrants mid-August–early November.

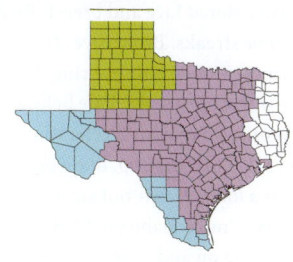

Olive Sparrow

(Arremonops rufivirgatus)

IUCN RED LIST STATUS (2021): Least Concern
POPULATION TREND: Stable

Olive Sparrow is one of the Texas specialties that birders come here to see. A skulking resident of South Texas, they have a wide tropical and subtropical distribution but are found in North America only in South Texas. **LENGTH:** 6.25 inches. **WINGSPAN:** 7.75 inches. **ADULT:** Pale gray face and breast. Brown eye stripe and crown stripes. Pale split eye ring. Dull olive on back. Belly and undertail are yellowish-olive color. **JUVENILE:** (Apr.–Aug.) Olive-colored face and breast. Breast has pale olive streaks. Brown eye stripe and crown stripes. Broken pale eye ring. Back is dull olive with dull streaks. Pale belly and undertail. **VOICE:** Song is an accelerating series of *TSIP-TSIP-TSIP-tsip-tsip*, like a bouncing ball. Call is a long, drawn-out sizzle: *tseeeeeepp*. **BEHAVIORS:** Similar habits to towhees; spends most of its time on and near the ground. Flies low across open gaps in habitat. **HABITAT:** Dense thorn-scrub. **STATUS:** Common resident throughout the South Texas Brush Country north to the Edwards Plateau, west to Devils River, and east to Calhoun County.

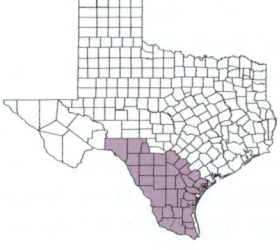

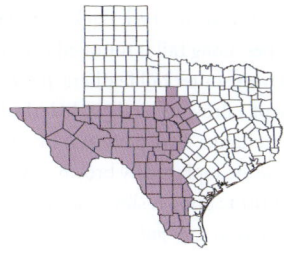

Black-throated Sparrow

(Amphispiza bilineata)

IUCN RED LIST STATUS (2021): Least Concern
POPULATION TREND: Decreasing

The Black-throated Sparrow is perhaps the easiest sparrow of the "little brown jobs"—or LBJs, as birders are fond of calling them—to identify. Its bold black-and-white pattern with a singular bold white eye stripe or supercilium can't be confused with any other regular sparrow in Texas.

LENGTH: 5.5 inches. **WINGSPAN:** 7.75 inches. **ADULT:** Dark gray head with white supercilium and white malar stripe. Thin lower eye arc. Bold black throat. Breast and belly are unmarked pale. Back is brownish-gray. Tail is dark with white corners. **JUVENILE:** (June–Oct.) Brownish-gray above and pale below. Breast band of indistinct streaks. Bold white supercilium and malar. Lower eye arc. **VOICE:** Song is a musical *swit swit sweeeee de de de de de de de*. Call is a high-pitched *sit sit*. **BEHAVIORS:** Hops on the ground. Flights between shrubs are short and direct. Males sing from a perch. **HABITAT:** Prefers semi-open habitat with evenly spaced shrubs. **STATUS:** Common to abundant resident in the Trans-Pecos to the western Edwards Plateau and south through South Texas Brush Country.

Lark Sparrow

(Chondestes grammacus)

IUCN RED LIST STATUS (2021): Least Concern
POPULATION TREND: Decreasing

As agricultural lands regrow to forest or are urbanized, Lark Sparrow's range is contracting.

LENGTH: 6.5 inches. **WINGSPAN:** 11 inches. **ADULT:** Bold harlequin face pattern. Brown crown with white central strip. White supercilium with dark eye stripe and brown ear or auricular patch. Bold white arc under eye. Dark malar stripe. Throat and breast are pale, almost white, with a bold central spot. Flanks are brown-washed. Back is brown with dark stripes. Long tail is rounded with white tips and white outer tail feathers. **JUVENILE:** (June–Sept.) Brown overall with dark-streaked breast and flanks. Face pattern is like adults with the white replaced by brown. Bold lower eye arc. **VOICE:** Song is slow-paced with a choppy rhythm: *zeer puk treeeee chido chido kreet-kreet-kreet-kreet trrrrr.* **BEHAVIORS:**
Often flies long distances with an undulating pattern. Males perform a strutting display with the tail upright, flashing the tail spots. **HABITAT:** An edge habitat specialist, often using very disturbed sites. **STATUS:** Common to uncommon migrant and summer resident statewide; generally absent from the most arid parts of the Trans-Pecos. Rare to locally uncommon during summer in the open parts of the Pineywoods. Rare to locally uncommon winter resident in the southern Oaks and Prairies, coastal prairies, and South Texas Brush Country. Spring migration is late March–early May, fall mid-August–early October.

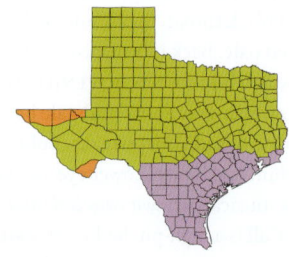

Lark Bunting

(Calamospiza melanocorys)

IUCN RED LIST STATUS (2021): Least Concern
POPULATION TREND: Decreasing

Lark Bunting is endemic to the grasslands of North America. It is one of the few species where the male molts from a cryptic winter plumage into a completely different, bold breeding plumage.
LENGTH: 7 inches. **WINGSPAN:** 10.5 inches.
ADULT BREEDING: (Mar.–July) Male is all black with a large white wing patch. Tip of the tail is white with central tail feathers all dark. Female is grayish-brown above with low-contrast brown streaking. Thin pale eye ring. Small white ear or auricular patch. A wide dark malar stripe on a white throat. White below with dark brown streaking. Tail is white-tipped. Large but poorly defined wing patch. **NON-BREEDING:** (Aug.–Mar.) Male is like breeding female but streaking is blackish, not brown. Female is like breeding female. **VOICE:** Song is four parts: *rheet-rheet-rheet sep-sep-sep-sep-sep wret-wret-wret-wret te-te-te-te-te-te-te we-we*. Call is a soft *heew*. **BEHAVIORS:** Hops when foraging on the ground. Flies low over weeds. **HABITAT:** Breeds in the grasslands of the High Plains. Winters in grasslands and fallow agricultural fields. **STATUS:** Uncommon to rare summer resident in the Panhandle. Abundant to uncommon migrant and winter resident in the western half of the state and south to the central coast and the Lower Rio Grande Valley. Rare migrant and winter visitor to the upper Texas coast. Spring migration is late March–early May, fall early July–late October.

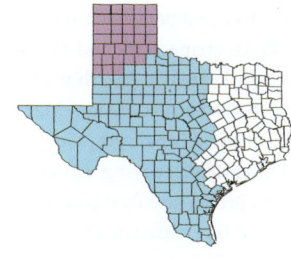

Chipping Sparrow

(Spizella passerina)

IUCN RED LIST STATUS (2021): Least Concern
POPULATION TREND: Decreasing

Chipping Sparrows are one of the most abundant and familiar of North America's songbirds. The moniker songbird seems misplaced when one hears its mechanical, unmusical trill. Chipping Sparrows are familiar to many as they visit feeders in large numbers, particularly in winter.

LENGTH: 5.5 inches. **WINGSPAN:** 8.5 inches.
ADULT BREEDING: (Apr.–Aug.) Chestnut crown and bold white supercilium. Dark eye stripe. Gray face and breast. Throat is white with gray malar stripe. Back and wings are brown with dark streaks. Two thin white wingbars. **NONBREEDING:** (Aug.–Mar.) Like breeding but throat is pale gray. Supercilium is buffy. Thin broken eye ring. **JUVENILE:** (June–Sept.) Brown-streaked crown and gray, finely streaked face with pale throat. Thin dark eye stripe. Breast is buffy with brown streaks. Thin broken eye ring. **VOICE:** Song is a long, dry, mechanical trill. Call is a sharp chip. **BEHAVIORS:** Flights are short and direct in close habitat. **HABITAT:** Variable, but generally open, grassy coniferous forests and edges. In winter, more tied to riparian than coniferous forests. **STATUS:** Common resident on the Edwards Plateau and Davis and Guadalupe mountains. Uncommon resident in the Pineywoods. Common to abundant migrant and winter resident in all parts of the state. Fall migration is mid-August–mid-October, spring mid-March–mid-May.

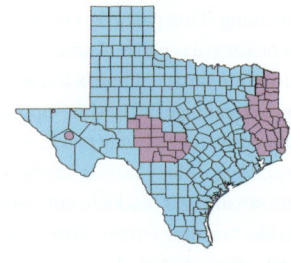

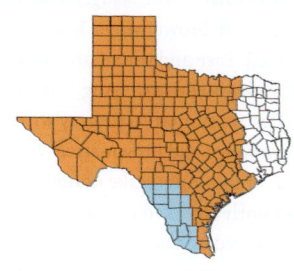

Clay-colored Sparrow

(Spizella pallida)

IUCN RED LIST STATUS (2021): Least Concern
POPULATION TREND: Decreasing

Clay-colored Sparrows do not breed in Texas but migrate through the state and winter in its southern region. They often form flocks with related species, like Brewer's Sparrows and Chipping Sparrows.
LENGTH: 5.5 inches. **WINGSPAN:** 7.5 inches. **ADULT BREEDING:** (Apr.–Aug.) Boldly marked head. Dark brown crown with white supercilium and brown face. Dark eye stripe. Grayish malar and white throat. Clean gray nape. Pale gray below. Back is brown with dark streaks. Two narrow white wingbars. Rump is brownish. **NONBREEDING:** (Aug.–Mar.) Like breeding but face is browner. Supercilium and breast are buffy. **JUVENILE:** (June–Sept.) Brown crown and finely streaked face. Breast is buffy with dark streaks. **VOICE:** Song is a series of two to five monotone, raspy buzzes: *zheee-zheee-zheee*. Call is a sharp, high-pitched *tsip*. **BEHAVIORS:** Sings from perches close to the ground. Hops and does not run on the ground. Flies and lands on the ground when disturbed. **HABITAT:** Open shrubby grasslands. **STATUS:** Common to uncommon migrant in the center of the state. Uncommon to rare migrant in the Trans-Pecos. Rare migrant in the eastern third of the state. Rare winter resident in the western half of the South Texas Brush Country. Fall migration is early September–early November, spring mid-March–mid-May.

Female

Male

Black-chinned Sparrow

(Spizella atrogularis)

IUCN RED LIST STATUS (2021): Least Concern
POPULATION TREND: Decreasing

Black-chinned Sparrows are common in their expected Texas range, but the mountains of Texas are some of the least accessible habitat to birders. Only a few of the sites where they are found are accessible, notably the Chisos Mountains in Big Bend National Park, the Franklin Mountains in El Paso County, and Guadalupe Mountans National Park. **LENGTH:** 5.75 inches. **WINGSPAN:** 7.75 inches. **ADULT BREEDING:** (Apr.–Aug.) Male is slate gray overall with brown wings. Back is brown and streaked. Face and throat are black with pink bill. **NONBREEDING:** (Aug.–Apr.) Males and females are similar: slate gray with pink bill. Brown wings and brown-streaked back. **JUVENILE:** (June–Aug.) Like nonbreeding. Breast is faintly streaked in gray. **VOICE:** Song is a slurred and accelerating trill of *slur-slur-slur-su-su-su-su-su*. Call is a sharp, high-pitched *tsip*. **BEHAVIORS:** Flight is low and direct. Males sing from a perch. **HABITAT:** Arid brush on steep slopes. **STATUS:** Uncommon summer resident in the mountains of the Trans-Pecos. In winter, the population moves to lower elevations and becomes much less common, suggesting that part of the Texas population moves out of state. Birds are found at lower elevations from late August to late March.

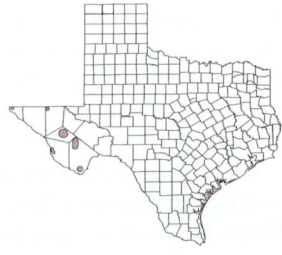

Field Sparrow

(Spizella pusilla)

IUCN RED LIST STATUS (2021): Least Concern
POPULATION TREND: Decreasing

With its plain face, white eye ring, and pink bill, Field Sparrow is one of the easiest sparrows to identify. Even though they look quite different, the dark-gray-and-black Black-chinned Sparrow may actually be Field Sparrow's closest relative. **LENGTH:** 5.75 inches. **WINGSPAN:** 8 inches. **ADULT:** Rufous crown and auricular patch. Bold white eye ring. Bill is pink. Single white wingbar. Back is rufous with dark streaks. Breast and belly are unmarked pale rufous to pale grayish. **JUVENILE:** (May–Oct.) Gray head with pale gray bill. White eye ring. Breast is pale buffy with dark streaks. Two indistinct wingbars. **VOICE:** Song is whistled, accelerating like a bouncing Ping-Pong ball: *su-su-su-su-su su-su.* **BEHAVIORS:** Hops on the ground; flights are direct. **HABITAT:** Breeds in older, successional fields and edges. Winters in similar habitat. **STATUS:** Uncommon resident in the northern half of the Pineywoods, the eastern half of the Panhandle, the eastern Rolling Plains, and the Edwards Plateau. Also in Brooks and Kenedy counties. Common to uncommon migrant and winter resident in Texas, except for the Trans-Pecos.

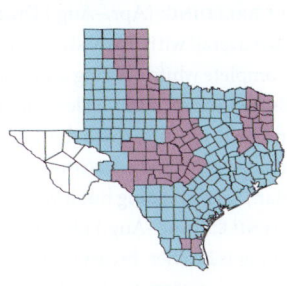

Brewer's Sparrow

(Spizella breweri)

IUCN RED LIST STATUS (2021): Least Concern
POPULATION TREND: Decreasing

Brewer's Sparrow is one of the drabbest, least showy-looking sparrows. While it does not breed in Texas, it has one of the most stand-out songs of all the sparrows—a long series of trills and buzzes that defies description. Brewer's Sparrow was named by John Cassin in 1856 in honor of Thomas M. Brewer, a Boston physician and ornithologist. **LENGTH:** 5.5 inches. **WINGSPAN:** 7.5 inches. **ADULT BREEDING:** (Apr.–Aug.) Drab gray-brown overall with finely streaked crown. Complete white eye ring and small bill. Indistinct supercilium. Pale malar mark. Breast and belly are unmarked dingy gray. Back and nape are streaked. **NONBREEDING:** (Aug.–Mar.) Like breeding but lower contrast. **JUVENILE:** (June–Aug.) Like breeding but streaking is heavier. Breast is buffy with heavy streaking. **VOICE:** Call is a loud *tsip tsip.*

BEHAVIORS: Hops on the ground and occasionally runs. Flight is direct and rarely more than 3 feet above the shrubs. Flies readily to shrubs when disturbed. **HABITAT:** Sagebrush shrublands and brushy desert habitat dominated by saltbush and creosote. **STATUS:** Common to uncommon migrant in the western third of Texas. Rare to abundant winter resident in the Trans-Pecos and irregular winter visitor along the Rio Grande south to the Lower Rio Grande Valley. Fall migration is late August–early November, spring early April–mid-May.

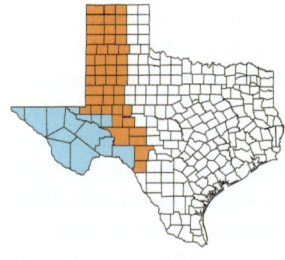

Fox Sparrow

(Passerella iliaca)

IUCN RED LIST STATUS (2021): Least Concern
POPULATION TREND: Decreasing

Fox Sparrow is one of the most complex species of the New World sparrows. As many as eighteen subspecies have been recognized and placed into four subspecies groups: Sooty Fox Sparrow, Thick-billed Fox Sparrow, Slate-colored Fox Sparrow, and Red Fox Sparrow. These four groups may represent four different species, based on DNA analysis. Only Slate-colored Fox Sparrow and Red Fox Sparrow have been recorded in Texas, and only Red Fox Sparrow is regular here. **LENGTH:** 7 inches. **WINGSPAN:** 10.5 inches. **ADULT:** Bright rufous-brown above with gray supercilium and nape. Back is streaked with gray. Rump is gray. White below with heavy rufous streaks and spots on breast and flanks. **VOICE:** Call is a loud, thrasher-like *smack*. **BEHAVIORS:** Typically hops with both legs. Flights are often close to the ground.

HABITAT: Prefers thick cover, thickets, and dense brushy tangles on the edges of woods.
STATUS: Uncommon to common migrant and winter resident from the eastern Panhandle south to the central Edwards Plateau and east to the central coast in the eastern half of Texas. Fall migration is mid-October–late November, spring late February–mid-April.

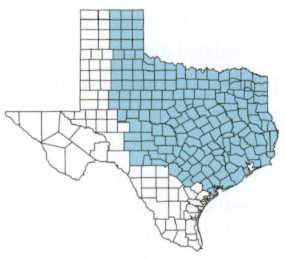

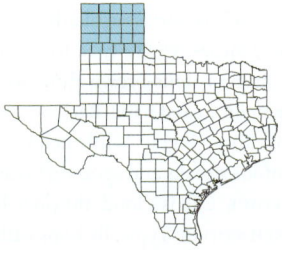

American Tree Sparrow

(Spizelloides arborea)

IUCN RED LIST STATUS (2021): Least Concern
POPULATION TREND: Decreasing

American Tree Sparrow's winter range has been contracting over the last several decades. In the 1950s, they could be found in the winter as far south as the northern Edwards Plateau. They now only regularly occur in the Panhandle, mostly its northern part, and are very rare south of the Panhandle.

LENGTH: 6.25 inches. **WINGSPAN:** 9.5 inches.
ADULT: Rufous crown and small bicolored bill. Broken white eye ring. Face is gray with rufous eye stripe. Breast and belly are whitish with an isolated dark spot on the breast. Rufous patches on sides of breast. Wings are black and brown. Back has dark stripes. Two white wingbars. **JUVENILE:** (May–Oct.) Brown above with brown eye stripe on gray face. Streaky brown auricular, or ear patch. Breast is whitish with dark brown streaks. Back is dark streaked. **VOICE:** Cheerful *chee-deep*.

BEHAVIORS: Hops on the ground and along branches. Flight is direct with rapid wingbeats. **HABITAT:** Fields, hedgerows, and open woodlands. **STATUS:** Locally uncommon winter visitor in the Panhandle. Fall birds arrive late October to early November. Spring migration is from early March to early April.

Pink-sided

Oregon

Gray-headed

Slate-colored

Dark-eyed Junco

(Junco hyemalis)

IUCN RED LIST STATUS (2021): Least Concern
POPULATION TREND: Decreasing

Until the 1970s, the Dark-eyed Junco was split into five separate species. These forms are all distinct from each other in plumage. As one moves from east to west in Texas, the diversity of Dark-eyed Junco's forms increases. In East Texas, most will be of the slate-colored form; in the mountains of the Trans-Pecos, one can often find three or four forms in a single foraging flock.

LENGTH: 6.25 inches. **WINGSPAN:** 9.25 inches. **ADULT:** Variable. Generally, all forms have pale to pinkish bills. Gray to black hood and pale belly. Tail and back are dark with white outer tail feathers. Most variable is the amount of rufous on sides and back. **VOICE:** Song is a musical, slow-paced trill: *swee-swee-swee-swee-swee.* Call is a rapid, piercing *stit-stit-stit-stit-stit.* **BEHAVIORS:** Forages on the ground in flocks. Hops on the ground.

HABITAT: Open forest areas with little understory. **STATUS:** Uncommon to abundant migrant and winter resident in the northern two-thirds of Texas. The Red-backed subspecies is an uncommon resident in the higher elevations of the Guadalupe Mountains. Fall migration is early October–mid-November, spring early March–mid-April.

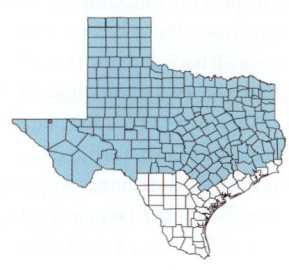

First winter

Male

White-crowned Sparrow

(Zonotrichia leucophrys)

IUCN RED LIST STATUS (2021): Least Concern
POPULATION TREND: Stable

Often described as elegant, White-crowned Sparrow is one of the most recognizable of the wintering songbirds in Texas. Its bold and distinctive head pattern and habit of perchng prominently in the sun on cold days make it easy to observe.

LENGTH: 7 inches. **WINGSPAN:** 9.5 inches.
ADULT: Black crown with white central crown stripe. Dull white supercilium and black eye stripe. Pink bill. Gray face with dark malar stripe. Unmarked gray-brown below. Brownish back and upper wings with dark back stripes. Two white wingbars. **FIRST WINTER:** (Aug.–Apr.) Like adults but brown and eye stripe are brown. Face and supercilium are buffy. **JUVENILE:** (June–Aug.) Brown above with well-defined dark crown stripes. Breast is whitish with heavy dark streaking. Hint of a pale eye stripe. **VOICE:** Song is a clear

whistle followed by a trill and buzzes: *suuuuu swee-swee-swee zwhee-zwhee-zwhee*. **BEHAVIORS:** Hops on the ground while foraging. Sings often on sunny winter days. **HABITAT:** Forms mixed flocks in winter. Usually stays close to cover but forages on the ground. **STATUS:** Abundant to uncommon migrant and winter resident in the northern two-thirds of Texas. Uncommon southward to the Lower Rio Grande Valley. Fall birds arrive in mid-September to mid-November. Spring migration is from late March to mid-May.

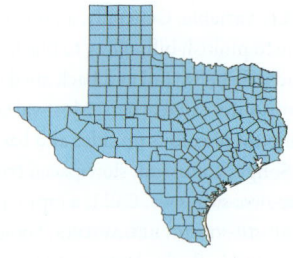

Harris's Sparrow

(Zonotrichia querula)

IUCN RED LIST STATUS (2020): Near Threatened
POPULATION TREND: Decreasing

Designated as a common species in steep decline by Partners in Flight, the Harris's Sparrow population has declined more than 50% in the last thirty years. The reason is unclear, as it is one of the least studied species in North America.

LENGTH: 7.5 inches. **WINGSPAN:** 10.5 inches.
ADULT BREEDING: (Mar.–Aug.) Black crown and face, extending down onto breast. Face is gray with dark crescent on cheek. Bill is pink. Brown on back and upper wings. Back is dark streaked. Two narrow wingbars. Belly is white. Brown flanks with dark streaks. Rump is grayish and tail is dark with pale corners. **NONBREEDING:** (Aug.–Mar.) Like breeding but face is brownish with brown crescent on cheek. Less black on crown and front of face. **FIRST WINTER:** (Aug.–Apr.) Like nonbreeding with dark wash on forward part of crown. Throat is white with dark, streaky breast band. **VOICE:** Song is a clear but raspy two-part whistle: *seeeeeee seeeee.* Call is a squeaky *cheek cheek.* **BEHAVIORS:** Hops on the ground but often perches high in trees when not active or disturbed. **HABITAT:** Hedgerows and brushy areas. Forest edges but not forest interiors. **STATUS:** Uncommon to locally common migrant and winter resident in the central part of Texas, including the eastern Panhandle, south to Edwards Plateau. Fall migrants arrive late October–early December. Spring migration is early March–mid-April.

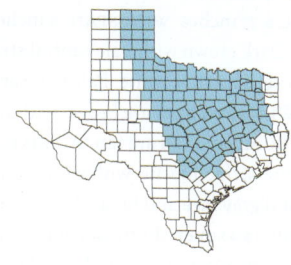

White-throated Sparrow

(Zonotrichia albicollis)

IUCN RED LIST STATUS (2021): Least Concern
POPULATION TREND: Decreasing

White-throated Sparrows come in two color morphs but have no subspecies. Some birds are white-striped with a bright white super-cilium; others are tan-striped with a tan su-pecilium. These color morphs are maintained by disassortative mating, in which each mates with the opposite morph. White-striped males sing more and are more likely to engage in extra-pair copulation. Tan-striped morphs of either sex provide more parental care. **LENGTH:** 6.75 inches. **WINGSPAN:** 9 inches. **ADULT:** Dark crown with pale central stripe. Yellow lore. Throat is white. Pale belly and tan flanks. Wings and back are brown with tan streaks. Rump and tail are tan. Adults range from white-striped forms with white super-cilium and gray face and breast to tan-striped forms with tan supercilium and gray-brown breast. **FIRST WINTER:** (Aug.–Mar.) Like tan-striped adults. Breast is streaked. **VOICE:** Song is a whistled *sooo seeeeee deedee deedee*, with a cadence like "Old Sam Peabody Peabody." Call is a very high-pitched *stink stink*. **BEHAVIORS:** Hops on the ground and on branches. Makes short flights between branches when foraging. **HABITAT:** Favors thick cover such as hedgerows, dense shrubs, and cattails. **STATUS:** Uncommon to common migrant and winter resident in the eastern half of Texas south to Corpus Christi. Uncommon to rare migrant and winter resident in the rest of the state. Fall migration is mid-October–mid-November, spring migration mid-March–early May.

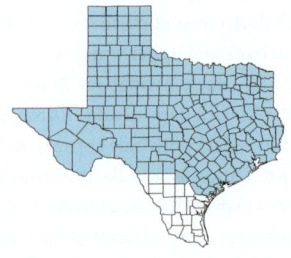

Sagebrush Sparrow

(Artemisiospiza nevadensis)

IUCN RED LIST STATUS (2021): Least Concern
POPULATION TREND: Stable

Sagebrush Sparrows winter in the western part of the Trans-Pecos region of Texas. They are not particularly rare but prefer some of the most inhospitable habitat in the state, where if a plant species doesn't have thorns, it's likely not a native species! Combine that with a preference to flee on the ground and they become one of the most cryptic birds in Texas.
LENGTH: 6 inches. **WINGSPAN:** 8.25 inches.
ADULT: Gray head with complete white eye ring. Small white loral spot. Narrow dark malar strips on white throat. Dark gray breast spot. Whitish on belly and undertail. Back and wings are olive-gray. Narrow but distinct streaks on back. Tail is dark with thin white outer feathers. **VOICE:** Call is a high-pitched *stit stit.* **BEHAVIORS:** Hops or walks on the ground. Runs across open areas back to cover; prefers to run when disturbed.

In flight, flies low over shrubs. **HABITAT:** Desert washes, sagebrush, creosote bush, and arid grasslands. **STATUS:** Uncommon migrant and local winter resident in the western part of the Trans-Pecos. Fall migrants arrive in mid-October and are present until mid-March.

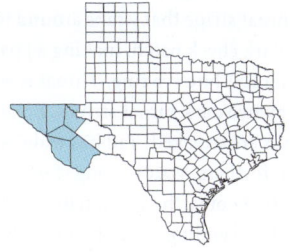

Vesper Sparrow

(Pooecetes gramineus)

IUCN RED LIST STATUS (2021): Least Concern
POPULATION TREND: Decreasing

Vesper Sparrow got its name because its song is supposedly sweetest in the evening, or at the time of vesper prayers. This species is one of the few that requires bare ground during the breeding season, and so benefited when forest was cleared for agriculture. Now that farms are reverting back to forest, it is in dramatic decline in eastern North America.
LENGTH: 6.25 inches. **WINGSPAN:** 10 inches.
ADULT: Brownish gray with pale belly. White throat stripe that wraps around to frame a dark cheek patch, making a J pattern on face. White eye ring. Throat is white framed by two dark throat stripes. Breast and flanks are streaked. Tail has wide, white outer tail feathers. **VOICE:** Song starts with two whistled notes, then slow trills accelerating and descending in pitch: *sueeeee sueeee chidle-childle-childle chidle-childle*. Chip call is a *chit chit chit*. Tink call is a short, high-pitched *stit stit*. **BEHAVIORS:** Strong flier but runs on the ground rather than flies. When flushed, flies a short distance. **HABITAT:** Grasslands and weedy fields. Arid subtropical scrub but avoids scrublands dominated by creosote bush. **STATUS:** Uncommon to common migrant and winter resident throughout most of Texas. Rare winter resident in the southern Panhandle but absent from the northern Panhandle as a winter resident. Fall migration is mid-August–early November, spring early March–early May.

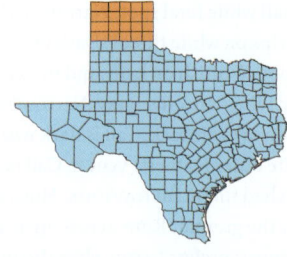

LeConte's Sparrow

(Ammospiza leconteii)

IUCN RED LIST STATUS (2021): Least Concern
POPULATION TREND: Decreasing

LeConte's Sparrow was named for John L. LeConte, a noted entomologist who collected the first specimen in Georgia in 1790. LeConte's Sparrow is so secretive it took more than 40 years until another specimen was collected in the early 1830s in North Dakota. When Arthur Bent published the species account in his life history series in 1968, fewer than fifty nests had been found.
LENGTH: 5 inches. **WINGSPAN:** 6.5 inches.
ADULT: Bright yellow face with grayish cheek patch and dark auricular patch. Dark crown with white central crown stripe. Yellow-buff on breast with fine dark streaks. Belly is mostly white. Flanks are yellow-buff with bolder streaks. Brownish collar is finely streaked with purplish spots. Back is brown with bold pale stripes. **VOICE:** Song is a hissing, unmusical buzz: *stit stziiiiiiiiiiiiiiiiiiiii*. **BEHAVIORS:**

More mouse than bird in habits; runs on the ground to flee if possible. When forced to fly, flight is short, and dives with a sharp turn into the grass. **HABITAT:** Fields with tallgrass, waist-high at least. Often lots of bluestem species (genus *Andropogon*). **STATUS:** Uncommon to common winter resident in the eastern third of the state, south to Kleberg County. Fall migration is mid-September–late November, spring mid-February–mid-April.

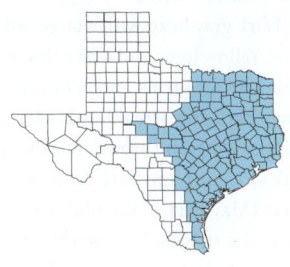

Adult

Immature

Seaside Sparrow

(Ammospiza maritima)

IUCN RED LIST STATUS (2021): Least Concern
POPULATION TREND: Stable

Seaside Sparrow is listed as a species of least concern by the IUCN Red List, but it is a specialist of a very limited zone of tidal marsh that has declined by almost 90% in Texas. Because tidal marshes occur only on the coast, they are one of the habitats most endangered by development. Additionally, the new threat of sea level rise could accelerate the loss of tidal habitat.

LENGTH: 6 inches. **WINGSPAN:** 7.5 inches.
ADULT: Dark gray head with a large bill for a sparrow. Yellow lores and white throat with dark malar stripe. Breast and belly are buffy with heavy dark gray streaking. Undertail is buffy. Upper wings are brown. Back is gray with dark streaking and dark dusky wash.
JUVENILE: (May–Oct.) Like adults but lacks the dark malar stripes. Belly is whitish. Breast is buffy with dark streaks. **VOICE:** Song is a high-pitched *"drink your teaaaa" dee dee deeeeee*. Call is a *tup tup*. **BEHAVIORS:** Perches just below the grass tops to sing from partial concealment. Climbs on grass. Flies low over grass tops for long distances. **HABITAT:** Tidal marshes dominated by cordgrass. **STATUS:** Uncommon to locally uncommon resident in the tidal marshes south to Calhoun County. Rare to uncommon south to the Rio Grande.

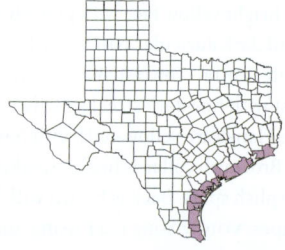

Nelson's Sparrow

(Ammospiza nelsoni)

IUCN RED LIST STATUS (2021): Least Concern
POPULATION TREND: Increasing

Until 1995, Nelson's Sparrow and Saltmarsh Sparrow were considered one species, the Sharp-tailed Sparrow. It was named for Edward Nelson, a naturalist and Chief of the USDA Bureau of Biological Survey from 1916 to 1927. Nelson's Sparrow breeds in interior grasslands in the northern-central United States and central Canada. It winters in coastal cordgrass marshes in the tidal zones of the Gulf and Atlantic coasts. Preservation of these two highly endangered habitats are key to future populations of this species.
LENGTH: 5 inches. **WINGSPAN:** 7 inches. **ADULT:** Dark brown crown and bright buffy-orange face. Nape is gray. Breast is buffy-orange with fine dark streaking and clean, well-defined white belly. Rufous-brown on back with whitish streaking. **VOICE:** Song is a hissing *tee-sheeeeeee-ugh*. **BEHAVIORS:**

Commonly moves on the ground by walking, hopping, or running, depending on circumstances. Perches high to survey surroundings. **HABITAT:** Winters in coastal cordgrass marshes. **STATUS:** Uncommon to locally common winter resident in tidal marshes south to Nueces County. Rare south of Nueces County to the Rio Grande. While it migrates through the easternmost quarter of Texas, it is rarely detected inland. Fall migration is late September–early November, spring late April–mid-May.

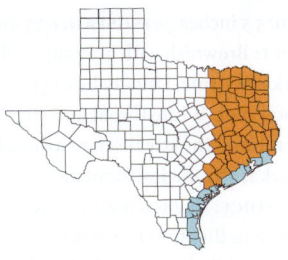

Savannah Sparrow

(Passerculus sandwichensis)

IUCN RED LIST STATUS (2021): Least Concern
POPULATION TREND: Decreasing

Savannah Sparrow is the most abundant and widely distrbuted of wintering sparrows in Texas. It is found statewide during migration and in all but the northern part of the Panhandle in winter. A sage observer of sparrows once gave the following advice to birders: "All roadside sparrows are Savannah Sparrows until proven otherwise." This is good advice. Mowed shortgrass along country roads is its favorite habitat.

LENGTH: 5.5 inches. **WINGSPAN:** 8.75 inches. **ADULT:** Brownish with gray supercilium and dark eye stripe. Usually has a yellow supraloral spot. White throat and streaked breast; at times the streaks may form a breast spot. Back is streaked without any scaled pattern. **VOICE:** Call is a high-pitched *sipt*, often given in flight. **BEHAVIORS:** Walks on the ground while foraging. Flight is low and direct when flushed. **HABITAT:** Open fields, roadsides, coastal marshes, and shortgrass edge habitats in open country. **STATUS:** Abundant to uncommon migrant statewide. Abundant to uncommon winter resident south of the southern Panhandle. Fall migration is mid-August–early November, spring mid-March–early May.

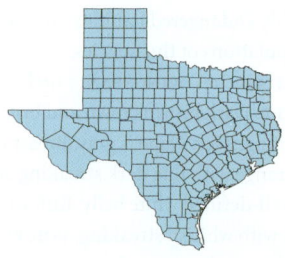

Song Sparrow

(Melospiza melodia)

IUCN RED LIST STATUS (2021): Least Concern
POPULATION TREND: Decreasing

The widespread and relatively common Song Sparrow is the champion of variation, as twenty-four subspecies are currently recognized. While these vary from the large, very dark birds of the Aleutian Islands to the small, very light birds of southeast Arizona, only two subspecies have been documented for Texas. Most of the state's Song Sparrows are from the eastern subspecies, and there are rare records of the Rocky Mountain subspecies. **LENGTH:** 6.25 inches. **WINGSPAN:** 8.25 inches. **ADULT:** Brown-colored with gray face and dark brown auricular patch. Supercilium is gray. White throat framed with thick, bold malar stripes. Breast is grayish with coarse dark brown streaks that form a dense spot in the middle. Flanks are buffy with coarse streaks. Belly is whitish. Back is streaked with gray. **VOICE:** Song is a three-part *zoot zoot zreeee zreeee chi-chi-chi-chi-chi-chi*. The most common call is a loud *chimp chimp*. Also makes a piercing *teenk teenk*. **BEHAVIORS:** Flights are short between perches. Primarily walks on the ground to forage or hops on rough ground. **HABITAT:** Wide range of forest and shrub habitats but always close to water. Can be numerous in cattails. **STATUS:** Common to uncommon migrant and winter resident statewide. Rare to very rare in the southern coastal prairies, South Texas Brush Country, and Lower Rio Grande Valley. Fall migration is late September–mid-November, spring early March–early May.

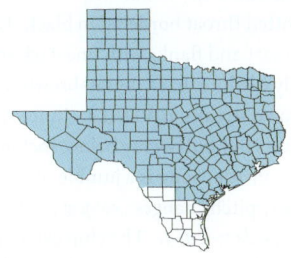

Lincoln's Sparrow

(Melospiza lincolnii)

IUCN RED LIST STATUS (2021): Least Concern
POPULATION TREND: Stable

Lincoln's Sparrow was named by John James Audubon during his expedition to northeastern Canada in 1833. It was named after Thomas Lincoln, the young man who collected the specimen. While crisply marked and easily recognized, it is generally secretive and has a reputation for being a skulker.
LENGTH: 5.75 inches. **WINGSPAN:** 7.5 inches.
ADULT: Brown crown with broad gray supercilium. Broad buffy malar strip with white fine-spotted throat bordered in black stripes. Buffy breast and flanks with fine dark streaking. Belly is white. Back is gray-brown with dark brown streaks. **JUVENILE:** (June–Aug.) Like adults but buffy-gray below, including the belly. **VOICE:** Song is a jumble of trills with many pitch changes: *jew-jew-jew jeeeeeeee de-de-de-de-de te-te-te-te*. The chip call is a *cheet cheet cheet*. **BEHAVIORS:** Walks on the ground while foraging. Flies short distances between shrubs. **HABITAT:** Wide variety of habitats with low dense cover. **STATUS:** Common to uncommon migrant statewide. Uncommon to rare winter resident in the Panhandle east across north-central Texas to the Pineywoods. Common to abundant winter resident in the remainder of Texas. Fall migration is late August–early November, spring mid-March–mid-May.

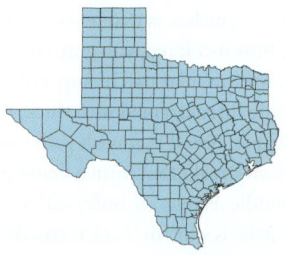

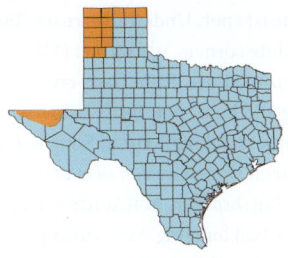

Swamp Sparrow

(Melospiza georgiana)

IUCN RED LIST STATUS (2021): Least Concern
POPULATION TREND: Increasing

Swamp Sparrow is an appropriately named bird. Both in breeding and in overwintering habitats, it is rarely found far from water. A skulker, it is almost always seen on the ground, emerging briefly from cover.
LENGTH: 5.75 inches. **WINGSPAN:** 7.25 inches.
ADULT: Brown crown with gray central stripe. Face is gray with dark line and auricular patch behind the eye. White to buffy malar stripe. White throat bordered in black. Breast is gray with faint blurry streaks. Belly is pale gray. Back is brown with dark bold stripes. Wings are brown with no wingbars. **FIRST WINTER:** (Sept.–Mar.) Like adults but face is tinged with olive. Flanks are more heavily streaked. **VOICE:** Chip call is an emphatic *CHEAT CHEAT*. Gives a soft *tink* call. When agitated, gives a raspy call: *RHEET RHEET RHEET*. **BEHAVIORS:** Walks on the ground,

especially on the edge of shallow water. Rarely flies long distances outside of migration.
HABITAT: Along watercourses, wet areas, and coastal marshes. **STATUS:** Common migrant through the eastern half of Texas. Local westward through the Trans-Pecos. Uncommon winter resident in most areas of the state. Most common in the Pineywoods and the upper and central coasts. Fall migration is late September–early November, spring early March–early May.

Canyon Towhee

(Melozone fusca)

IUCN RED LIST STATUS (2021): Least Concern
POPULATION TREND: Decreasing

Canyon Towhee is not a showy bird but seems to get along well with man outside of dense urban settings. They seem to be always present in parking lots, rural settlements, and cattle pens. Adaptable, they occupy many open and semi-open spaces in West Texas. **LENGTH:** 9 inches. **WINGSPAN:** 11.5 inches. **ADULT:** Pale gray-brown. Rusty crown. Buffy loral spot and eye ring. Thick rusty eye line. Buffy throat with "necklace" of dark spots. Dark breast spot. Undertail is rusty. Tail has small white corners. **JUVENILE:** (May–Aug.) Gray-brown above. Faint rusty crown. Buffy eye ring and loral spot. Buffy below with faint thick streaking. **VOICE:** Song is a slow trill: *swee-ut swee-ut swee-ut swee-ut.* Call is a hoarse *chep chep chep.* **BEHAVIORS:** Hops on the ground when foraging. Very curious, exploring cars in parking lots and campsites. **HABITAT:** Open arid habitats with bare ground for foraging and brush for cover. **STATUS:** Common to uncommon resident in the Trans-Pecos east to the southern Panhandle, South Plains, western Rolling Plains, and the western Edwards Plateau. Rare and local on the eastern Edwards Plateau.

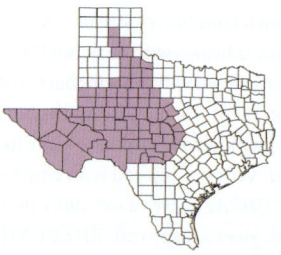

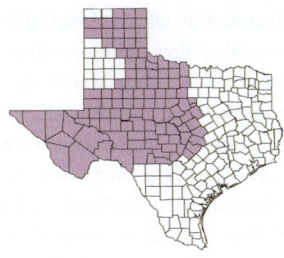

Rufous-crowned Sparrow

(Aimophila ruficeps)

IUCN RED LIST STATUS (2021): Least Concern
POPULATION TREND: Decreasing

Rufous-crowned Sparrow is a widely distributed species in the southwestern United States and Mexico. It is nonmigratory. Secretive and preferring rocky, brush-covered slopes, it can be locally common but difficult to observe.

Pairs generally nest on the ground, and nests are carefully hidden. Rufous-crowned Sparrows seldom wander far from cover. Due to its stealthy behavior and remote habitat, little is known about its life history.
LENGTH: 6 inches. **WINGSPAN:** 7.75 inches.
ADULT: Gray face with rufous crown and eye stripe. White eye ring. Pale buffy malar stripe and a pale gray throat framed by black stripes. Gray below. The back is rufous with gray streaking. **JUVENILE:** (May-Aug) Like a paler version of the adults. **VOICE:** The song is a mechanical trill of *chep-chep-chep-chep*. The call is a nasal *beer beer beer beer* or single nasal *cherp*. **BEHAVIORS:** Spends most of the time on or close to the ground. Stays out of sight. Flights are floppy and rapid. **HABITAT:** Semiarid grassy shrublands and open woodlands on rocky slopes. **STATUS:** Common to uncommon resident in the western two-thirds of Texas east through the Rolling Plains to Tarrant and Johnson counties and south to the southern edge of the Edwards Plateau. Mostly absent from the western Panhandle and the South Plains.

Green-tailed Towhee

(Pipilo chlorurus)

IUCN RED LIST STATUS (2021): Least Concern
POPULATION TREND: Stable

Green-tailed Towhee, with its rusty crown and green plumage, is the most distinct of the towhee species in Texas. Most of the year it is seldom seen, though its presence is often given away by its distinctive double foot scratching and ascending, catlike, mewing call.
LENGTH: 7.25 inches. **WINGSPAN:** 9.75 inches.
ADULT: Gray head with rufous crown. Bright white throat and white malar stripe. Eye is dark red. Breast is gray with paler belly and yellowish undertail. Olive-colored back with bright greenish-yellow edging on wings and tail. **FIRST WINTER:** (Aug.–Mar.) Like adults but lacking the bright greenish-yellow edging on the wings and tail. Rufous crown is reduced. **JUVENILE:** (June–Aug.) Grayish-green with fine dark streaking on head. Back and breast are heavily streaked with dark. Belly is pale. Some greenish-yellow on tail and wings.

VOICE: Call is a mewing *skew skew skew*. **BEHAVIORS:** Hops on the ground when foraging. Flights between perches and foraging are low. **HABITAT:** Low weedy brush and desert grasslands with scrub. **STATUS:** Rare to uncommon summer resident in the higher elevations of the Davis and Guadalupe mountains. Common to uncommon migrant and winter resident in the western half of the state from the western two-thirds of the Panhandle to Starr and Zapata counties. Fall migration is early September–early November, spring early April–mid-May.

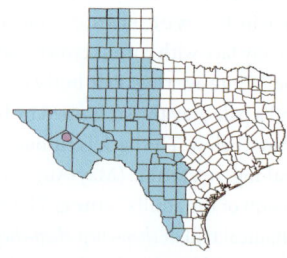

Male

Female

Spotted Towhee

(Pipilo maculatus)

IUCN RED LIST STATUS (2021): Least Concern
POPULATION TREND: Stable

The Spotted Towhee and Eastern Towhee were until recently classified as one species under the name Rufous-sided Towhee. This was based on hybridization of populations primarily along rivers on the Great Plains that allowed the western Spotted Towhee to come in contact with the eastern species then known as Rufous-sided Towhee. In 1995 the American Ornithological Union reconsidered this and split the species.
LENGTH: 8.5 inches. **WINGSPAN:** 10.5 inches.
ADULT: Male has black hood extending to the breast. Bright red eye. Pale belly with white rufous flanks. Undertail is rufous. Back and upper wings are dark charcoal-gray with white spots. Tail and rump are charcoal-gray with white tail corners. Female is like male but hood is dark brown. **JUVENILE:** (May–Aug.) Dark brown on head, breast, and belly with heavy streaking. Eye is red. Upper wings are dark brown with white spots. Undertail is rufous. **VOICE:** Song is a loud *cheat-cheat-cheat treeeeeeeeee*. Call is a screeching *hewwwww*.
BEHAVIORS: Hops on the ground; hops backwards when scratching with both feet.
HABITAT: Shrubs and thickets. **STATUS:** Uncommon to locally abundant summer resident in the mountains of the Trans-Pecos. Common to uncommon migrant and winter resident in most of the state, except the upper Texas coast and the Pineywoods. Fall migration is late September–late October, spring late March–early April.

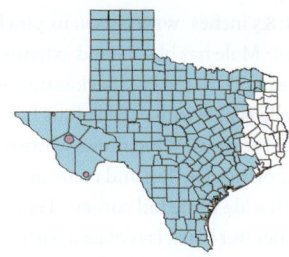

Eastern Towhee

(Pipilo erythrophthalmus)

IUCN RED LIST STATUS (2021): Least Concern
POPULATION TREND: Decreasing

Eastern Towhees come in pale-eyed and red-eyed subspecies. The birds that breed further north have red eyes and are migratory. The southern birds, particularly in Florida, have pale straw-colored eyes and are nonmigratory. The subspecies that occurs in Texas is red-eyed, and even though singing territorial males are regularly found in East Texas, there is only one confirmed nesting in the state from 1914.

LENGTH: 8.5 inches. **WINGSPAN:** 10.5 inches. **ADULT:** Male has black hood extending to the breast. Dark red eye. Back is uniform black. Belly is pale with wide rufous flanks. Undertail is rufous. Wing shows white patch at base of primaries. Tail and rump are black with wide white tail corners. Female is like males but black is replaced with dark brown. **VOICE:** Song is often described as "drink your tea": *suit-seer-treeeeee*. Call is a piercing *SUR-rheee*. **BEHAVIORS:** Hops on the ground. Hops backward to scratch leaf litter with both feet. **HABITAT:** Generally, edge habitat of small trees with dense shrubs and well-developed litter layer for foraging. **STATUS:** Uncommon to rare migrant and winter resident in the eastern third of Texas south to the central Texas coast. Territorial males have been regularly found in the Pineywoods, but no nesting has been confirmed. Fall migration mid-October–mid-November, spring early March–mid-May.

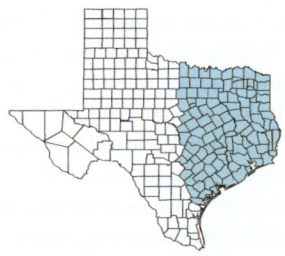

ICTERIIDAE
(YELLOW-BREASTED CHAT)

The Yellow-breasted Chat is an avian enigma. Recent molecular DNA analysis has shown that it is not, as was once speculated, one of the largest of the wood-warblers. Instead the mystery is still being coaxed out of its DNA. Some studies have shown it to be a sister family to the blackbirds and orioles. Other studies suggest it is sister family to a larger super-family of species that includes blackbirds, orioles, wrenthrushes, and Cuban Warblers. For now, at least, it is in its own monotypic family of one species.

Yellow-breasted Chat

(Icteria virens)

IUCN RED LIST STATUS (2021): Least Concern
POPULATION TREND: Decreasing

Phylogenetic studies using DNA have placed Yellow-breasted Chat close to the families of wood-warblers, New World orioles and blackbirds, and sparrows and buntings. However, it has been placed in a monotypic family by itself, *Icteriidea*. **LENGTH:** 7.5 inches. **WINGSPAN:** 9.75 inches. **ADULT:** Plain olive above. Bold white spectacles. Throat and breast are bright yellow. Belly is white with dusky olive sides. Undertail is white. **JUVENILE:** (June–Aug.) Like adults but lacks white spectacles. Has only partial eye crescents. Lower underpart feathers are filamentous, or "fluffy." **VOICE:** Mimics other species but also has a variety of distinct calls, like a raspy *reep* and *chur-chur-chur rhet-rhet-rhet-rhet*. **BEHAVIORS:** Flight is usually direct and steady, often though dense vegetation. When crossing open areas, often flies low. Males during breeding season have a fluttering display flight with the head held up. **HABITAT:** Shrubby open-canopy forest, abandoned fields, clear cuts, and utility corridors. **STATUS:** Common to uncommon summer migrant and summer resident in the Trans-Pecos, Edwards Plateau, and Pineywoods. Uncommon and local summer resident in the western part of the South Texas Brush Country. Uncommon to common migrant in the eastern two-thirds of Texas, becoming rare in the west. Rare winter resident in the Lower Rio Grande Valley north to the central coast. Spring migration is early April–late May, fall mid-August–late October.

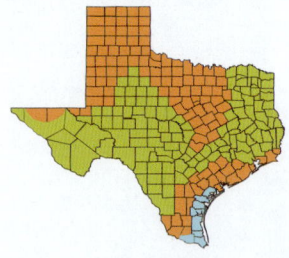

ICTERIDAE
(ORIOLES AND BLACKBIRDS)

The orioles and blackbirds include the meadowlarks, grackles, and oropendolas. So diverse is this group that ornithologists tire of naming the many groups and simply call them the ictrids. Ictrids are found in the New World from the arctic circle to the tip of South America at Tierra del Fuego. They mostly occupy edge and open habitats, grasslands, marshes, forest edges, open forests, savannahs, and even urban parks. Many species have adapted well to human-altered habitats.

Most species are monogamous, but there are polygynous breeders and colonial breeders. The cowbirds have evolved to be nest parasites—they lay their eggs in other species' nests. Blackbirds build cup nests. Meadowlarks nest on the gound and cover the nest with a dome. Orioles and oropendolas build large hanging nests. Cowbirds build no nest at all.

There are 106 species of ictrids. Fifteen are of conservation concern. The marsh-nesting Slender-billed Grackle of Mexico went extinct in the 20th century, as Mexico City swallowed up its habitat. The Montserrat Oriole lost a good portion of its habitat to a recent volcanic event. Twenty-two species of ictrids have been recorded in Texas. Three—Black-vented Oriole, Streak-backed Oriole, and Shiny Cowbird—are rare visitors.

Female

Male

Yellow-headed Blackbird

(Xanthocephalus xanthocephalus)

IUCN RED LIST STATUS (2016): Least Concern
POPULATION TREND: Increasing

A male Yellow-headed Blackbird is unmistakable. Females and first-year males are less showy but still show a bold yellow chest. Birders search for them in large flocks of blackbirds in winter.

LENGTH: 9.5 inches. **WINGSPAN:** 15 inches. **ADULT:** Male has mustard-yellow head and breast. Black lores. Back, wings, and underparts are black. White primary coverts form a white wing patch. Undertail coverts have a small yellow patch. Female has yellow breast and face. Throat is white. Crown, wings, and back are brown. Belly is brown, and there is a small yellow patch on the undertail. **FIRST-YEAR MALE:** Like female with diffuse white patch on primary coverts. **JUVENILE:** (June–Aug.) Overall creamy-colored with darker auricular patch. Wings are dark brown with two bold wingbars. **VOICE:** Call is a harsh, raspy *skraaa skraaa skraaa*. **BEHAVIORS:** Walks when foraging on the ground. Climbs and slides on vegetation to find a perch. Flight is undulating. **HABITAT:** Open agricultural areas and emergent wetlands. Flocks around grain silos. **STATUS:** Common to uncommon migrant in the western half of Texas, becoming less common to rare eastward to the Pineywoods and upper Texas coast. Locally common summer resident in the Panhandle. Rare and irregular winter resident statewide, except the western Trans-Pecos, where they can be abundant in El Paso and Hudspeth counties. Fall migration is mid-July–early November, spring mid-March–mid-May.

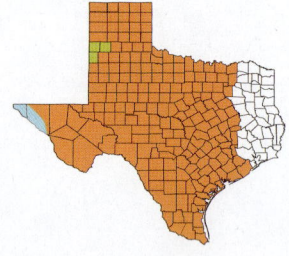

Bobolink

(Dolichonyx oryzivorus)

IUCN RED LIST STATUS (2016): Least Concern
POPULATION TREND: Decreasing

Historically, Bobolinks fed on wild rice during their southbound migration. This is likely the source of an old common name for Bobolink: Rice Bird. Though the Texas rice harvest is timed well for Boblinks, they are not commonly reported in Texas rice fields.

LENGTH: 7 inches. **WINGSPAN:** 11.5 inches.
ADULT BREEDING: (Mar.–Aug.) Male has black crown and face with straw-yellow nape. All black below. Rump and back are white, with two wide black stripes on back. Tail is black. Female has yellow-buff head with pale nape and dark crown. Dark line extends behind the eye. Throat and breast are pale buffy. Fine streaking on flanks. Back and rump are buffy. Back is streaked black. Wings are dark with buffy edges on flight feathers. **NON-BREEDING:** (Aug.–Mar.) Like breeding female but streaking on sides is bolder. Stripe behind eye is thinner and less pronounced. Lores are pale. **VOICE:** Song is a jumbled mechanical series of notes, like a toy robot. **BEHAVIORS:** Walks on the ground. Flights are low and direct. Seems to just disappear into tallgrass. **HABITAT:** In migration, often found in marshes and moderate to tallgrass fields. **STATUS:** Uncommon to rare spring migrant and casual fall migrant in the eastern half of the state west to the Oaks and Prairies region. Spring migration is early April–late May, fall late August–mid-October.

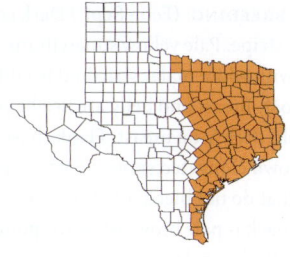

Chihuahuan Meadowlark

(Sturnella lilianae)

IUCN RED LIST STATUS (2022): Not Assessed
POPULATION TREND: Unknown

Chihuahuan Meadowlark was split from
Eastern Meadowlark in 2022. This split was
first proposed in 2016 but was not accepted.
Later work published in 2021 was accepted as
evidence that Chihuahuan Meadowlark is a
distinct species. As a result, though this split
was long anticipated by Texas birders, the status and range of the Chihuahuan Meadowlark
in the state is still unclear.
LENGTH: 9.5 inches. **WINGSPAN:** 14 inches.
ADULT BREEDING: (Feb.–Sept.) Dark crown
and eye stripe. Pale yellow supercilium. Cheek
is gray with white malar strip and bright yellow throat. Breast and belly are bright yellow
with bold dark V on breast. Flanks have bold,
dark brown elongated spots on off-white
flanks that do not bleed into the yellow of the
breast. Back is pale brownish with spotted
pattern. Back is pale and streaked. Tail is dark

with four white outer tail feathers, distinctly
whiter than an Eastern Meadowlark. **NON-
BREEDING:** (Oct.–Jan.) Like breeding but the
supercilium, throat, and flanks are buffy. Eye
stripe and crown have less contrast. The V on
the breast is faded. **VOICE:** A pure whistled
searrooo-searreeee. Call is a rapid *chi-chi-chi-chi-
chi*. **BEHAVIORS:** Walks or runs on the ground
when foraging. Flights alternate periods of
gliding with periods of rapid, stiff wingbeats.
HABITAT: Desert grasslands. **STATUS:** Common resident in the central Trans-Pecos and
southern High Plains west to El Paso County.

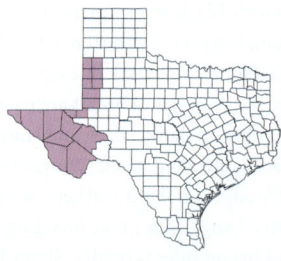

Adult

Immature

Eastern Meadowlark

(Sturnella magna)

IUCN RED LIST STATUS (2020): Near Threatened
POPULATION TREND: Decreasing

Eastern Meadowlark is synonymous with open country and grasslands in eastern North America.
LENGTH: 9.5 inches. **WINGSPAN:** 14 inches.
ADULT BREEDING: (Feb.–Aug.) Dark crown and eye stripe. White supercilium with little or no yellow in front of the eye. Cheek is gray with white malar strip and bright yellow throat. Breast and belly are bright yellow with bold dark V on breast. Flanks have bold dark brown streaks. Back is brownish with spotted pattern. Back is streaked. Tail is dark with white outer tail feathers. **NONBREEDING:** (Sept.–Jan.) Like breeding but supercilium is all yellow. Eye stripe and crown have less contrast. The V on the breast is faded. **VOICE:** Song is a clear, whistled *seeeaaawooo seeee-aaa-weeee*. Call is a mechanical *chp-chp-chp-chp-chp-cho*. **BEHAVIORS:**

Walks or runs on the ground when foraging. In flight, alternates periods of gliding with periods of rapid, stiff wingbeats. **HABITAT:** Native grasslands, alfalfa fields, weedy crop borders, and roadsides. **STATUS:** Uncommon to common summer resident from the eastern Panhandle south through central Texas to the eastern edge of the Edwards Plateau, the coastal prairies, and the southeastern South Texas Brush Country. Common to rare migrant and winter resident from the eastern two-thirds of the Panhandle south through the Rolling Plains, eastern Edwards Plateau, and South Texas Brush Country. Fall migration is early September–late October, spring late February–early April.

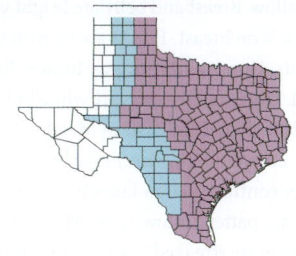

Western Meadowlark

(Sturnella neglecta)

IUCN RED LIST STATUS (2018): Least Concern
POPULATION TREND: Decreasing

Superficially, Western and Eastern Meadowlark appear nearly identical, but there are subtle differences. Western Meadowlarks' flanks are whitish while the flanks of an Eastern Meadowlark are buffy, and the face of a Western Meadowlark has much less contrast than an Eastern's. **LENGTH:** 9.5 inches. **WINGSPAN:** 14.5 inches. **ADULT BREEDING:** (Feb.–Aug.) Gray-brown crown and face. White supercilium behind eye and yellow in front of eye. Throat and malar are bright yellow. Breast and belly are bright yellow with dark V on breast. Flanks are whitish with bold spots. Upper parts are gray-brown. Back is streaked. Outer tail feathers are white but there is less white than in an Eastern Meadowlark. **NONBREEDING:** (Sept.–Jan.) Like breeding but lower contrast; yellow fades to a buffy yellow. Face pattern is low-contrast. Whitish flanks are more streaked than spotted. **VOICE:**

Song is a clear *sleep-wee-err-weedle-weep*. Call is a raspberry-like *chur-brbrbrbrbrbrbbr-brt*. **BEHAVIORS:** Walks or runs on the ground when foraging. Flights alternate periods of gliding with periods of rapid, stiff wingbeats. **HABITAT:** Uses a wide range of grasslands, both native and converted croplands. Less abundant where grass is tall and dense. **STATUS:** Common to uncommon resident in the Panhandle and Rolling Plains. Uncommon to rare migrant and winter resident west to the Pineywoods and south through the South Texas Brush Country. Fall migration is late September–mid-November, spring migration mid-February–early April.

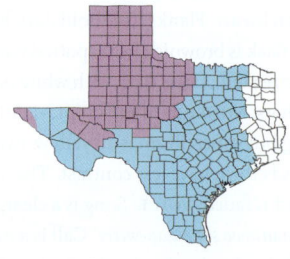

Female

First-summer male

Male

Orchard Oriole

(Icterus spurius)

IUCN RED LIST STATUS (2016): Least Concern
POPULATION TREND: Stable

Orchard Oriole, our smallest oriole, favors open-canopy forests, hedgerows, and shrubby lands. It does well even in large stands of cane on the coast of Texas. The semi-open habitat it favors allows it to interact with humans, and it can be found in suburban and park habitats close to people.
LENGTH: 7.25 inches. **WINGSPAN:** 9.5 inches.
ADULT: Male has black hood and back. Breast and belly are chestnut. Rump is chestnut, tail is black. Wings are dark with a lower, thin white wingbar and a wide chestnut upper wingbar or chevron. Female is yellow-green overall, brighter yellow below. Wings are grayish with two well-defined wingbars.
FIRST-SUMMER MALE: (Feb.–Aug.) Like female with dark face and throat. Wings have more contrast than a female's. **VOICE:** Song is a sweet, bubbling jumble of calls. Call is a piercing *yeert yeert*. **BEHAVIORS:** Flight is light and buoyant. **HABITAT:** Bushy trees, often near water and in fresh and brackish marshes in cattails and tall reeds. **STATUS:** Uncommon to locally common summer resident in the eastern two-thirds of Texas south to the South Texas Brush Country, where they are absent. Uncommon summer resident in the southern Trans-Pecos and along the Rio Grande River. Spring migration is late March–mid-May, fall early July–late September.

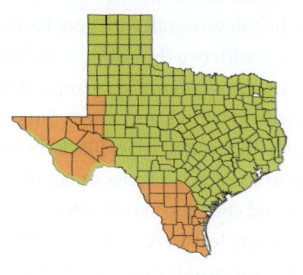

Female

Male

Hooded Oriole

(Icterus cucullatus)

IUCN RED LIST STATUS (2016): Least Concern
POPULATION TREND: Increasing

Hooded Oriole is named for the male's bright orange hood. In Texas, this species is declining while its range is expanding.
LENGTH: 8 inches. **WINGSPAN:** 10.5 inches.
ADULT: Male has orange head and breast with black face and throat; the black face extends around the eye. Rump is orange. Wings, back, and tail are black. Median coverts make a wide white bar or chevron on the wing. Thin white wingbar below the chevron. Female is grayish-yellow overall with gray flanks. Gray wings with two thin white wingbars.
FIRST-SUMMER MALE: (Feb.–Aug.) Yellowish overall with gray back and wings. Single wide wingbar with serrated edge. Black throat and lores. **VOICE:** Song is a series of rapid, short whistles and chatters. **BEHAVIORS:** Forages for insects among leaves; searches leaf undersides and may hang upside down from branches.

Flits across open spaces. **HABITAT:** Breeds in areas of scattered trees; palm, mesquite, and Texas ebony are favorites. **STATUS:** Uncommon to local migrant and summer resident on the southwestern Edwards Plateau, western and southern South Texas Brush Country, Lower Rio Grande Valley, and north to Nueces County on the coastal prairies. Rare in El Paso County and along the Rio Grande River from Presidio County to Terrell County. Rare winter resident in the Lower Rio Grande Valley. Spring migration is late March–mid-May, fall mid-August–early October.

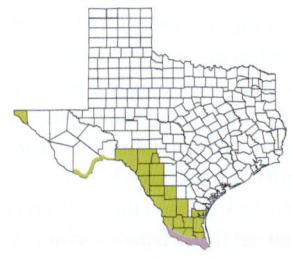

Female (left) and male

Female

Immature male

Bullock's Oriole

(Icterus bullockii)

IUCN RED LIST STATUS (2016): Least Concern
POPULATION TREND: Stable

Bullock's Oriole and Baltimore Oriole overlap in range on the Great Plains.
LENGTH: 9 inches. **WINGSPAN:** 12 inches.
ADULT: Male has black crown, nape, and back. Face is orange with black eye line. Throat is black. Wings are black with wide white wing panel. Rump and tail are orange. Central tail feathers and tip of tail are black in a T pattern. Female has yellow face and throat with dusky crown and dusky shadow of an eye line. Belly is pale gray. Back and wings are gray. Two white wingbars. Tail and rump are yellow with dusky-pale central feathers. **FIRST-YEAR MALE:** Brighter yellow than female with black throat and eye line. Back and wings are dark. **VOICE:** Song is a series of short, chatty whistles and chirps. Call is a raspy series of *check-chack-chack-chack.* **BEHAVIORS:** Makes short hops in trees. Hangs upside down. Flight is strong and direct. **HABITAT:** Areas with mature mesquite. Nests in mesquites, willows, cottonwoods, and pecans. Trees are usually well spaced. **STATUS:** Common to uncommon summer resident in western half of Texas from the Panhandle to the central Rolling Plains, central Edwards Plateau, and the South Texas Brush Country. Common to uncommon migrant in the western half of the state, becoming an uncommon to rare migrant east to the Pineywoods. Spring migration is late March–mid-May, fall late July–late September.

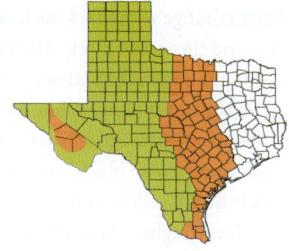

Altamira Oriole

(Icterus gularis)

IUCN RED LIST STATUS (2020): Least Concern
POPULATION TREND: Stable

Altamira Oriole is one of the Texas specialties that birders come here to see. If you want to see it in the United States, you have to visit Texas. It is a relative newcomer to the state; the first record was in 1938, and it was not documented as nesting in Texas until 1951. The population began to grow in the 1960s, but some signs indicate the population in Texas is now in decline.
LENGTH: 10 inches. **WINGSPAN:** 14 inches.
ADULT: Bright orange with black back and wings. Lores and throat are black. Median coverts are orange and make a wide wingbar or chevron on the upper wing. White wingbar below the orange bar. Tail is black. **FIRST YEAR:** Yellow-orange with black lores and throat. Back is dusky yellow-orange. Blackish wings. Yellow wingbar above white wingbar. **JUVENILE:** (June–Sept.) Like first-year

birds but lacks the black face and throat.
VOICE: Song is a series of clear whistles: *two-tweee-too-too-too-too-two*. Call is a raspy *greep-greep-greep-greep*. **BEHAVIORS:** Flight is quick and jerky. Often perches in exposed locations. **HABITAT:** Thorn forest, but revegetated and remnant. **STATUS:** Uncommon resident in the Lower Rio Grande Valley north to Zapata County.

Audubon's Oriole

(Icterus graduacauda)

IUCN RED LIST STATUS (2020): Least Concern
POPULATION TREND: Decreasing

Audubon's Oriole is another of the Texas specialties, one of those species you must see in Texas to see in the United States. This secretive species is not well studied—it even sings from hidden perches. It seems apparent that the population is declining in Texas, both from habitat loss and from cowbird parasitism.

LENGTH: 9.5 inches. **WINGSPAN:** 12 inches. **ADULT:** Yellow with yellow-green back and black wings. Head and upper breast are hooded in black. Tail is black. Greater secondary coverts are tipped in white and make a narrow white wingbar. **JUVENILE:** (June–Aug.) Drab yellow overall. Wings are dark gray with two wingbars. Head and upper breast have dusky wash. As it matures, head and breast are mottled with black. **VOICE:** Song is a series of clear whistles rising and falling, like someone learning to whistle. Call is a series of raspy *raaap-raaap-raaap*. **BEHAVIORS:** Hops and climbs along branches. Flight is strong but short. **HABITAT:** Varied habitats with mature trees: riparian forest, thorn forest, and live oak motts. **STATUS:** Uncommon to rare resident in the South Texas Brush Country north to Goliad and Uvalde counties. Absent from Cameron and the eastern half of Hidalgo County.

Male

Immature

Baltimore Oriole

(Icterus galbula)

IUCN RED LIST STATUS (2018): Least Concern
POPULATION TREND: Stable

Baltimore Orioles are one of the species Texas birders are most likely encounter in a fallout, when large numbers of migrating birds are grounded after a well-timed cold front passes the Texas coast.
LENGTH: 8.75 inches. **WINGSPAN:** 11.5 inches.
ADULT: Male is bright orange below. Black hood and back. Rump and tail are orange. Base of the rump and center of the tail are dark in a Y-like pattern. Upper wings have orange median coverts, making a wide orange bar on the wing. Below that is a white wingbar. Female is dull orange below and variable in the amount of darkness on the hood. Two white wingbars. Tail and rump are dull orange.
FIRST FALL: (Aug.–Mar.) Dull grayish-orange. Dark wings with two white wingbars. **VOICE:** Song is a series of rapid whistles: *twedi twedi twedi tew tidew*. Call is a harsh *chep-chep*

chep-chep-chep. **BEHAVIORS:** Makes short hops using the wings when foraging. Hangs upside down. Strong flier. **HABITAT:** Breeds in open deciduous woodlands, often streamside. In migration, favors open woodlands, edges, hedgerows, and parklands. **STATUS:** Uncommon summer resident in the eastern third of the Panhandle and uncommon to rare summer resident eastward across the northern Rolling Plains and north-central Texas. Common to uncommon migrant in the eastern half of Texas. Spring migration is early April–late May, fall late July–mid-November.

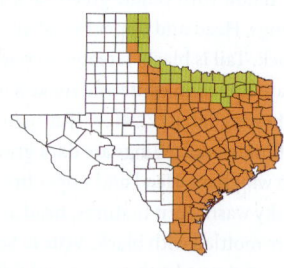

Male

Scott's Oriole

(Icterus parisorum)

IUCN RED LIST STATUS (2019): Least Concern
POPULATION TREND: Decreasing

Scott's Oriole was named by General Darius Couch, who named it for his superior officer during the Mexican War, General Winfield Scott. Scott's Orioles are one of earliest arriving migrants in Texas, and many pairs complete two broods, with at least some attempting a third.
LENGTH: 9 inches. **WINGSPAN:** 12.5 inches.
ADULT: Black hood and breast. Back is black. Underwing coverts, belly, and undertail coverts are yellow. Rump is yellow, tail is black. Lesser coverts are yellow, making a yellow patch on the shoulder. Two white wingbars. Female is like males but variable amounts of black on face and breast. Head, back, and rump are olive. **FIRST YEAR:** Olive overall. Wings are dark with two jagged wingbars. **VOICE:** Song is a series of gurgling, clear whistles: *wheato-wheato-wheato weedle wheat.*

Call is a raspy *chep chep.* **BEHAVIORS:** When foraging in trees and bushes, moves along limbs by climbing, not using its wings. Flies long distances to forage low across the landscape. **HABITAT:** Elevated arid slopes and desert-facing slopes of mountains. **STATUS:** Uncommon to common migrant and summer resident in the Trans-Pecos east to the southern Edwards Plateau. Spring migration is early February–mid-May, fall late August–early October.

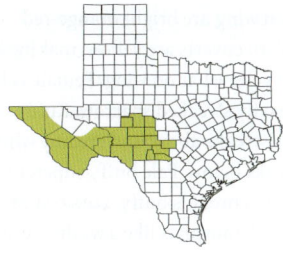

Male

Female

Red-winged Blackbird

(Agelaius phoeniceus)

IUCN RED LIST STATUS (2018): Least Concern
POPULATION TREND: Decreasing

Red-winged Blackbird is one of the most abundant species in Texas. Expected statewide, they are well known for the huge roosts they form, often with other blackbird species. The damage these flocks can do to crops has resulted in considerable efforts at blackbird control, so much so that humans have become a major source of mortaility for the species. **LENGTH:** 8.75 inches. **WINGSPAN:** 13 inches. **ADULT:** Male is dull black overall. Lesser coverts on wing are bright orange-red and median coverts are yellow, making a red-over-yellow bar on wing. Female is like a large, heavily streaked sparrow. Rufous-brown with pale streaks above and creamy white with dark streaks below. Pale buffy supercilium and malar. Throat is buffy. **FIRST-SUMMER MALE:** Dark morph is like a washed-out version of adult males. Lacks the red and yellow coverts on the wing. Hint of a pale supercilium and malar. Light morph is like adult female but has a darker background color. Pale red-over-yellow patch on wing. Whitish supercilium and malar. **VOICE:** Song is a raspy, emphatic *koo-kaa reeeeeeee*. Call is *chup chup*. **BEHAVIORS:** Forages on the ground; slides on small branches. Strong flier. Forms large flocks outside of breeding. **HABITAT:** Breeds in a variety of wetlands habitats. **STATUS:** Abundant to locally uncommon migrant and resident in Texas. Fall migration is late August–mid-November, spring late February–early May.

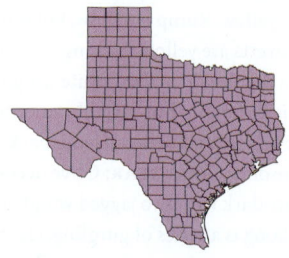

Male

Bronzed Cowbird

(Molothrus aeneus)

IUCN RED LIST STATUS (2017): Least Concern
POPULATION TREND: Decreasing

The Bronzed Cowbird, like the closely related Brown-headed Cowbird, is a brood parasite. Females lay their eggs in other species' nests to be reared. More than 100 species have been parasitized by Bronzed Cowbirds. Being larger than Brown-headed Cowbirds, they select larger hosts to parasitize and seem to favor orioles as hosts.

LENGTH: 8.75 inches. **WINGSPAN:** 14 inches.
ADULT: Glossy black with bronzy sheen. Eye has bright red iris. Wings are a dark, glossy blue that contrasts with the body. Nape has a thick ruff that is often puffed, giving a helmeted look. Bill is heavy, almost grosbeak-like. Eastern female is dull bronzy-black overall with red iris. Western female is dull olive-clay-colored with faint scalloped pattern in the breast. Iris is red. Wings have a slight blue sheen and contrast with the body. **VOICE:** Song is a whistled, high-pitched, rapid trill. Call is a rapid, rattled *cheb-cheb-cheb-cheb-cheb*.
BEHAVIORS: Usually walks on the ground. **HABITAT:** Open habitats, open fields, pastures, agricultural areas. **STATUS:** Common to uncommon summer resident in the southwestern half of Texas. Locally common to uncommon resident in the South Texas Brush Country and coastal prairies north to Galveston Bay. Spring migration is mid-March–mid-May, fall mid-July–early September.

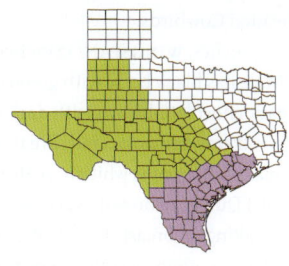

Male

Brown-headed Cowbird

(Molothrus ater)

IUCN RED LIST STATUS (2018): Least Concern
POPULATION TREND: Decreasing

Brown-headed Cowbird is the best-known brood parasite in North America. Evolved to follow North American bison on the shortgrass prairie, man's development of North America literally opened up vast new areas and new hosts for Brown-headed Cowbirds. Females wander widely and may lay as many as forty eggs in a season in other species' nests. In Texas, both the endangered Golden-cheeked Warbler and Black-capped Vireo are regular hosts of Brown-headed Cowbirds.
LENGTH: 7.5 inches. **WINGSPAN:** 12 inches.
ADULT: Male has black body with green gloss. Head is dark chocolate-brown with dark eyes. Female is dull gray-brown overall. Breast is faintly streaked. Throat is whitish. **JUVENILE:** (June–Sept.) Dull gray-brown overall with distinct streaking on breast. Back feathers have light fringes, giving a scaled appearance.

VOICE: Song is a gurgling note followed by a high-pitched whistle: *glug-glip screeeeeeeeeee*. Call is a piercing whistled *screeeeeeee*. **BEHAVIORS:** Usually feeds by walking on the ground. Flight is quick, matching grackles it often flocks with. **HABITAT:** Habitats with low or scattered trees in grasslands. In winter, associates in large numbers with other blackbird species in flocks. **STATUS:** Common to abundant summer resident throughout the state. Uncommon in the South Texas Brush Country. In winter, mostly withdraws from the Trans-Pecos, Edwards Plateau, and Pineywoods. Spring migration is mid-February–mid-April, fall late August–early October.

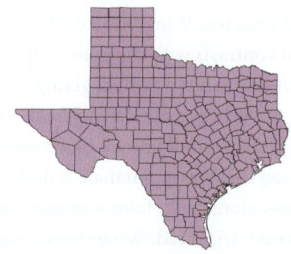

Rusty Blackbird

(Euphagus carolinus)

IUCN RED LIST STATUS (2020): Vulnerable
POPULATION TREND: Decreasing

Rusty Blackbirds are one of North America's most rapidly declining species. In the 1950s, roosts up to 5000 birds were reported wintering in Texas. Now reports of more than a few dozen birds are noteworthy. Loss of wooded wetlands where they winter, poisoning of other blackbirds on winter roosts, and contaminants on breeding grounds have all been identified as contributing factors in the decline.

LENGTH: 9 inches. **WINGSPAN:** 14 inches.
ADULT BREEDING: (Jan.–Aug.) Male is somewhat glossy black overall, with purplish glossy head. Iris is pale. Bill is thin and slightly decurved. Female is brown overall with pale throat and supercilium. Dark lores and dusky auricular patch. Tail is dark. **NONBREEDING:** (Aug.–Mar.) Dark feathers have bronzy-rusty tips. Mottled rusty overall in varying amounts.

Female is like breeding female but lighter overall. Obvious gray rump and tail. **VOICE:** Song is a soft gurgling followed by a piercing whistle: *chur-chur TEEEEEE*. Call is *chep chep*. **BEHAVIORS:** Forages on the ground walking and running. Wades in water sometimes up to the belly. **HABITAT:** Swamps, wet woodlands, and pond edges. Usually does not associate with other blackbirds. **STATUS:** Uncommon to rare migrant and winter resident in the eastern third of Texas. Rare migrant and winter resident in the Rolling Plains and eastern Panhandle. Fall migration is mid-November–mid-December, spring early February–mid-March.

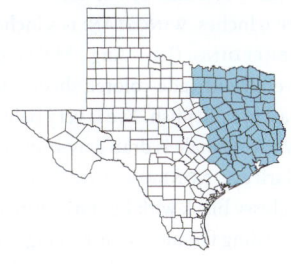

Male

Female

Brewer's Blackbird

(Euphagus cyanocephalus)

IUCN RED LIST STATUS (2016): Least Concern
POPULATION TREND: Decreasing

Brewer's Blackbird has been known locally by some as the Satin Blackbird, which captures perfectly the glossy, silky look of the males. Prior to the early 1900s, Brewer's Blackbirds were not known to breed east of Minnesota. Beginning about 1914 and continuing for about the next four decades, their breeding range expanded about 1000 miles eastward as human modification of the landscape allowed the species to colonize new areas.
LENGTH: 9 inches. **WINGSPAN:** 15.5 inches.
ADULT BREEDING: (Jan.–Aug.) Male is glossy black overall with jade-colored sheen. Head and breast are purplish. Eye is very pale yellow to white. Female is drab gray-brown. Most have a dark eye. **NONBREEDING:** (Aug.–Jan.) Male is glossy black with light iris. Female is like breeding females. **VOICE:** Song is a screeching *ska-reeeeeeee*. Call is a *check check*.

BEHAVIORS: Walks with head bobbing forward and back. Flight is an undulating series of flaps and glides. **HABITAT:** In Texas, usually is in flocks and found in open areas, lawns, agricultural fields, and feedlots. **STATUS:** Common to locally abundant migrant and winter resident throughout Texas. Uncommon to rare in the Pineywoods, South Texas Brush Country, and the Lower Rio Grande Valley. Fall migration is mid-September–mid-November, spring late March–mid-May.

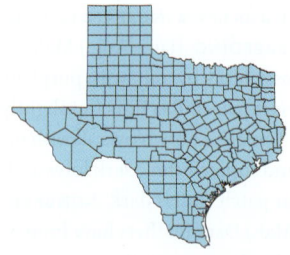

Adult

Immature

Common Grackle

(Quiscalus quiscula)

IUCN RED LIST STATUS (2018): Near Threatened

POPULATION TREND: Decreasing

Common Grackle is a bird that loves edge habitats and therefore does well around humans. Agricultural development has allowed it to become one of the most abundant species in North America. Common Grackles inflict one of the greatest economic impacts on agriculture, by consuming agricultural grains, particularly corn. Its conservation status is confusing. The IUCN lists it as near threatened due to the eastern population dropping rapidly, apparently due to control measures. **LENGTH:** 12.5 inches. **WINGSPAN:** 17 inches. **ADULT:** Male has blackish body with bronze iridescence. Head has purple iridescence. Iris is pale. Female is like males but less glossy and browner in color. **JUVENILE:** (June–Sept.) Dull dark brown overall. **VOICE:** Song is a squeaky *shir-reeek*. Call is a harsh *check check check*.

BEHAVIORS: Walks on the ground. Flight is level without undulation. **HABITAT:** Wide variety of open and partially open habitats. Does not use mature woodlands. **STATUS:** Common to uncommon summer resident in the eastern two-thirds of the state west to the High Plains and central Edwards Plateau, south to the Guadalupe River drainage on the coastal prairies. Common to abundant migrant and winter resident in the eastern third of the state, becoming less common west to the Rolling Plains and Edwards Plateau. Fall migration is mid-September–early November, spring mid-February–late May.

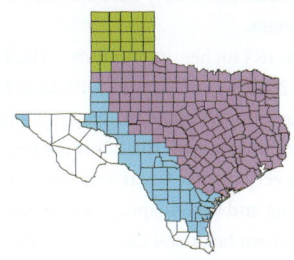

Male

Female

Boat-tailed Grackle

(Quiscalus major)

IUCN RED LIST STATUS (2016): Near Threatened
POPULATION TREND: Decreasing

Boat-tailed Grackle and Great-tailed Grackle were long considered one species, though the more western population of Great-tailed Grackle was isolated from the more eastern Boat-tailed Grackle. By the 1970s, Great-tailed Grackles had expanded their range to come in contact with Boat-tailed Grackles on the Texas coast. The two forms maintained reproductive isolation and eventually were elevated to separate species.
LENGTH: 16.5 inches. **WINGSPAN:** 23 inches.
ADULT: Male is dark overall, with bluish iridescence on the body. In Texas, all birds have dark brown eyes. Head is more rounded than the very similar Great-tailed Grackle. Tail is long and keel-shaped. Female has rich rufous-brown head and dark wings. Tail is relatively flat compared to males. **VOICE:**

Song is an emphatic *ria-ria-ria-ria-REET*.
BEHAVIORS: Gregarious year-round. Walks on the ground. Flight is direct and does not undulate. **HABITAT:** Mostly found in fresh and brackish marshes. Rarely found far from the coast. **STATUS:** Uncommon to abundant resident from Jefferson to Aransas County on the coastal prairies. Most common on the upper Texas coast, it is becoming less common on the central coast.

Male

Female

Great-tailed Grackle

(Quiscalus mexicanus)

IUCN RED LIST STATUS (2018): Least Concern
POPULATION TREND: Stable

Long considered a subspecies of Boat-tailed Grackle, the Great-tailed Grackle was elevated to full species in 1973. Texas is one of the few places the two coexist. Luckily, they are easily separated in the field. On the US East Coast, Boat-tailed Grackles all have pale eyes like Great-tailed Grackles. On the western Gulf Coast, Boat-tailed Grackle have dark eyes in contrast to the light yellow eyes of Great-tailed Grackles.

LENGTH: 18 inches. **WINGSPAN:** 23 inches.
ADULT: Male has uniform purple-blue iridescence on body. Head is relatively flat compared to Boat-tailed Grackles. Iris is bright yellow. Tail is large and keel-shaped. Female is dark brown above and paler brown below. Iris is yellow. Tail is flat compared to male tails. **JUVENILE:** (May–Aug.) Like females but iris is dark, usually becoming yellow

by October, but a dark iris can be retained for the first year. **VOICE:** Sound is a raucous *aaaa-reeep aaaa-reeep*. Call is a *chep-chep chep-chep*. **BEHAVIORS:** Walks on the ground. Flight is direct and not undulating. **HABITAT:** Wide variety of open areas with scattered trees. Forms large roosts in urban areas during winter. **STATUS:** Abundant resident in the southern half of the state. Uncommon to rare in the Pineywoods, primarily in urban areas. Abundant but limited to urban areas in the northern half of the state.

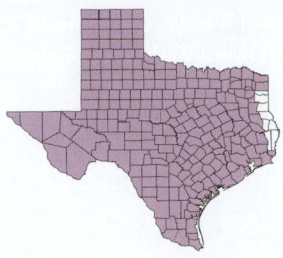

PARULIDAE
(NEW WORLD WARBLERS)

The New World warblers, often shortened to just warblers, are a New World–exclusive family. Warblers are found from the arctic circle to the tropics of South America. It appears that the family originated in North America and spread through Central America to South America. Some species, like the Blackpoll Warbler make one of the longest migrations of any songbird, from the boreal forest to central South America. Early ornithologists named the family after the Old World warblers because they thought them similar, but Old World warblers are much duller in appearance than their New World counterparts and are not closely related.

The warblers have been called the butterflies of the bird world—no family is more prized by birdwatchers. Many make long-distance migrations, passing over the Gulf of Mexico to make landfall on the Gulf Coast of the United States, and Texas in particular. With a fair southern tailwind and blue skies, most pass over the coast unobserved and disperse inland to rest and refuel after a nonstop transgulf flight of close to twenty-four hours. On those occasions when a flight of warblers encounters a cold front and stormy north wind, many will barely reach the coast, where they make an emergency stop, stream into small patches of wooded habitat known as migrant traps, and delight birders in groups of sometimes hundreds of warblers—including twenty or even thirty species—in a few hours.

There are 115 species of warblers. Twenty-four are of conservation concern. Two, Bachman's Warbler and Semper's Warbler, are already considered extinct. Fifty-three species have been recorded in Texas, nearly half of the total species of warblers. Seven—Crescent-chested Warbler, Connecticut Warbler, Gray-crowned Yellowthroat, Fan-tailed Warbler, Rufous-capped Warbler, Golden-crowned Warbler, and Slate-throated Redstart—are considered rare visitors to Texas.

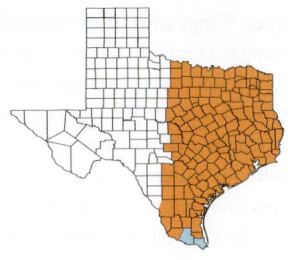

Ovenbird

(Seiurus aurocapilla)

IUCN RED LIST STATUS (2021): Least Concern
POPULATION TREND: Stable

Ovenbirds are a common migrant in the eastern half of Texas, but even though there are nearby populations in Oklahoma and Arkansas, it is not known to breed in Texas. In our state it is most often encountered in the coastal migrant traps during migration. **LENGTH:** 6 inches. **WINGSPAN:** 9.5 inches. **ADULT:** Olive-brown above with bold white eye ring. Two black crown stripes bordering a tawny crown. White below with black spots. **VOICE:** Two-syllable repeating phrase, rising in volume: *chertee chertee CHERtee CHERtee CHERTEE CHERTEE.* Chip note is a loud, wet *CHET CHET CHET.* **BEHAVIORS:** Usually walks on the ground. Tail isn't bobbed up and down but often held high with the wings drooping. **HABITAT:** Generally found in places with broadleaf forest and understory but bare ground to walk on. **STATUS:** Uncommon to common migrant through the eastern half of the state. Mostly encountered on the coast in migration. Rare to uncommon winter resident in the Lower Rio Grande Valley, becoming rare northward along the coast. Spring migration is late March–mid-May, fall early August–early October.

Worm-eating Warbler

(Helmitheros vermivorum)

IUCN RED LIST STATUS (2021): Least Concern
POPULATION TREND: Increasing

Worm-eating Warbler is primarily a migrant but an uncommon nesting bird in the eastern third of Texas. Worm-eating Warblers winter on the eastern slope of Central America and in the Caribbean. They are well known for probing clumps of dead leaves for caterpillars, which are their primary food and were once called "worms"—hence their common name. **LENGTH:** 5.25 inches. **WINGSPAN:** 8.5 inches. **ADULT:** Buffy-olive overall. Black and buffy head stripes. Wings and tail are unmarked. **VOICE:** Song is a rapid, flat, insect-like buzz. Chip note is a soft *tsip tsip*. **BEHAVIORS:** Hops and flits from branch to branch. Occasionally creeps up trunks and branches or hops on the ground without walking. Flight is direct in the subcanopy. Known for probing in clumps of dead leaves for caterpillars. **HABITAT:** Primarily nests in tracts of mature deciduous forest with hillsides. In migration, mostly encountered in coastal migrant traps. **STATUS:** Very rare and local summer resident in the central and southern Pineywoods. Uncommon to common migrant in the eastern third of Texas but mostly encountered on the coast. Spring migrants found mid-March–early May. Fall migration is late August–early November.

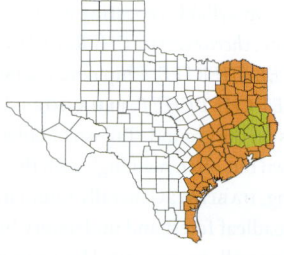

Louisiana Waterthrush

(Parkesia motacilla)

IUCN RED LIST STATUS (2021): Least Concern
POPULATION TREND: Increasing

The Louisiana Waterthrush is one of the earliest-arriving wood-warblers in Texas, with some birds arriving in March. While its plumage is crypic and hard to spot, its ringing call carries long distances on the breeding ground. Because it depends on clear running streams, surveys of Lousiana Waterthrushes have been used to measure water quality in a watershed. **LENGTH:** 6 inches. **WINGSPAN:** 10 inches. **ADULT:** Dark brown above with bold white supercilium that usually flares at the rear. Wide dark eye stripe and lighter cheek. Throat is usually white and unspotted. Usually white below with brown streaks. Rear flanks are buffy. **VOICE:** Sound is usually three descending notes followed by a couple of jumbled ones: *SUEEW SUEEW SUEEW sweeto sweeto.* Chip note is a wet *chet chet.* **BEHAVIORS:** Mostly walks while constantly bobbing its tail.

HABITAT: Breeds along clear flowing perennial streams in closed-canopy deciduous forest. **STATUS:** Rare to locally uncommon summer resident in the Pineywoods, eastern part of the Blackland Prairies, and the Brazos River drainage north to Hood County and west to the central Edwards Plateau. Uncommon migrant through the eastern two-thirds of Texas. Spring migration is early March–late April, fall migration early June–late September.

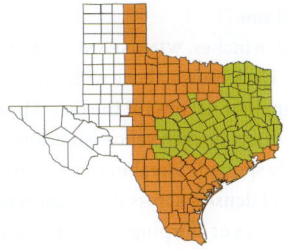

Northern Waterthrush

(Parkesia noveboracensis)

IUCN RED LIST STATUS (2021): Least Concern
POPULATION TREND: Increasing

Separating the Northern Waterthrush from the very similar Louisiana Waterthrush is one of the quintessential problems for Texas birders. The field marks separating the two species are subtle and take some practice in judging the shape of the supercilium and how spotted the throat is. Even though they have different migration windows, there is considerable overlap in their presence in the state, and birdwatchers are often left unsure which species they just saw.

LENGTH: 6 inches. **WINGSPAN:** 9.5 inches.
ADULT: Dark brown above with white supercilium that narrows at the rear. Most are slightly dingy or even yellowish below. Throat is usually spotted. Streaking and spotted below and densest across the breast. **VOICE:** Song is series of chirping notes, accelerating at the end: *sweet sweet sweet swear swear swear swit swit swit*. Chip note is a clear *chink chink chink*. **BEHAVIORS:** Walks on the ground; constantly bobs tail. **HABITAT:** In migration, found in thick cover along streams, in marshes, and by stagnant pools. In winter, found in wooded habitats along the coast with standing water. **STATUS:** Common to uncommon migrant in the eastern half of the state. Uncommon to rare west through the Trans-Pecos. Rare winter visitor along the coast. Spring migration is early April–late May, fall mid-August–mid-October.

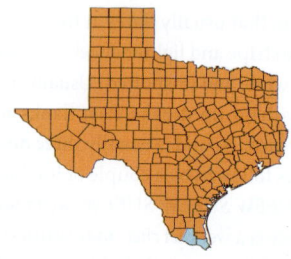

Golden-winged Warbler

(Vermivora chrysoptera)

IUCN RED LIST STATUS (2018): Near Threatened
POPULATION TREND: Decreasing

Golden-winged Warbler has declined rapidly in the southern part of its range. In the past, agricultural land-use practices favored it, and its population and range expanded. In recent decades, however, forest has regenerated and a close relative, the Blue-winged Warbler, has moved in. In many parts of the Golden-winged Warbler's former range, the Blue-winged Warbler has completely replaced it.

LENGTH: 4.75 inches. **WINGSPAN:** 7.5 inches. **ADULT:** Male has black throat and auriculars on white face. Crown is yellow to the front and gray to the rear. Back is gray. Pale gray below with bright golden-yellow wing panel. Large white corners on the gray tail. Female is like males but face is patterned in gray, not black. Crown is greenish in front. **VOICE:** Song is a high, buzzing *zeee zaa-zaa*; the first note is higher. Chip note is a *chet*. **BEHAVIORS:** Forages with head high in smaller trees, often hanging upside down to get at prey in clumps of leaves. **HABITAT:** Migration habitat is poorly documented, but generally forest edge and tall second-growth trees. **STATUS:** Uncommon to rare spring migrant and rare to very rare fall migrant through the eastern third of Texas. Golden-winged Warblers are most often encountered in the coastal migrant traps. Spring migration is mid-April–mid-May, fall late August–mid-October.

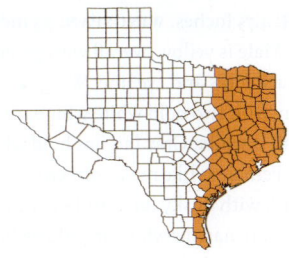

Blue-winged Warbler

(Vermivora cyanoptera)

IUCN RED LIST STATUS (2021): Least Concern
POPULATION TREND: Decreasing

Blue-winged Warblers are closely related to the near-threatened Golden-winged Warbler. Both expanded their range into the northeast United States, but the Blue-winged has faired better, often replacing Golden-winged Warblers. The two species often hybridize and have fertile offspring. Two of the common hybrids were for a time considered their own species: Lawrence's Warbler and Brewster's Warbler.

LENGTH: 4.75 inches. **WINGSPAN:** 7.5 inches.
ADULT: Male is yellow overall with greenish back. Black lores and eye line. Wings are blue-gray with two bold white wingbars. Undertail is white. Gray tail has bold white corners. Female is like males but crown is greenish with dusky lores and eye line. Wingbars are narrow white or yellowish.
VOICE: Song is a buzzing *zeee-brrrrrrrrr*. Call

is *chep chep*. **BEHAVIORS:** More deliberate and vireo-like than other wood-warblers. Sometimes hangs upside down. Short, flitting flights from branch to branch. **HABITAT:** In migration, uses scrub, thorn, and lowland riparian forests. **STATUS:** Common to uncommon spring migrant and uncommon fall migrant in the eastern third of the state. Spring migration is late March–mid-May, fall mid-August to mid-October.

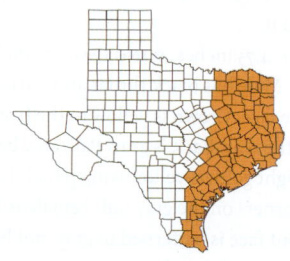

Female

Male

Black-and-white Warbler

(Mniotilta varia)

IUCN RED LIST STATUS (2021): Least Concern
POPULATION TREND: Stable

The charismatic Black-and-white Warbler is one of the most distinctive of the wood-warblers. Its habit of foraging on the trunks and large branches of trees makes it easy to observe. The genus name *Mniotilta* means "moss-plucking," and is a reference to this foraging behavior.

LENGTH: 5.25 inches. **WINGSPAN:** 8.25 inches. **ADULT:** Male is black-and-white-streaked overall. Throat is unevenly black. Auriculars are also black. Undertail is spotted black. Female is black-and-white-streaked overall with pale throat and paler collar. Auriculars are pale. Gray-streaked below with black-spotted undertail. **FIRST WINTER:** (Aug–Mar.) Like females but pale buffy below. **VOICE:** Song is a very high-pitched *weeza weeza weeza weeza*. Chip note is a loud *cheet cheet*. **BEHAVIORS:** Hops rather than walks. Moves methodically over trunks and branches while foraging, moving easily in all directions including heading down the trunk and upside down on branches. **HABITAT:** Mature and second-growth deciduous and mixed-deciduous forests. **STATUS:** Rare to common summer resident in the Pineywoods, Oaks and Prairies, and west through the southern Edwards Plateau where it is local. Uncommon to common migrant through most of the state. Uncommon winter resident on the coastal prairies and the Lower Rio Grande Valley. Spring migration is early March–late May, fall mid-July–late October.

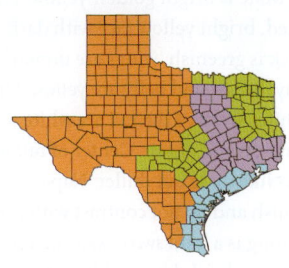

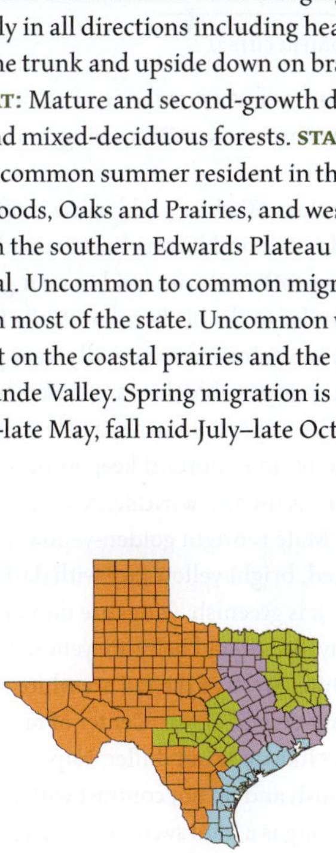

With orange crown (few)

Prothonotary Warbler

(Protonotaria citrea)

IUCN RED LIST STATUS (2021): Least Concern
POPULATION TREND: Decreasing

Prothonotary Warbler's odd name is often attributed to the bird having been named for the Vatican's prothonotaries, said to wear golden robes similar to the bird's color. While this makes a good story, they actually wear purple or black. It is possible the name refers to some other office of prothontaries, a title that is used to refer to important keepers of records. **LENGTH:** 5.5 inches. **WINGSPAN:** 8.75 inches. **ADULT:** Male is bright golden-yellow. Plain, unmarked, bright yellow face with dark eyes. Back is greenish, wings are unmarked blue-gray. Breast and belly are yellow. Undertail is white. Webs of the tail are white with dark band on end of tail. Center of tail is gray. Female is like male but duller. Nape and crown are greenish and do not contrast with the back. **VOICE:** Song is a loud *sweet sweet sweet so-sweet*. Chip note is a loud *chit chit chit*. **BEHAVIORS:** Hops along branches and twigs in trees and shrubs, on logs, and on the ground. Flies within trees. **HABITAT:** Usually near water in wooded areas with cavities for nesting. Nest is often over water. **STATUS:** Rare to common summer resident in the eastern half of the state south to the coastal prairies. Uncommon to common migrant in the eastern half of the state. Spring migration is mid-March–late May, fall late July–late October.

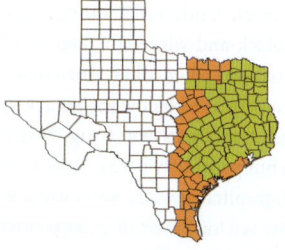

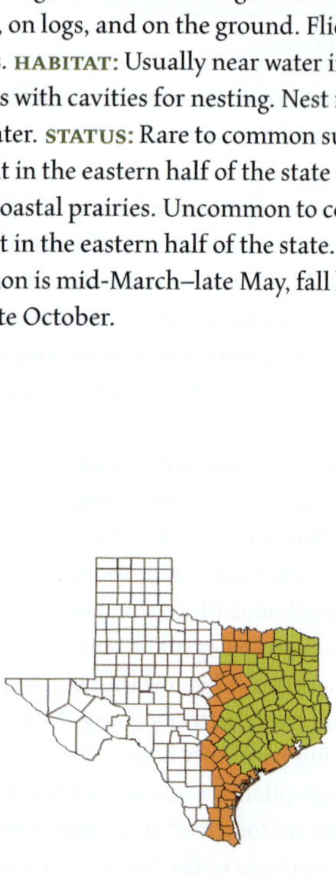

Swainson's Warbler

(Limnothlypis swainsonii)

IUCN RED LIST STATUS (2021): Least Concern
POPULATION TREND: Increasing

Swainson's Warbler is one the most secretive bird species in North America. Its preference for dense, shaded understory makes it difficult to observe and study. It was named in 1833 by John James Audubon for ornithologist William Swainson. Evidence suggests that the orginal discoverer of the species was John Abbot, a naturalist from Georgia who made drawing and watercolors of Swainson's Warbler as early as 1801.

LENGTH: 5.5 inches. **WINGSPAN:** 9 inches.
ADULT: Brown above and pale dusky below. Face is pale with dark eye line. Undertail is white. Crown is rufous-tinted. Bill is long.
VOICE: Song is a clear *swee swee-SIS sis-sis sew*. Chip note is a spaced *chit chit*. **BEHAVIORS:** Steps rapidly on the ground. May "patter" or rapidly stomp feet to flush prey. Flies direct from perch to perch; may cross its territory in a single flight. **HABITAT:** Variety of habitats with shaded and dense understory, abundant leaf litter, and little herbaceous ground cover. **STATUS:** Rare to locally uncommon summer resident in the Pineywoods; rare and local south to Aransas and Victoria counties and west to Bastrop County. Rare spring migrant in the eastern third of the state. Uncommon migrant at coastal migration sites. Spring migration is late March–mid-April, fall late August–mid-October.

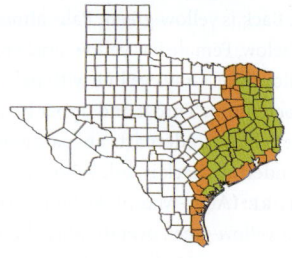

Female

Male

Tennessee Warbler

(Leiothlypis peregrina)

IUCN RED LIST STATUS (2021): Least Concern
POPULATION TREND: Decreasing

Tennessee Warbler was named for the location the first specimen was collected, along the Cumberland River in Tennessee. This bird was a migrant, though, because Tennessee Warbler nests entirely in the boreal forest of Canada and the most northern parts of the United States.
LENGTH: 4.75 inches. **WINGSPAN:** 7.75 inches.
ADULT BREEDING: (Apr.–Aug.) Male has gray head with gray eye line and pale supercilium. Back is yellow-green. Pale, almost white, below. Female has yellow-gray crown and yellow face. Gray eye line with pale yellow supercilium. Throat and upper breast are pale yellow. Underparts are pale gray with white undertail. Back is dull yellow-green.
IMMATURE: (Aug.–Mar.) Like female but brighter yellow-green overall. May show pale narrow wingbars. Undertail, very pale yellow

to white, is always lighter than the breast.
VOICE: Song is a high-pitched *tipit tipit tipit tipit tipit swit swit swit titititititititititit*. Chip call is a high-pitched *cheat cheat cheat*. **BEHAVIORS:** Primarily forages by hopping within foliage. Uses both flight and hops to access leaf tips. Sometimes hawks and hovers while foraging. **HABITAT:** All types of woodlands during migration. **STATUS:** Uncommon to common spring migrant and uncommon to rare fall migrant in the eastern half of Texas. Sometimes abundant on the upper Texas coast and Pineywoods. Spring migration is late March–mid-May, fall late August–late October.

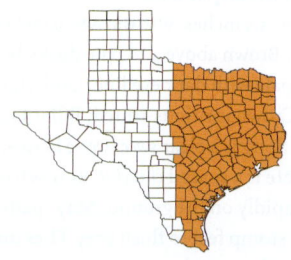

Orange-crowned Warbler

(Leiothlypis celata)

IUCN RED LIST STATUS (2021): Least Concern
POPULATION TREND: Decreasing

Many birders consider Orange-crowned Warbler one of the most misnamed birds in North America. Few have ever seen its namesake orange crown patch. It is also often called "the bird with no field marks." It does have field marks, but they are subtle and better appreciated with experience. Within a narrow range, field marks vary greatly, with the base color ranging from a light brownish to yellow to an almost light green.

LENGTH: 5 inches. **WINGSPAN:** 7.25 inches.
ADULT: Pale olive overall with faint, slightly darker streaking on breast. Faint eye line. Pale broken eye ring. Adults have an orange crown patch that is difficult to see and almost always concealed. Undertail coverts are always yellowish. **IMMATURE:** (July–Mar.) Like adults but generally grayer. Eastern birds often have a gray head, while birds from the west are more yellow. **VOICE:** Song is a clear trill: *tititititititi-wilwilwilwil.* Chip note is *chit chit chit.* **BEHAVIORS:** Hops within the foliage to forage. Flights are direct. Will hover or hawk for insects. **HABITAT:** Wide variety of brushy and dense foliage, and deciduous trees. **STATUS:** Uncommon to common winter resident in most of Texas. Casual north of Lubbock in the Panhandle as a winter resident. Uncommon to abundant migrant statewide. Fall migration is late August–early October, spring late March–mid-May.

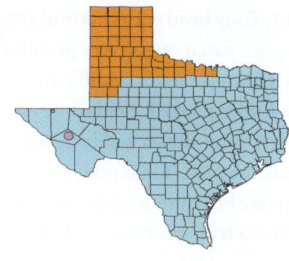

Note chestnut crown

Note yellow undertail

Colima Warbler

(Leiothlypis crissalis)

IUCN RED LIST STATUS (2020): Least Concern
POPULATION TREND: Stable

The Colima Warbler was discovered to be breeding in the Chisos Mountains in Big Bend National Park in 1932, and this is still their only known breeding site in Texas and the United States. Birders who want to see this bird in the United States must make the long hike up into the Chisos Mountains. It occurs over a larger range in the mountains of north-central Mexico.

LENGTH: 5.5 inches. **WINGSPAN:** 7.75 inches. **ADULT:** Gray head with chestnut crown patch. Crown patch can be very prominent when the crown is raised, or difficult to see when the bird is relaxed. Bold white eye ring. Back is olive-gray. Breast and belly are gray, flanks are brownish. Tail is gray. Undertail and rump are bright mustard-yellow. **VOICE:** Song is a long trill, often with a slurred note at the end: *tetetetetetetetetetete-sew*. Chip note is a soft *cheep cheep*. **BEHAVIORS:** Hops both on the ground and in tree branches while foraging. Often wags tail while foraging. **HABITAT:** Breeds in mixed oak-piñon-juniper forest with bunch grasses. **STATUS:** Extremely local and restricted to the Chisos Mountains in Big Bend National Park. Spring migrants arrive in the Chisos Mountains in mid-April and depart by early September.

Lucy's Warbler

(Leiothlypis luciae)

IUCN RED LIST STATUS (2021): Least Concern
POPULATION TREND: Increasing

Lucy's Warbler is named for Lucy Hunter Baird, daughter of ornithologist Spencer Baird. It is one of only two North American wood-warblers that breed in tree cavities. It is also the smallest of the warblers, only slightly larger than the very similar-looking Blue-gray Gnatcatcher.
LENGTH: 4.25 inches. **WINGSPAN:** 7 inches. **ADULT:** Male is gray overall with rusty cap. Pale whitish lores and dark eye on plain face. Rump is dark rusty. Undertail is white. Female is like males but less rusty on cap. Face is plainer. Underparts are tinged with buffy. **JUVENILE:** (May–July) Like adult females but lacking the rusty colors. Rump is pale tawny. Faint wingbars. **VOICE:** Call is a rapid, clear trill: *titititititititititi-swe-swe-swe-swe-swe*. Call is a *tep tep tep*. **BEHAVIORS:** Hops between branches while foraging. Flights are short and direct. **HABITAT:** Breeds in dense lowland riparian mesquite woodlands. Trees must be large enough for nesting cavities. **STATUS:** Rare to locally uncommon summer resident in arid woodlands along the Rio Grande River from Hudspeth County to western Brewster County. Spring migrants arrive in mid-March and are present until late August.

Nashville Warbler

(Leiothlypis ruficapilla)

IUCN RED LIST STATUS (2021): Least Concern
POPULATION TREND: Stable

Nashville Warbler is named for a specimen collected near Nashville, Tennessee in 1811. There are two breeding populations in North America, an eastern and a western. The western subspecies was once regarded as a separate species and was known as Calavera's Warbler. **LENGTH:** 4.75 inches. **WINGSPAN:** 7.5 inches. **ADULT:** Gray head with chestnut crown patch. Bold white eye ring. Throat and underparts are bright yellow. Undertail is yellow. Back is green. Rump is yellow and tail is yellowish-gray. Female is like males with less contrast. Throat is pale yellow. **VOICE:** Song is a sweet *sweeta-sweeta-sweeta-sweeta titititititi*. Call is a *chet chet*. **BEHAVIORS:** Flicks and wags its tail while foraging. **HABITAT:** The eastern population uses deciduous trees and shrubs in open mixed forests. The western group uses drier habitat, often in mountains. **STATUS:** Uncommon to abundant migrant throughout most of the state. More common in fall than in spring. They are circum-Gulf migrants and are less common from the central coast northward on the immediate coast. Uncommon to rare winter resident in the Lower Rio Grande Valley. Spring migration is mid-March–early May, fall mid-August–early November.

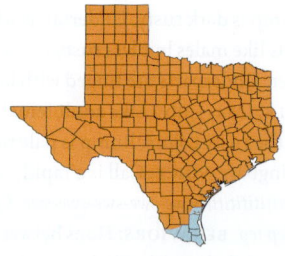

Virginia's Warbler

(Leiothlypis virginiae)

IUCN RED LIST STATUS (2021): Least Concern
POPULATION TREND: Decreasing

Virginia's Warbler was named for Virginia Anderson, wife of W.W. Anderson, who collected the first specimen in 1858. It is one of the last species discovered in North America and only made it as a footnote in the 1860 *Birds of North America* by Baird, Cassin, and Lawrence. **LENGTH:** 4.75 inches. **WINGSPAN:** 7.5 inches. **ADULT:** Grayish overall with bold white eye ring. Chestnut crown patch is often concealed. Throat and breast are yellow. Undertail coverts are yellow. Back is gray, rump is yellow. **FIRST YEAR:** Like adults but the amount of yellow is variable in the breast and may be absent all together. **VOICE:** Song is a *swee-swee-swee-swee-se-se-se*. Call is a slurred *chert chert*. **BEHAVIORS:** Moves both on the ground and in vegetation with short hops. **HABITAT:** Typically found in piñon-juniper and oak woodlands. **STATUS:** Uncommon to rare migrant in the Trans-Pecos. Rare to uncommon summer resident in the upper elevations of the Davis and Guadalupe mountains. Spring migrants arrive late March–mid-May. Fall migration is mid-August–early October.

MacGillivray's Warbler

(Geothlypis tolmiei)

IUCN RED LIST STATUS (2021): Least Concern
POPULATION TREND: Decreasing

MacGillivray's Warbler was named for William MacGillivray, a Scottish naturalist and ornithologist, by his friend John James Audubon, though credit for the discovery of the species in 1839 goes to John Townsend. MacGillivray's Warbler is closely related to the Mourning Warbler, and in the small part of its range where the two species both breed they readily hybridize. Some experts now suggest they are cospecific, or one species.
LENGTH: 5.25 inches. **WINGSPAN:** 7.5 inches.
ADULT: Male has dark charcoal-gray hood with black lores. Bold white eye arcs above and below the eyes. Bright yellow below, including undertail with greenish flanks. Back and tail are green. Female is like male but hood is light gray, not dark gray. Lacks the black lores. **FIRST WINTER:** (July–Mar.) Like females. **VOICE:** Song is a rhythmic series: *seat-seat-seat-seat breet breet breet*. Chip note is a *chit chit*. **BEHAVIORS:** Flicks tail sideways while foraging. Primarily hops along the ground and among twigs. Flights are short, moving rapidly in and out of vegetation. **HABITAT:** Dense shrubs and well-shaded habitats in migration. Riparian habitats are important during migration. **STATUS:** Uncommon to rare migrant through the western half of the state. Rare and very local summer resident in the highest elevations of the Davis Mountains. Spring migration is early April–early June, fall early August–mid-September.

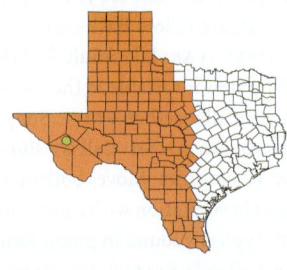

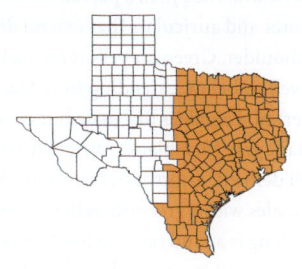

Mourning Warbler

(Geothlypis philadelphia)

IUCN RED LIST STATUS (2021): Least Concern
POPULATION TREND: Decreasing

Mourning Warbler is a relatively common but not often observed migrant species in Texas. Because it breeds in the boreal forest, it is a later migrant than many of the wood-warblers, and the bulk of the population passes through Texas late in spring and early in fall. These are the seasons where Texas birders become less active due to swarming insects and heat. **LENGTH:** 5.25 inches. **WINGSPAN:** 7.5 inches. **ADULT:** Male has dark gray hood with no eye ring. Face is dark. Black crescent on breast. Yellow belly and undertail. Green back. Female has light gray hood with no eye ring. Belly and undertail are yellow. Back is green. **FIRST WINTER:** (July–Mar.) Greenish head with yellow throat. Narrow broken eye ring. Belly and undertail are yellow. Back is green. **VOICE:** Song is a short, rhythmic *churree churree churree twee twee twee*. Chip note is *cheet cheet cheet*. **BEHAVIORS:** Hops on the ground and from branch to branch in dense vegetation. Flights are short, darting in and out of vegetation. **HABITAT:** Migration habitat is poorly known, generally dense undergrowth with vining plants like poison ivy and giant ragweed. **STATUS:** Uncommon to rare spring migrant and uncommon fall migrant in the eastern half of Texas. Spring migration is late April–early June, fall late August–late October.

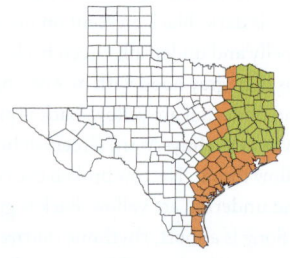

Kentucky Warbler

(Geothlypis formosa)

IUCN RED LIST STATUS (2021): Least Concern
POPULATION TREND: Decreasing

Kentucky Warbler is more often heard than seen. With a song not unlike a Carolina Wren or Ovenbird, Kentucky Warblers have great stamina when singing and often hold perfectly still while singing in dense vegetation, making it a challenge to get a glimpse of the bird. **LENGTH:** 5.25 inches. **WINGSPAN:** 8.5 inches. **ADULT:** Male has blackish crown flecked with gray. Bold yellow supercilium that wraps around behind the eye in a partial eye ring. Black lores and auriculars that extend down to the shoulder. Greenish-yellow on back and bright yellow below, extending from throat to undertail. Female is like male but crown is more dark gray than black. Auricular patch less well defined. **FIRST WINTER:** (July–Mar.) Like females with a greenish-yellow crown. **VOICE:** Song is a loud *tweet-aa tweet-aa tweet-aa tweet-aa tweet*. Chip note is *chipt chipt chipt*.

BEHAVIORS: Walks and hops on the ground. Moves with hops and short flights when foraging in low branches. **HABITAT:** Bottomland woodlands near streams with dense understory. Non-sloping land is important in habitat selection. **STATUS:** Rare to locally uncommon summer resident in the eastern quarter of Texas, west to Grayson and Bastrop counties and south to Hardin County. Uncommon to common migrant in the eastern third of Texas. Spring migration is late March–mid-May, fall mid-August–late September.

Male

Female

Common Yellowthroat

(Geothlypis trichas)

IUCN RED LIST STATUS (2021): Least Concern
POPULATION TREND: Decreasing

Common Yellowthroat is one of the most widespread species of wood-warblers in the Americas. There are currently thirteen sub-species recognized **LENGTH:** 5 inches. **WINGSPAN:** 6.75 inches. **ADULT:** Male has bold black face mask with gray border between mask and crown. Crown, back, and wings are grayish-green. Throat is bright yellow. Belly and undertail are light yellow. Female's head, back, and wings are grayish-green. Throat is bright yellow. Belly is gray, undertail is light yellow. **FIRST WINTER:** (July–Mar.) Male is like adult females with a vague dark mask. Female is like adult females but throat is pale creamy-yellow. **VOICE:** Song is somewhat variable regionally but is always a form of *wichety wichety wichety*. Chip note is a loud *chet chet*. **BEHAVIORS:** Hops among branches. May walk on the ground. Flights are usually short and direct. **HABITAT:** Uses a wide range of habitats in wetlands and prai-ries, and some pine woodlands with dense undergrowth. **STATUS:** Rare to common sum-mer resident in the northeastern part of the state south to the upper coast and the coastal prairies south to Kleberg County. Uncommon summer resident along the length of the Rio Grande, Pecos, and Devils rivers. Common to uncommon winter resident in the south half of Texas. Common to uncommon migrant statewide. Spring migration is mid-March–mid-May, fall migration early September–early November.

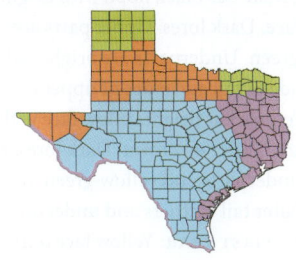

Male

Female

Hooded Warbler

(Setophaga citrina)

IUCN RED LIST STATUS (2021): Least Concern
POPULATION TREND: Increasing

Hooded Warbler is a showy species that never disappoints when it appears by the dozens in coastal migrant traps. An edge specialist formerly dependent on tree-fall gaps in the forest to create nesting habitat, this species has done well with humans. Utility rights-of-way, roads, and even homesteads create the kinds of gaps it can exploit.

LENGTH: 5.25 inches. **WINGSPAN:** 7 inches.
ADULT: Male has black hood with bright yellow face. Dark lores. Upper parts are yellow-green. Underparts are bright yellow. Underside of tail and sides of upper tail are white. Female has bright yellow face with wide dark border on crown. Dark lores. Bright yellow underparts and yellow-green upper parts. Outer tail feathers and underside of tail are white. **FIRST YEAR:** Yellow face outlined in yellow-green. Lores are dark. Body and

tail are like adults. **VOICE:** Song is *sa-wheat sa-wheat sa-wheat tse-sue*. Chip note is *che che che*. **BEHAVIORS:** Hops through vegetation and on the ground, flicking open its tail. Flight is direct and acrobatic while pursuing prey. **HABITAT:** Mature forests with shrub understory for nesting, usually light gaps. **STATUS:** Uncommon to abundant summer resident in the eastern quarter of Texas and local population in Bastrop, Colorado, and Matagorda counties. Uncommon to common migrant in the eastern third of Texas east of the Balcones Escarpment. Spring migration is mid-March–mid-May, fall mid-August–late October.

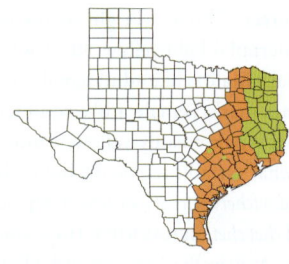

Male

Female

American Redstart

(Setophaga ruticilla)

IUCN RED LIST STATUS (2021): Least Concern
POPULATION TREND: Increasing

The American Redstart is conspicuous when present. Males are a flashy black-and-orange that catch any observers' eyes. Both males and females frequently flash their large tail spots when foraging. More aboreal than most warblers, they often acrobatically pursue prey in the air.
LENGTH: 5.25 inches. **WINGSPAN:** 7.75 inches.
ADULT: Black head, breast, and wings. Sides of breast are orange. Belly is white. Wings have a wide, bold orange bar. Upper portion of tail is orange with black center. Wide black band on end of tail. Female has gray head. Underparts are dingy white. Sides of breast are yellow. Wings are greenish-gray. Back is olive. Wide yellow wingbar. Upper portion of tail is yellow with dark gray center. End of the tail has a wide gray band. **FIRST YEAR:** Like females with less yellow. **VOICE:** Song

is *tsee tsee tsee tsee tee-sue*. Chip note is *chwip chwip*. **BEHAVIORS:** Moves rapidly when foraging, often flashing the bright patches on the tail. **HABITAT:** Breeds in moist, deciduous second-growth woodlands. Uses a wide variety of shrubby and wooded habitats in migration. **STATUS:** Rare to uncommon resident in the forests of East Texas. Uncommon to common migrant in the eastern half of Texas and rare to uncommon migrant in the western half. Spring migration is late March–mid-May, fall mid-August–end of October.

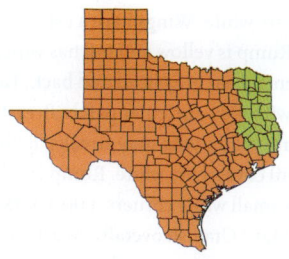

Cape May Warbler

(Setophaga tigrina)

IUCN RED LIST STATUS (2021): Least Concern
POPULATION TREND: Increasing

Cape May Warbler is a spruce budworm specialist that nests in the boreal forest. Most of the population winters in the West Indies. **LENGTH:** 5 inches. **WINGSPAN:** 8.25 inches. **ADULT:** Male has dark crown with yellow face. Auriculars are chestnut-colored. Thin dark eye line. Face pattern forms a yellow neck band. Olive on back with dark streaks. Throat and breast are yellow with dark streaks that converge on throat. Belly and undertail coverts are white. Wings have a diffuse white panel. Rump is yellow and tail has white corners. Female has gray head and back. Face is yellow with gray auricular patch. Breast and throat are yellow with gray streaks. Belly and undertail coverts are white. Rump is yellow. Tail has small white corners. **FIRST WINTER:** (Aug.–Mar.) Grayish overall. Breast has blurry streaking. Rump is greenish. **VOICE:** Song is a thin, high-pitched *seet seet seet seet*, rising in volume. Call is a high *tseet tseet*. **BEHAVIORS:** Walks rapidly along branches. Hops within foliage. Flight is direct and rapid. Hovers when gleaning insects. **HABITAT:** Variety of forest, woodlands, scrub, and thickets. Ornamental flowering trees and shrubs. **STATUS:** Rare spring migrant and very rare fall migrant along the coastal prairies. Rarely encountered away from the immediate coast. Spring migration is late March–mid-May, fall early September–late October.

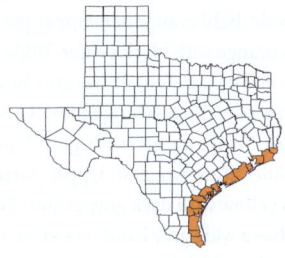

Male Female

Cerulean Warbler

(Setophaga cerulea)

IUCN RED LIST STATUS (2021): Near Threatened

POPULATION TREND: Decreasing

The Cerulean Warbler is one of the longest-distance migrating wood-warblers. It breeds in the northeastern United States and spends its nonbreeding time in the South American Andes Mountains. Because of this long-distance migration, it migrates sooner than other wood-warblers. Once an abundant breeder, it has been in steep decline for most of a century.

LENGTH: 4.75 inches. **WINGSPAN:** 7.75 inches. **ADULT:** Blue above with paler blue supercilium. White below with narrow, dark blue breast band and dark blue streaks on flanks. Two white wingbars. White patches in the tail corners above a dark terminal band. Female is like a blue-green male. **FIRST YEAR:** Green above, yellow-green on breast. Broad yellow-green supercilium and dark eye line.

Two white wingbars and white undertail. **VOICE:** Song is a buzzy *zeep zeep zeep zeep ze-ze-ze-ze-ze*. Chip note is a sharp *chert chert*. **BEHAVIORS:** Hops on small branches and twigs in forest canopy. **HABITAT:** Mostly encountered in the canopy of second-growth trees. **STATUS:** Uncommon to rare spring migrant and very rare fall migrant in the eastern third of Texas. Mostly encountered in the coastal migrant traps. Spring migration is late March–early May, fall late August–mid-October.

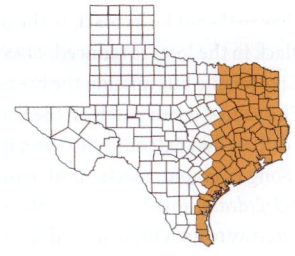

Female   Male

Northern Parula

(Setophaga americana)

IUCN RED LIST STATUS (2021): Least Concern
POPULATION TREND: Increasing

Northern Parula is one of the smallest of the wood-warblers. It moves through the tree canopy much like a chickadee does. **LENGTH:** 4.5 inches. **WINGSPAN:** 7 inches. **ADULT:** Male is blue-gray above with yellow-green patch on back. Dark lores and two short eye arcs. Throat and breast are bright yellow with a black-over-rufous band. Belly and undertail coverts are white. Two white wingbars. Tail has white corners. Female is like male but lacks black in the breast band. Black in the lores is reduced. **FIRST YEAR:** Like female but missing the breast band. The yellow-green patch on the back is more diffuse and extends onto the nape. **VOICE:** Song is a rapid, mechanical, rising trill: *zidededededededede-zup*. Chip note is *sipt sipt*. **BEHAVIORS:** Often hops; does not walk or climb. **HABITAT:** Found in canopy and subcanopy of Spanish moss–festooned bottomland forests, usually around water. **STATUS:** Uncommon to common summer resident in the eastern half of Texas and south to the San Antonio River on the coastal prairies. Summer resident along the major rivers on the southern Edwards Plateau. Common to uncommon migrant in the eastern three-quarters of Texas, becoming less common in the west. Spring migration is late February–mid-May, fall late August–early November.

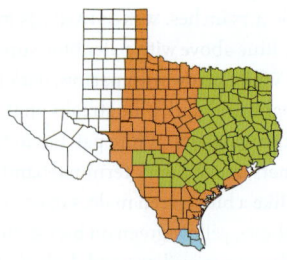

Tropical Parula

(Setophaga pitiayumi)

IUCN RED LIST STATUS (2020): Least Concern
POPULATION TREND: Decreasing

Tropical Parula was formerly known as Sennett's Warbler, named for a Pennsylvania businessman and ornithologist who made an extensive study of the birds of the Lower Rio Grande Valley in the late 1800s. While widespread in Mexico, Central America, and South America, Tropical Parula only regularly makes it into North America in South Texas, and so is one of the Texas specialties for birders.

LENGTH: 4.5 inches. **WINGSPAN:** 6.25 inches.
ADULT: Male has blue-gray head with black face. Yellow-green back. Yellow below with orange breast. Two bold white wingbars. Blue-gray rump. Belly and undertail coverts are white. White corners in the tail are smaller compared to a Northern Parula. Female is like male but lacks orange on breast. Face has a reduced amount of black compared to a male. **VOICE:** Essentially identical to Northern Parula. **BEHAVIORS:** Moves by creeping or hopping. **HABITAT:** In Brooks and Kenedy counties, found in live oak woodlands with abundant epiphytes. In the Lower Rio Grande Valley, cedar elm, Mexican ash, and hackberry woodlands with epiphytes. **STATUS:** Rare to uncommon resident in the live oak woodlands of the Coastal Sand Plains in Kenedy and Brooks counties. Rare to very rare in the Lower Rio Grande Valley. Rare to locally uncommon summer resident along the Devils and Pecos rivers in Val Verde County. Migrants arrive in late March and are present until mid-September.

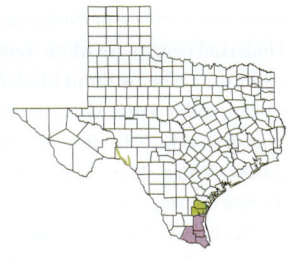

Nonbreeding

Breeding male

Magnolia Warbler

(Setophaga magnolia)

IUCN RED LIST STATUS (2021): Least Concern
POPULATION TREND: Increasing

The first specimen of Magnolia Warbler was found in a magnolia tree, though these boreal forest breeders have little affinity for magnolia trees.

LENGTH: 5 inches. **WINGSPAN:** 7.5 inches.
ADULT BREEDING: (Mar.–Aug.) Male has gray crown with white supercilium over black face mask. Lower white eye arc. Nape and back are black. Rump is yellow. Tail is dark with white band. Wings have a white wing panel. Yellow below with necklace of thick black streaks. Undertail coverts are white. Female is like breeding male but without black face mask. Dark auricular patch. Grayish-yellow back with dark streaks. Two white wingbars.
NONBREEDING: (Aug.–Mar.) Male has gray head and white eye ring. Yellow throat and breast with thick dark streaks on sides. Two white wingbars and grayish-yellow back.

FIRST WINTER: (Aug.–Mar.) Female has gray head with greenish back. White eye ring. Throat and breast are yellow with narrow gray neckband. White undertail with wide dark band at the tip. **VOICE:** Song is a simple, musical *sweeter sweeter sweeter SWEETEST*. Call is a very soft *zeep*. **BEHAVIORS:** Hops while feeding on arboreal surfaces. **HABITAT:** In migration, found in low trees, shrubs, and forest edges. **STATUS:** Uncommon to common migrant in the eastern half of the state. Encountered most often on the coast during migration. Spring migration is mid-April–late May, fall late August–late October.

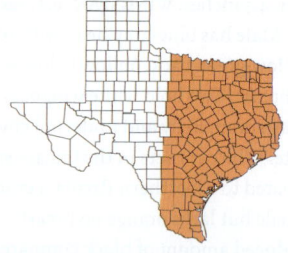

Male

Female

Bay-breasted Warbler

(Setophaga castanea)

IUCN RED LIST STATUS (2021): Least Concern
POPULATION TREND: Increasing

Bay-breasted Warbler is another of the boreal forest–breeding wood-warblers dependent on spruce budworms. The spruce budworm is declining due to control efforts, and Bay-breasted Warbler and others are declining with it. Bay-breasted Warbler is very distinctive in breeding plumage, but fall and winter birds closely resemble fall and winter Blackpoll and Pine Warblers, often confusing birders.

LENGTH: 5.5 inches. **WINGSPAN:** 9 inches.
ADULT BREEDING: (Apr.–Aug.) Bay-colored crown. Black face mask. Bay-colored throat. Sides of neck are pale creamy-yellow. Breast and belly are whitish with bay-colored sides. Undertail coverts are pale. Wings and back are dark slate-colored with darker streaks. Two white wingbars. White outer tail corners. Female has pale head with faint dark eye line.

Pale throat and breast with bay-colored wash on sides. Dark wings with two white wingbars. **NONBREEDING:** (Aug.–Apr.) Male is like breeding female. **FIRST-WINTER FEMALE:** (Aug.–Apr.) Like breeding females but lacks bay-colored flanks. Back is yellowish. **VOICE:** Song is a thin, high-pitched *see-ah see-ah see-ah see-oo*. Call is a soft *tswip tswip*. **BEHAVIORS:** Moves slowly and deliberately among branches. **HABITAT:** Found in a wide variety of habitats in migration, usually among higher branches. **STATUS:** Uncommon to common spring migrant and rare fall migrant in the eastern half of the state. Spring migration is late April–mid-May, fall early September–late October.

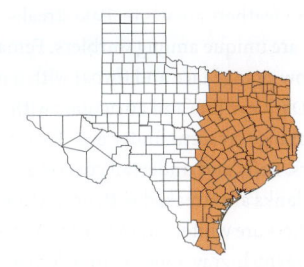

Female

Male

Blackburnian Warbler

(Setophaga fusca)

IUCN RED LIST STATUS (2021): Least Concern
POPULATION TREND: Increasing

Blackburnian Warbler are forest interior specialists and do not use edge habitats. **LENGTH:** 5 inches. **WINGSPAN:** 8.5 inches. **ADULT:** Male has dark crown with yellow-orange central stripe. Flame-orange face and throat with dark eye line and auriculars. Orange crescent under the eye. Pale yellow on breast with black streaks on flanks. White undertail coverts. Two bold white wingbars almost merge into a single wing panel. Outer tail feathers are white. Pale streaks on back are unique among warblers. Female has yellow-orange face and throat with dark crown. Dark eye line and auriculars with yellow-orange crescent under eye. Two white wingbars. Belly and undertail coverts are white. Flanks are streaked with gray. Outer tail feathers are white. **IMMATURE:** (Aug.–Mar.) Greenish-gray above with pale yellow supercilium and throat. Pale-yellow arc under eye. Belly and undertail coverts are white with gray streaks on flanks. Pale streaks on back. **VOICE:** Song is a sharp series, rising to very high notes: *sept sept sept sept sept see-see-see*. **BEHAVIORS:** Not often on the ground but hops exclusively. **HABITAT:** In migration, uses all woody habitats. **STATUS:** Uncommon to locally common spring migrant in the eastern half of the state. Rare fall migrant in the eastern third of the state. Most birds are encountered in the coastal migrant traps. Spring migration is early April–mid-May, fall early September–late October.

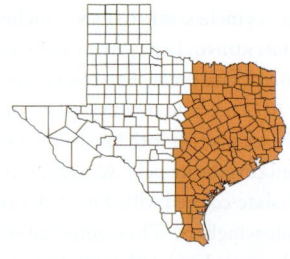

Male Female

Mangrove Warbler

Yellow Warbler

(Setophaga petechia)

IUCN RED LIST STATUS (2021): Least Concern
POPULATION TREND: Decreasing.

The eastern form of Yellow Warbler was once a nesting species in Texas, but the last recorded nesting was in 1956. In 2004 a breeding population of the chestnut-headed subspecies of Yellow Warbler, known as the Mangrove Warbler, was discovered breeding in the mangroves that grow in the Laguna Madre in South Texas. This is a tropical subspecies that was formerly a rare visitor to Texas. This population has continued to expand and is now an uncommon breeder at several sites in Cameron County.
LENGTH: 5 inches. **WINGSPAN:** 8 inches.
ADULT: Male is bright yellow-green above, bright yellow below. Face is bright yellow. Bright chestnut streaks on breast. Female is bright yellow overall. **FIRST YEAR:** Variable from pale yellow to brownish to grayish with slightly darker wings. **VOICE:** Song is a clear, high-pitched *sweet-sweet-sweet si si SWEET*. Chip note is *chi chi*. **BEHAVIORS:** Does not walk but hops on branches. **HABITAT:** Mainly scrub/shrub and second-growth forest, often in association with wetlands. **STATUS:** Common to abundant migrant statewide, more abundant in fall. The only breeding Yellow Warbler population in Texas is the distinct Mangrove Warbler, a chestnut-headed subspecies that breeds in the mangroves around the Laguna Madre in South Texas, particularly in Cameron County. Spring migration is late March–mid-May, fall late July–mid-October.

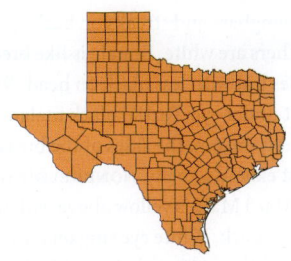

Breeding male

First winter

Chestnut-sided Warbler

(Setophaga pensylvanica)

IUCN RED LIST STATUS (2021): Least Concern
POPULATION TREND: Decreasing

Chestnut-sided Warbler is drawn to scrubby second growth and forest edges. While it has declined since the 1960s, it still maintains a healtly population.
LENGTH: 5 inches. **WINGSPAN:** 7.75 inches.
ADULT BREEDING: (Mar.–Aug.) Male has bright yellow crown, black supercilium, and bright white cheeks. Black malar and white throat. White underparts with chestnut sides. Back and upper wings are dark. Two pale yellowish wingbars and streaks on back. Outer tail feathers are white. Female is like breeding males but with less black on head. White crescent under eye. Partial black malar. The chestnut is only on the sides of the breast, does not extend to belly. **NONBREEDING:** (Aug.–Mar.) Male is yellow above with dark streaks on back. White eye ring on gray face, cheek, and throat. Gray below with chestnut on lower flanks. Two whitish wingbars. Female is yellow above with white eye ring. Pale gray below. Two white wingbars. **VOICE:** Song is a high-pitched *swee-swee-swee-swee WHEAT-too.* Chip note is a soft *twip twip.* **BEHAVIORS:** Hops on the ground and in foliage. Searches the underside of leaves for prey. **HABITAT:** In migration, uses a wide variety of habitats, from shrubbery to deep forests. **STATUS:** Uncommon to common spring migrant and uncommon to rare fall migrant in the eastern third of Texas. Spring migration is mid-April–late May, fall late July–mid-October.

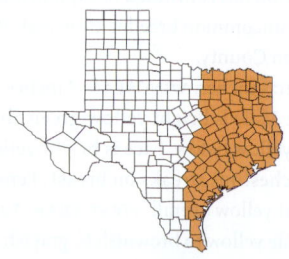

Blackpoll Warbler

(Setophaga striata)

IUCN RED LIST STATUS (2018): Least Concern
POPULATION TREND: Decreasing

Blackpoll Warbler is one of the largest North American wood-warblers.
LENGTH: 5.5 inches. **WINGSPAN:** 9 inches.
ADULT BREEDING: (Apr.–Aug.) Male has black cap and white lower cheek. Thin black malar stripe. White below with thin dark streaking on flanks. Undertail coverts are clear white. Back is gray with black streaks. Two white wingbars. Tail has white corners. Yellowish legs. Female is gray-green above with fine dark streaks on crown and dark streaks on back. Two white wingbars. Underparts are grayish with fine dark streaks. Indistinct thin dark malar stripe. Undertail is clear white. White outer tail corners. **NONBREEDING:** (Aug.–Apr.) Male is yellowish-green above with darker wings. Dark streaks on crown and back. Yellow face with dark eye stripe and dusky auriculars. Yellowish wash on breast and flanks with dark streaks. Dark malar streak. Female is yellow-green with pale whitish belly. Two pale wingbars. White undertail coverts. Pale-yellow face with dark eye stripe. Faint streaks on flanks. **VOICE:** Very high-pitched *sisisisisiSISISISsisi*. Call is a soft *chip chip*. **BEHAVIORS:** During migration, forages in the outer branches of trees. **HABITAT:** Can be found in almost all habitats in migration. **STATUS:** Uncommon to rare spring migrant and rare fall migrant along the coast. Rare to very rare migrant inland in the eastern half of Texas. Spring migration is mid-April–late May, fall early August–late October.

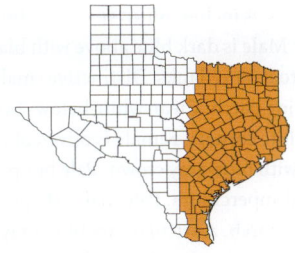

Black-throated Blue Warbler

(Setophaga caerulescens)

IUCN RED LIST STATUS (2021): Least Concern
POPULATION TREND: Increasing

Black-throated Blue Warbler males and females are so different looking that many
early ornithologists thought they were two
species. Most migrate east of the Appalachian
Mountains and only a very small part of the
population passes through Texas. A bird of the
forest interior, its population declined as its
habitat was cleared for farming in the 18th and
19th centuries. As farms return to forest, the
population is on the rise again.

LENGTH: 5.25 inches. **WINGSPAN:** 7.75 inches.
ADULT: Male is dark blue above with black
face, throat, and flanks. Distinctive small
white wing patch. Clean white below. Tail has
small white corners. Female is olive-colored
overall with blue-gray crown. Pale but pronounced supercilium. Pale malar stripe. Pale
under-eye arch. Auriculars are blue-gray.
Undertail coverts are whitish. Small white
wing patch at base of primaries. **FIRST YEAR:**
Like breeding female but some birds lack the
white wing patch. **VOICE:** Song is a buzzy *zwoo
zwoo zheeeeeee*. Chip note is a loud *chirt chirt*.
BEHAVIORS: Hops on the ground. Flits rapidly
among perches and foliage. **HABITAT:** There
is little data on migration habitat, but usually
encountered in the interior of larger coastal
migrant traps. **STATUS:** Rare migrant along the
Texas coast; it is more common in fall. Spring
migration is mid-April–late May, fall early
September–late October.

Breeding

Nonbreeding

Palm Warbler

(Setophaga palmarum)

IUCN RED LIST STATUS (2021): Least Concern

POPULATION TREND: Increasing

The Palm Warbler, despite the name, is one of the most northerly nesting of the wood-warblers. Only Blackpoll Warblers nest farther north. Palm Warbler aquired its name because the first specimen was collected from a palm thicket on Hispaniola in the Caribbean. **LENGTH:** 5.5 inches. **WINGSPAN:** 8 inches. **ADULT BREEDING:** (Apr.–Aug.) Eastern form has extensive rufous crown. Yellow-olive above and bright yellow below. Yellow supercilium. Grayish auricular patch. Yellow malar and unmarked yellow throat. Rufous streaks on breast. Bright yellow undertail. White tail corners. Western form has rufous crown and pale supercilium. Greenish-gray above. Pale yellow supercilium and bright yellow throat. Gray auricular patch and dark eye line. Undertail coverts are bright yellow. White tail corners. **NONBREEDING:** (Aug.–Mar.) Eastern form is yellow-olive overall. Yellow supercilium and dark eye stripe. Undertail coverts are yellow. Western form is dull brownish-yellow overall. Dark eye line and pale yellow supercilium. Undertail coverts are yellow. **VOICE:** Song is a buzzy trill: *zwe-zwe-zwe-zwe-zwe-zwe.* Chip call is *tswip tswip.* **BEHAVIORS:** Walks on the ground often. Nearly continually bobs its tail up and down. **HABITAT:** Variety of habitats, secondary-growth edges, and thickets. **STATUS:** Uncommon winter resident on the upper and central Texas coast. Rare to uncommon migrant in the eastern two-thirds of Texas. Spring migration is mid-April–mid-May, fall late August–late October.

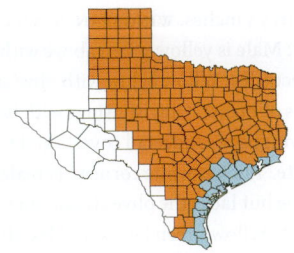

Male

Female

Pine Warbler

(Setophaga pinus)

IUCN RED LIST STATUS (2021): Least Concern
POPULATION TREND: Increasing

The Pine Warbler is nearly unique among North American wood-warblers in that it breeds and winters almost entirely in North America. Rarely found away from pine trees, they can be abundant, particularly in winter when the northern migrants join flocks in the south. It is able to move among the trees much like a Brown Creeper, so much so that early ornithologists referred to it as the Pine-creeping Warbler.

LENGTH: 5.5 inches. **WINGSPAN:** 8.75 inches. **ADULT:** Male is yellow-green above with yellow spectacles. Yellow below with olive streaks on sides of breast. Grayish wings with two white wingbars. Undertail coverts and belly are white. Tail has white corners. Female is like male but lacks the olive streaks on sides of breast. Yellow-green below and less distinct spectacles. **FIRST YEAR:** Drab olive-brown overall. Whitish undertail coverts. Two white wingbars on low-contrast wings. **VOICE:** Two distinct songs: a rapid, chipping, sparrow-like trill and a slower *whep-whep-whep-whep-whep.* **BEHAVIORS:** Hops exclusively both on the ground and in the trees. **HABITAT:** Upland pine and pine-hardwood forests. **STATUS:** Common resident in the pine forests of East Texas. Found south to the "Lost Pines" of Bastrop and Caldwell counties in isolated patches of pine and on the coastal prairies. Uncommon winter resident in the eastern third of Texas. Winter birds arrive in early October and are present until late March.

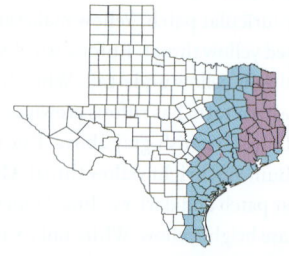

Yellow-rumped "Myrtle" Warbler

(Setophaga coronata coronata)

IUCN RED LIST STATUS (2017): Least Concern
POPULATION TREND: Unknown

Yellow-rumped Warbler comes in two forms
in Texas: Myrtle Warbler and Audubon's
Warbler. There is a third form resident in
Mexico, known as Goldman's Warbler.
LENGTH: 5.5 inches. **WINGSPAN:** 9.25 inches.
ADULT BREEDING: (Apr.–Aug.) Male has
small yellow crown patch. Blue-gray above.
Face is black with thin white supercilium and
thin arc under the eye. Back has dark streaks.
Rump is yellow and tail has white corners.
Throat is white and breast has a necklace of
dense streaks. Sides of breast are yellow. Undertail coverts are white. Female is like males
but face mask is less distinct and streaking on
breast is less dense. No crown patch. **WINTER:**
(Aug.–Apr.) Sexes are alike. Brownish above
with dark streaks. Indistinct supercilium and
thin arc above and below the eye. Dirty white
below with dark streaks and yellow patches
on sides of breast. Rump is bright yellow. Two
white wingbars. **VOICE:** Song is a series of musical *sep sep sep sep sep*. Call is a loud *chep chep*.
BEHAVIORS: Mostly hops, both on the ground
and in vegetation. Makes short flights between
trees. **HABITAT:** Open forest. Often dense in
Chinese tallow stands. **STATUS:** Common to
locally abundant migrant and winter resident
in the eastern two-thirds of Texas. Uncommon
in the western third of the state. Fall migration is early September–mid-October, spring
mid-March–mid-May.

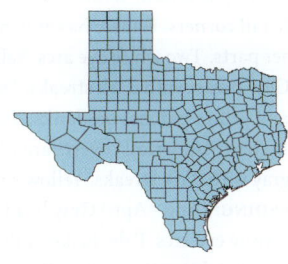

Yellow-rumped "Audubon's" Warbler

(Setophaga coronata auduboni)

IUCN RED LIST STATUS (2017): Least Concern
POPULATION TREND: Unknown

Audubon's Warbler is the western subspecies of Yellow-rumped Warbler in Texas. **LENGTH:** 5.5 inches. **WINGSPAN:** 9.25 inches. **ADULT BREEDING:** (Apr.–Aug.) Male has yellow crown patch. Black face with two thick eye arcs. Throat is bright yellow. Black breast. Belly and undertail coverts are white with black streaks and yellow patches on the sides. Blue-gray upper parts with dark streaks on back. Rump is yellow. White wing patches. White tail corners. Female has gray head and upper parts. Two white eye arcs. Yellow throat. Gray breast with dark streaks. Dark streaks and yellow patches on flanks. Two white wingbars. Undertail coverts are white. Back is gray with dark streaks. Yellow rump. **NONBREEDING:** (Aug.–Apr.) Gray head with two thin white eye arcs. Pale dusky underparts with blurry streaking. Pale yellow throat.

Yellow rump. **VOICE:** Song is a musical *sweeto sweeto sweeto swee swee swee*. Chip note is a sharp *chit chit*. **BEHAVIORS:** Mostly hops on the ground and in vegetation. **HABITAT:** Open areas and second growth, pine forest. **STATUS:** Uncommon to locally common migrant and winter resident in the Trans-Pecos and South Plains. Common to locally abundant migrant in the eastern two-thirds of the Panhandle and rare winter visitor in the entire Panhandle. Uncommon summer resident in the higher elevations of the Davis and Guadalupe mountains. Fall migration is early September–mid-October, spring mid-March–mid-May.

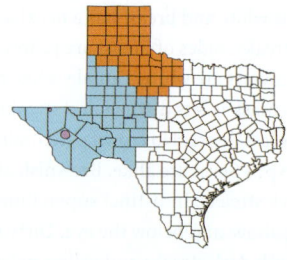

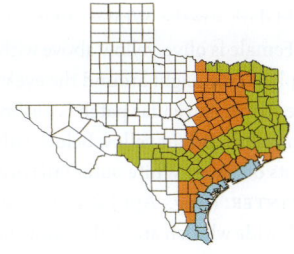

Yellow-throated Warbler

(Setophaga dominica)

IUCN RED LIST STATUS (2021): Least Concern
POPULATION TREND: Increasing

Yellow-throated Warbler is one of the earliest-arriving migrant wood-warblers in Texas. Its song will often be heard filtering down from high up in trees that still have not leafed out. It moves slower and more deliberately than many other wood-warblers as it forages almost creeper-like. For unknown reasons it retreated from the most northern parts of its breeding range in the early 20th century, but it is now expanding back into its former range. **LENGTH:** 5.5 inches. **WINGSPAN:** 8 inches. **ADULT:** Gray above with bold white supercilium. Front of crown is blackish. White neck spot and black auricular patch. White eye arc below eye. Bright yellow throat. White below with dark gray streaks on flanks. Two bold white wingbars. Tail has white corners. Female is like males but lacks black on the crown. Yellowish wash on flanks. **VOICE:** Song is *sweet sweet sweet sweet sue sue sue*. Chip note is *chi chi*. **BEHAVIORS:** Climbs trunks and large limbs like a Black-and-white Warbler or Brown Creeper. Moves horizontally by hopping. **HABITAT:** Bottomland forests, cypress swamps, and upland pines. **STATUS:** Uncommon to common summer resident in the Pineywoods and the forest belt through the southern Post Oak Savannah to the Edwards Plateau. Uncommon to rare summer resident in riparian corridors on the Edwards Plateau. Spring migration is late February–early April, fall late July–late September.

Prairie Warbler

(Setophaga discolor)

IUCN RED LIST STATUS (2021): Least Concern
POPULATION TREND: Decreasing

Since the 1970s, as North American forests are regenerating, the Prairie Warbler population has been in decline and its status is now a matter of concern.
LENGTH: 4.75 inches. **WINGSPAN:** 7 inches.
ADULT: Male. Olive-yellow above and bright yellow below. Arched yellow supercilium and white-yellow eye arc below eye. Yellow neck outlines an olive-yellow auricular patch. Rufous streaks on back. Sides of breast and belly have bold dark streaks. Whiter outer corners on tail. Female is olive-yellow above with a mostly plain face. Light around the eye with faint eye stripe. Auricular patch is outlined below in yellow. Bright yellow below with dark spots on sides. White outer tail corners.
FIRST WINTER: (Aug.–Apr.) Gray head and face with wide whitish arcs below and above a thin eye stripe. Olive-yellow above with faint yellowish wingbars. Bright yellow below with faint streaks on sides. White outer tail corners are not as extensive as adults. **VOICE:** Song is buzzy, rising in pitch: *zree-zree-zree-zree-zree*. Chip note is *chep chep*. **BEHAVIORS:** Hops instead of walking. **HABITAT:** Shrubby habitats. Most often found in young pines less than 20 feet high. **STATUS:** Rare to common summer resident in the northeastern corner of Texas, often in tracts of recently replanted pine trees. Uncommon migrant along the coast. Spring migration is mid-March–late April, fall mid-July–mid-September.

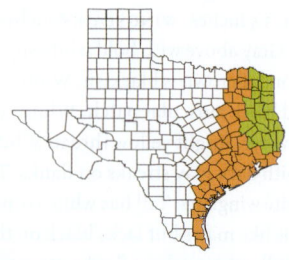

Grace's Warbler

(Setophaga graciae)

IUCN RED LIST STATUS (2021): Least Concern
POPULATION TREND: Decreasing

Grace's Warbler is a small-pine specialist. In North America it is a montane species, but resident populations in Central America use some coastal pine savannahs. Forestry practices of logging and fire suppression in North America have reduced the mature pine habitat it prefers, and its numbers have declined by more than 50% over the last 50 years.
LENGTH: 5 inches. **WINGSPAN:** 8 inches.
ADULT: Male is gray above with bright yellow supercilium and black stripe above supercilium, Bright yellow arc below eye. Thin dark loral line. Bright yellow throat and upper breast. Belly and undertail coverts are white. Flanks have dark streaks. Two white wingbars. Female is like males but blackish parts are instead gray. **FIRST WINTER:** (Aug.–Mar.) Gray-brown above with broad yellow supercilium and yellow lower eye arc. Throat and breast are yellow with yellow flanks and pale belly. Faint streaks on flanks. Two white wingbars. **VOICE:** Song is a slow trill: *tew tew tew tew tee tee tee tee*. Chip note is *chi chi chi*. **BEHAVIORS:** Hops and flits in the middle or upper branches, seldom on exposed perches. Sticks to smaller branches. **HABITAT:** Pine and pine-oak forests. **STATUS:** Uncommon to common summer resident in the higher elevation of the Davis and Guadalupe mountains. Migrants are rarely seen away from the mountains. Spring migrants arrive in mid-April and are present until mid-September.

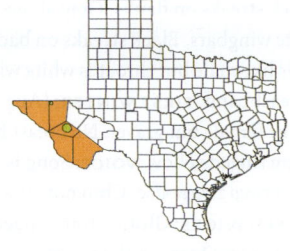

Black-throated Gray Warbler

(Setophaga nigrescens)

IUCN RED LIST STATUS (2021): Least Concern
POPULATION TREND: Decreasing

The Black-throated Gray Warbler breeds mostly west of the Rocky Mountains and winters in Mexico. Unlike most of the neotropical migrant songbirds, it seems not to have been affected by human activities.
LENGTH: 5 inches. **WINGSPAN:** 7.75 inches.
ADULT: Male has black crown with broad white supercilium. Black cheek with broad white stripe below it. Yellow loral spot in front of eye. Black throat. Clean white below with bold black streaks on flanks. Gray above with two white wingbars. Black streaks on back. Female is like male but throat is white with black breast band. **FIRST WINTER:** (Aug.–Mar.) Like female but grayer. No breast band. Flanks are tinged yellow. **VOICE:** Song is *zreep zreep zreep zreeeee-doo*. Chip note is a soft *chit chit*. **BEHAVIORS:** Gleans from foliage of trees. **HABITAT:** Open coniferous or mixed coniferous-deciduous woodland with brushy undergrowth. **STATUS:** Rare to uncommon migrant in the Trans-Pecos and High Plains. Rare to uncommon winter resident in the Lower Rio Grande Valley and rare winter visitor along the coastal prairies. Spring migration is late March–late May, fall early August–early November.

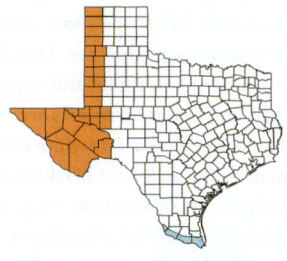

Female  Male

Townsend's Warbler

(Setophaga townsendi)

IUCN RED LIST STATUS (2021): Least Concern
POPULATION TREND: Decreasing

Townsend's Warbler was one of many species first collected by Thomas Nuttall and John Kirk Townsend during their expedition to the Pacific coast through the Rocky Mountains in 1834.

LENGTH: 5 inches. **WINGSPAN:** 8 inches.
ADULT: Male has black crown and yellow supercilium. Black lores and auriculars. Yellow arc under eye. Yellow cheek and malar area. Throat is black. Breast is yellow. Belly is white with black streaked flanks. Undertail coverts are white with black spots. Back is green with dark streaks. Wings are gray with two white wingbars. Tail is gray with white corners. Female has gray crown and yellow face. Gray lores and auricular patch with yellow arc below eye. Throat is whitish with dark patches on sides of yellow breast. Gray smudgy streaks on flanks. Belly and undertail coverts

are white. Wings are gray with two white wingbars. Tail is gray with white corners. **FIRST WINTER:** (Aug.–Mar.) Like females with grayish belly. **VOICE:** Song is a buzzy *zep zep zep zeee zeee zep*. Chip note is a soft *tip tip*. **BEHAVIORS:** Hops through foliage. Most flights are short with short glides. **HABITAT:** Variety of wooded habitats in migration. **STATUS:** Uncommon to common migrant in the Trans-Pecos, mostly at higher elevations. Rare to uncommon migrant in the western High Plains. Spring migration is early April–late May, fall early August–early November.

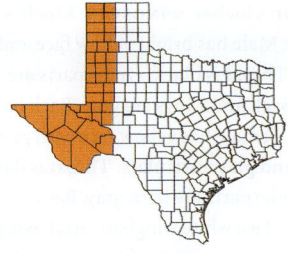

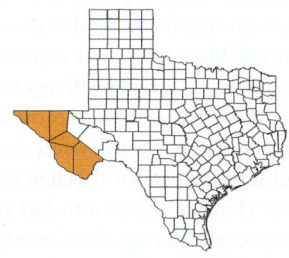

Hermit Warbler

(Setophaga occidentalis)

IUCN RED LIST STATUS (2021): Least Concern
POPULATION TREND: Stable

Hermit Warbler was first collected and described by John Kirk Townsend near Vancouver, Washington, in 1837. He named the species Hermit Warbler because of its apparent solitary and secretive behavior. Preferring to stay high in conifer trees, it is easy to hear but difficult to see. It is closely related to Townsend's Warbler, and hybrids are frequently found in a narrow zone where the two species meet in Washington and Oregon. **LENGTH:** 5 inches. **WINGSPAN:** 8 inches. **ADULT:** Male has bright yellow face and crown. Throat is black. Underparts are clean white. Back and wings are gray. Back has dark streaks. Female has yellow face with gray crown and gray auriculars. Throat is dark gray, underparts are light gray. Back and wings are gray. Two white wingbars. Back has gray streaks. **FIRST WINTER:** (Aug.–Mar.) Like female but face not as bright, with a pale yellow eye ring. Flanks have faint yellow wash. **VOICE:** Song is *ze-ta ze-ta ze-ta ze-ta zeee to*. Chip note is a very quiet *stit stit*. **BEHAVIORS:** Hops while foraging. Flies within and between trees. **HABITAT:** In migration, found in pine and pine-oak forest. **STATUS:** Rare migrant in the major mountain ranges of western Trans-Pecos. Most often found in the Chisos Mountains. Spring migration is early April–late May, fall late July–late September.

Golden-cheeked Warbler

(Setophaga chrysoparia)

IUCN RED LIST STATUS (2020): Endangered
POPULATION TREND: Decreasing

Golden-cheeked Warbler is the only breeding species endemic to Texas. It winters in southern Mexico, Guatemala, and Honduras. Its migration path is through the mountains of Mexico, not along the coast, and it enters Texas along the Rio Grande.

LENGTH: 5 inches. **WINGSPAN:** 7.75 inches.

ADULT: Male has bright golden cheeks. Black line through the eye from the lores to the crown. Black throat, crown, and back. Heavy black streaking on breast and flanks. Two white wingbars. Mostly light below. Female is similar to males but throat is pale yellow with thin dark malars. Black streaks on back and crown. Outer tail feathers are white.

FIRST WINTER: (Aug.–Mar.) Yellow face with distinct dark eye line. Pale throat and breast with gray streaks on flanks. Olive back with faint streaks. Pale below with white wingbars.

VOICE: Low-pitched, buzzy *zrr zoo zeedl zeeee twip* or *zeedl zeedl zeedl zweee tsip.* **BEHAVIORS:** Flight is often undulating. Flutters between branches when foraging; sometimes hovers and hangs upside down. **HABITAT:** Old growth and mature regrowth juniper-oak woodlands on limestone hills and canyons. **STATUS:** Breeding range restricted to canyon lands of the eastern Edwards Plateau and northward to Palo Pinto and Somervell counties. Very few records exist outside of the breeding range. Breeding birds begin to arrive in late February and migrate early, as soon as late June.

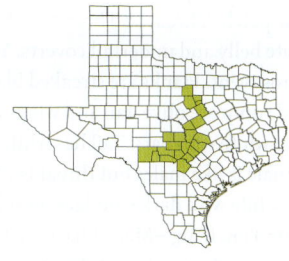

Male

Female

Black-throated Green Warbler

(Setophaga virens)

IUCN RED LIST STATUS (2021): Least Concern
POPULATION TREND: Stable

Black-throated Green Warbler is the eastern representative of the closely related
group of *Setophaga* warblers that includes
Golden-cheeked, Hermit, Townsend's, and
Black-throated Gray Warblers. These species
likely date from periodic episodes of isolation
caused by Pleistocene glaciations.
LENGTH: 5 inches. **WINGSPAN:** 7.75 inches.
ADULT: Male has bright yellow face with
olive crown and nape. Olive eye line. Olive
auricular patch. Throat and breast are black
with white belly and undertail coverts. Yellow
band across vent. Flanks are streaked black.
Back and rump are olive, tail and wings are
gray. Two white wingbars. Tail has white corners. Female is like males but throat is white.
Breast is white with dark smudges on sides.
FIRST WINTER: (Aug.–Mar.) Like female but
lacks dark smudges on breast. Faint streaking

only on sides of the breast. **VOICE:** Song is a
high, buzzy *zee zee zee zee zee zoo ze*. Chip note
is *chit chit*. **BEHAVIORS:** Hops exclusively.
Flies within and between trees. **HABITAT:** In
migration, uses all woody habitats including
forest edges. **STATUS:** Common to uncommon
migrant in the eastern half of Texas. Uncommon to rare winter resident in the Lower Rio
Grande Valley and on the lower coast. Spring
migration is mid-March–mid-May, fall early
August–mid-November.

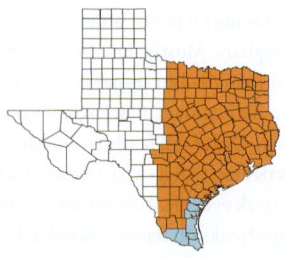

Male

Canada Warbler

(Cardellina canadensis)

IUCN RED LIST STATUS (2021): Least Concern
POPULATION TREND: Decreasing

Canada Warbler breeds mostly in the boreal forest. It is one of the last wood-warblers to pass through Texas during spring migration and one of the first to arrive on its long southbound journey to northern Central America. Canada Warbler was named for a specimen collected in Canada in 1760.

LENGTH: 5.25 inches. **WINGSPAN:** 8 inches.
ADULT: Male is gray above with black forecrown. Bright yellow eye ring and yellow lores on black face. Malar area and throat are yellow. Yellow below with bold necklace of black streaks on breast. Undertail coverts are white. Tail, wings, and back are gray. Female is like males without black on face and crown. Breast is streaked gray. **FIRST YEAR:** Gray above, dull yellow below. Face is gray with white eye ring. Lores are yellow. Malar and throat are yellow. Faint gray streaks on breast. Undertail coverts are white. **VOICE:** Song is a descending series of sputtering trills, highly variable. Chip note is a loud *che che.* **BEHAVIORS:** Capable flier in thick vegetation, through which it also hops and climbs. Often has tail cocked in low vegetation. **HABITAT:** Found in shrubbery, brushes, and vine tangles in migration. **STATUS:** Uncommon to rare spring migrant in the eastern half of Texas. Common to uncommon fall migrant in the eastern half of Texas. Spring migration is mid-April–May, fall late July–mid-October.

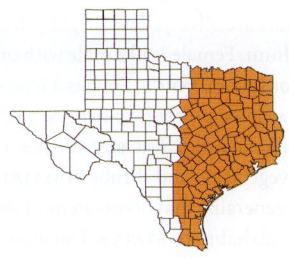

Male

Wilson's Warbler

(Cardellina pusilla)

IUCN RED LIST STATUS (2021): Least Concern
POPULATION TREND: Decreasing

Wilson's Warbler was first described by
Alexander Wilson from a specimen collected
in New Jersey in 1811. He gave it the specific name *pusilla*, which in Latin means "very
small." The name suits Wilson's Warbler well,
as it is one of the smallest, if not the smallest,
of the wood-warblers.

LENGTH: 4.75 inches. **WINGSPAN:** 7 inches.
ADULT: Male is olive-yellow above, bright yellow below. Distinctive black cap like a beanie.
Auriculars are olive-yellow. Bright yellow
supercilium. Female is like male with only a
partial or no cap. **VOICE:** Song is a repeating,
mellow *swip-swip-swip-swip-swip-swip*. Chip
note is a hoarse *chep chep*. **BEHAVIORS:** Hops
among vegetation and shrubs. **HABITAT:** A
habitat generalist that occurs in most deciduous shrub habitats. **STATUS:** Uncommon to
common migrant throughout Texas. Most

common in migration in the Trans-Pecos,
Panhandle, and South Plains. More common
in fall. Uncommon to rare winter resident
on the coastal prairies and in the Lower Rio
Grande Valley.

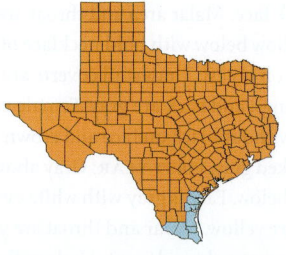

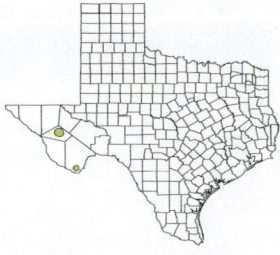

Painted Redstart

(Myioborus pictus)

IUCN RED LIST STATUS (2020): Least Concern
POPULATION TREND: Decreasing

Painted Redstart populations have fluctuated greatly in Texas. They have been regular for a long period in the upper elevations of the Chisos and Davis mountains and then absent for years at a time. In the last decade there have been a few wintering individuals in the Lower Rio Grande Valley and on the coastal prairies, much to the delight of birders who haven't yet been able to make the strenuous hike up to see this flashy black-and-red bird in West Texas.

LENGTH: 5.2 inches. **WINGSPAN:** 8.75 inches. **ADULT:** Dull blackish above, with red belly and gray undertail coverts. Bold white wing patch. Often appears crested. Thin but obvious white arc under eye. Outer three tail feathers are brilliant white. **JUVENILE:** (June–Aug.) Like adults but belly and undertail coverts are gray. **VOICE:** Song is *che-we che-we che-we wheata wheata wi wi*. Call is a clear *teed-woo teed-wo*. **BEHAVIORS:** Pivots and hops while foraging, swinging its long tail back and forth. Generally forages in shady habitats. **HABITAT:** Oak-juniper woodlands. **STATUS:** Rare summer resident in the Chisos Mountains of Big Bend National Park. Very rare summer resident in the Davis Mountains. Spring migrants arrive late March–mid-April. Fall migration is mid-September–early November.

CARDINALIDAE (CARDINALS AND ALLIES)

The cardinals as a group are sexually dimorphic; the males are much bolder and more colorful than the females. The plumages are bold in blues, reds, and yellows. They can be divided into two groups: the cardinals and grosbeaks with large, white, seed-crushing bills, and the buntings and tanagers with more delicate bills. Somewhat confusing for birders and birdwatchers is the fact that many birds called buntings and tanagers belong to other families of birds.

The cardinals, grosbeaks, and buntings are primarily fruit- and seed-eaters, while the tanagers are primarily insect- and fruit-eaters. Most of the family is centered in North America and most species are migratory.

The cardinal family is most likely a sister to the tanager family. Recent DNA studies have shown that the North American *Piranga* tanagers are actually part of the cardinal family, not the tanager family. The resemblence to the true tanagers is mostly due to evolutionary convergence.

There are fifty-one species of cardinals in the New World. Six are of conservation concern. The Carrizal Seedeater, first described to science in 2003, is critically endangered due to habitat loss from deforestation and flooding by the creation of hydroelectric dams in Venezuela. Eighteen species of cardinals have been recorded in Texas. Four are considered rare visitors: Flame-colored Tanager, Crimson-collared Grosbeak, Yellow Grosbeak, and Blue Bunting.

Female

Male

Hepatic Tanager

(Piranga flava)

IUCN RED LIST STATUS (2019): Least Concern
POPULATION TREND: Increasing

Hepatic Tanager has a small range in North America but is actually one of the most widely distributed of the *Piranga* tanagers, breeding all the way to southeastern South America. There are three subspecies that look quite different and could be elevated to separate species. The common name Hepatic Tanager referes to the liver-red color of the males in northern population, but the specific name *flava* means "yellow" and was assigned from the original specimen collected in Paraguay, where the subspecies females are bright yellow.
LENGTH: 8 inches. **WINGSPAN:** 12.5 inches. **ADULT:** Male is red-orange; the color is brightest on the crown and throat. Bill is dark. Dusky gray auriculars. Gray flanks. Back is grayish. Female is yellow-orange, brightest on crown and throat, with dark bill. Dark eye line and grayish auriculars. Flanks are grayish. **VOICE:** Clear, slow jumble of phrases, grosbeak-like. Call is a high clipped *chup*. **BEHAVIORS:** Movements are slow and deliberate. Forages in pairs or groups. **HABITAT:** Open pine-oak woodlands. **STATUS:** Uncommon to common summer resident in the Chisos, Davis, and Guadalupe mountains. Rare to very rare migrant in the rest of the western Trans-Pecos. Breeders arrive in mid-April and are present until mid-October.

First-spring male

Male

Female

Summer Tanager

(Piranga rubra)

IUCN RED LIST STATUS (2016): Least Concern
POPULATION TREND: Stable

Summer Tanager is one of the boldest of neo-tropical songbirds in Texas. Adult males are the only all-red bird in Texas—even the ubiquitous Northern Cardinal has a black face. Even young male Summer Tanagers display a striking mottled red on yellow that is unmistakable when you see one.

LENGTH: 7.75 inches. **WINGSPAN:** 12 inches. **ADULT:** Male is bright red overall. Often shows a slight crest. Bill is a light horn color. Female is yellow to orange-yellow. Bill is a light horn color. **FIRST-SPRING MALE:** (Mar.–July) Like females with variable amounts of blotchy red. **VOICE:** Song is a series of three-note, robin-like phrases. Call is *pituk pittuktuk* or *chebek*. **BEHAVIORS:** Hops between perches in vegetation and on the ground. Flights are swift and direct. **HABITAT:** Deciduous and pine-oak forests. **STATUS:** Rare to common summer resident in the eastern half of Texas and south through the Edwards Plateau and South Texas Brush Country, except for the Lower Rio Grande Valley. Uncommon summer resident in the Trans-Pecos. Uncommon to common migrant statewide. Spring migration is early April–late May, fall mid-August–late October.

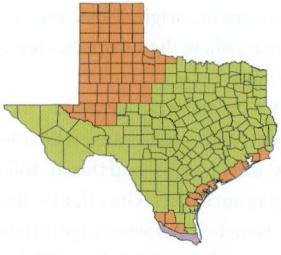

Male

Female

Scarlet Tanager

(Piranga olivacea)

IUCN RED LIST STATUS (2016): Least Concern
POPULATION TREND: Stable

Scarlet Tanager is the smallest of the *Piranga* tanagers. The birds in the genus are all still called tanagers, but the genus is no longer considered part of the "true tanagers" in the family Thraupidae. They are now placed in the family Cardinalidae, or the cardinals, grosbeaks, and buntings. All of the *Piranga* tanagers are colored in reds, oranges, and yellows. These colors are rare in the Thraupidae family, but common in the Cardinalidae family. **LENGTH:** 7 inches. **WINGSPAN:** 11.5 inches. **ADULT BREEDING:** (Mar.–Aug.) Brilliant red with black wings and black tail. Bill is gray. Female is yellow-green overall, with dark wings and olive upper wing coverts. Tail is dark. **NONBREEDING:** (Aug.–Mar.) Male is olive-yellow with black wings and tail. **VOICE:** Song is a rapid series of four or five raspy phrases. Call is *chik-brrr*. **BEHAVIORS:** Hops between perches and walks on branches and on the ground while foraging. **HABITAT:** Uses a variety of woodland habitats with tall trees in migration. **STATUS:** Rare to common spring migrant in the eastern third of Texas. Rare fall migrant in the eastern third of Texas. Rarely observed away from the coast. Spring migration is early April–mid-May, fall late August–late October.

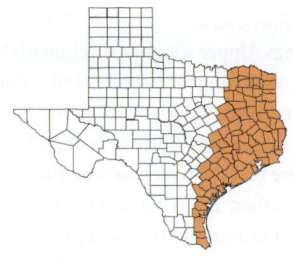

Male

Female

Western Tanager

(Piranga ludoviciana)

IUCN RED LIST STATUS (2016): Least Concern
POPULATION TREND: Increasing

Western Tanager's red male plumage differs from the red of the other *Prianga* tanagers, which produce their reds from yellow pigments. Western Tanager's red comes from a diet of insects that consume red pigment and pass it on. **LENGTH:** 7.25 inches. **WINGSPAN:** 11.5 inches. **ADULT BREEDING:** (Mar.–Aug.) Male has red head, yellow nape, and yellow body. Back is black. Black wings with yellow wingbar over white wingbar. Tail is black. Female is olive-yellow overall. Grayish back and dark gray wings. Upper wingbar is yellowish to white. Lower wingbar is white. Tail is dark. **NONBREEDING:** (Aug.–Mar.) Yellow head with reddish face. Yellow body with black back. Wings are black with yellow wingbar over white wingbar. Tail is black. **VOICE:** Song is a series of four or five repeated phrases. Call is *pri-deet*. **BEHAVIORS:** Hops on the ground and in trees. Jumps between branches. Flights are strong and direct. **HABITAT:** Breeds in open conifer and mixed conifer-deciduous woodlands. In migration, a variety of forest, woodland, scrub, and partially open habitats. **STATUS:** Uncommon to common summer resident in the Guadalupe and Davis mountains above 6000 feet. Uncommon to common migrant in the Trans-Pecos and High Plains. Rare migrant on the coastal prairies. Rare winter resident in the Lower Rio Grande Valley and coastal prairies. Spring migration is mid-April–late May, fall late August–early October.

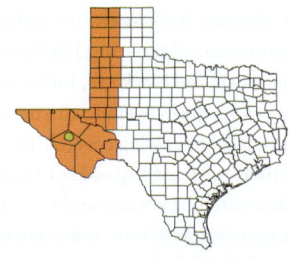

Female

Male

Northern Cardinal

(Cardinalis cardinalis)

IUCN RED LIST STATUS (2018): Least Concern
POPULATION TREND: Stable

The Northern Cardinal, often known colloquially as the Redbird in Texas, is familiar to almost all Texans as it is present in all but a very small part of the state. For many, seeing a Northern Cardinal represents a visit by a loved one who has passed away. Many have stories of being comforted by a cheery red Northern Cardinal when faced with the loss of someone special in their life.

LENGTH: 8.75 inches. **WINGSPAN:** 12 inches.
ADULT: Male is bright red overall with prominent red crest and red conical bill. Face and throat are black. Female is brownish overall with red flight feathers on wings and red tail. Tip of crest is red. Bill is red-orange. Face is dark gray. **JUVENILE:** (Apr.–Sept.) Brown overall with black bill that gradually changes to orange. Hints of red on wings and tail. **VOICE:** Song has many variations but typically

is *woit woit woit chew chew chew chew*. Call is a hard, sharp *tik tik*. **BEHAVIORS:** Hops on the ground and in vegetation. Males often sing from a high perch, but most birds forage close to or on the ground. **HABITAT:** Areas with shrubs and small trees, forest edges. **STATUS:** Common to abundant resident throughout most of the state, except El Paso County and the northern parts of Culberson and Hudspeth counties.

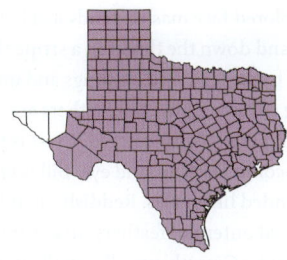

Male

Female

Pyrrhuloxia

(Cardinalis sinuatus)

IUCN RED LIST STATUS (2016): Least Concern
POPULATION TREND: Decreasing

Pyrrhuloxia is the Southwestern version of the closely related Northern Cardinal. The name Pyrrhuloxia is descriptive and was derived from two other genera of birds: *pyrrhula* from the bullfinches, referring to the flame color of the males, and *loxia* meaning "crooked," referring to the crooked bill.

LENGTH: 8.75 inches. **WINGSPAN:** 12 inches. **ADULT:** Male is gray with wine-colored crest and yellow, parrot-like hooked bill. Wine-colored face mask extends just behind the eye and down the throat in a stripe that extends to the legs. Red on wings and underwing coverts. Outer tail feathers are red. Female is grayish-clay-colored. Crest is tipped in wine-color. Dark around eye. Bill is yellow and rounded like males. Reddish flight feathers and red outer tail feathers. **JUVENILE:** (May–Sept.) Grayish overall. Small amount

of red on crest and wing. Rounded bill is dark. **VOICE:** Song is a repeated *tchew tchew tchew tchew*. Call is a rapid series of hoarse *chert-chert-chert-chert-chert*. **BEHAVIORS:** Hops on the ground. Flight is undulating. **HABITAT:** Scrubby mesquite, shrubby grasslands, and chaparral. **STATUS:** Common to abundant resident in the southwestern half of the state, east to the Guadalupe Delta, and west to the southern High Plains.

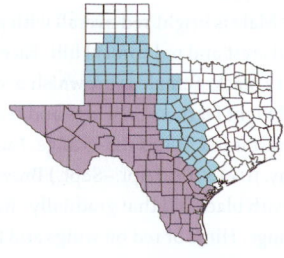

Male

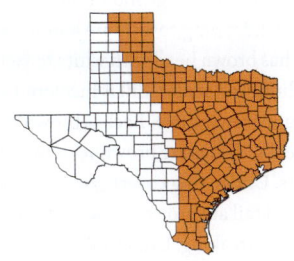

Female

Rose-breasted Grosbeak

(Pheucticus ludovicianus)

IUCN RED LIST STATUS (2017): Least Concern
POPULATION TREND: Stable

The male Rose-breasted Grosbeak is distinctive with with its black head, white wing patches, and rose-red breast pattern. **LENGTH:** 8 inches. **WINGSPAN:** 12.5 inches. **ADULT BREEDING:** (Mar.–Aug.) Male has black head and bright red breast. Bill is pale pink. White below. Back and wings are black with large white wing patch and white wingbar. Underwing coverts are red. Tail is black with white corners. Female has brown head with bold white supercilium and malar area. Back is brown with dark streaks. Wings are brown with white wingbars. Underparts are white with heavy brown streaks on breast and flanks. **NONBREEDING FEMALE:** (Aug.–Mar.) Like breeding females but with buffy wash across breast and flanks. **FIRST-WINTER MALE:** (Aug–Mar.) Brown head with white supercilium and malar. Buffy collar. Breast and flanks are buffy with faint thin streaks. Belly and undertail coverts are white. Back is brown with dark streaks. Wings are brown with white wingbars. **VOICE:** Song is a slow, whistling, robin-like warble. Call is a squeaking *chiit*. **BEHAVIORS:** Does not walk but hops. **HABITAT:** In migration, uses a wide variety of habitats, including shrubs and secondary forests. **STATUS:** Common to uncommon migrant in the eastern half of the state. Rare winter visitor along the coastal prairies in the Lower Rio Grande Valley. Spring migration is early April–late May, fall mid-September–early November.

Male

Female

Black-headed Grosbeak

(Pheucticus melanocephalus)

IUCN RED LIST STATUS (2016): Least Concern
POPULATION TREND: Increasing

The Black-headed Grosbeak, like the closely related Rose-breasted Grosbeak, has very different male and female plumages.
LENGTH: 8.25 inches. **WINGSPAN:** 12.5 inches.
ADULT BREEDING: (Mar.–Aug.) Male has black head with orange-rufous collar and breast. Back is black with orange-rufous streaks. Rump is orange-rufous. Tail is black with white corners. Wings are black with two wide white wingbars. In flight, the wings show bold white wing patches. Underwing coverts are lemon-yellow. Female has brown head with white to yellowish supercilium and malar. Bill is bicolored with darker upper mandible. Back is brown with pale streaks. Wings are brown with two white wingbars. Underwing coverts are lemon-yellow. Rump and tail are brown. Breast is buffy with fine sparse streaking. Undertail coverts are white. **NONBREEDING FEMALE:** (Aug.–Mar.)

Like breeding females but more rufous below and on nape. **FIRST-WINTER MALE:** (Aug.–Mar.) Dark brown head with a bold white supercilium and malar. Orange-rufous below and on nape. Brown on back with rufous streaks. **VOICE:** Song is a rapid, choppy, whistled warble. Call is a squeaky *PIK*. **BEHAVIORS:** Hops on the ground and on branches. **HABITAT:** Variety of woodlands. **STATUS:** Common migrant and summer resident in the mountains of the Trans-Pecos. Uncommon to rare migrant in the Trans-Pecos, High Plains, and western part of the Rolling Plains. Spring migration is early April–late May, fall mid-September–mid-November.

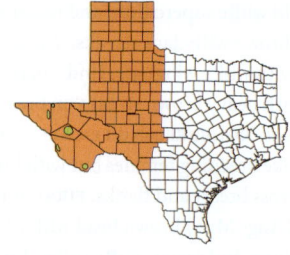

Male

Female

Blue Grosbeak

(Passerina caerulea)

IUCN RED LIST STATUS (2018): Least Concern
POPULATION TREND: Increasing

Blue Grosbeak is actually a bunting and member of the genus *Passerina*, not one of the grosbeaks in the genus *Pheucticus*. It is closely related to Indigo Bunting. While widely distributed in North America, its numbers are actually relatively low. This general scarcity of study subjects means virtually all aspects of its biology are poorly understood.
LENGTH: 6.75 inches. **WINGSPAN:** 11 inches. **ADULT:** Male is dark blue overall with heavy blue-gray bill. Black lores with rufous wingbars. Back has dark streaks. Nonbreeding and immature females are gray-brown overall with buff-rufous wingbars. Tail and rump are often bluish. **FIRST-SUMMER MALE:** (Mar.–Sept.) Gray-brown overall with variable amounts of mottled dark blue. Two rufous wingbars. **VOICE:** Song is a steady warble, very bunting-like. Call is a metallic *pink pink*.

BEHAVIORS: Flight is swift and low. Hops on the ground and in trees. **HABITAT:** Old fields, forest edge, open scrubby areas. **STATUS:** Locally common to uncommon summer resident statewide. Common to uncommon migrant statewide. Spring migration is early April–mid-May, fall mid-August–mid-October.

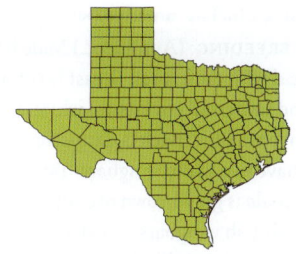

Lazuli Bunting

(Passerina amoena)

IUCN RED LIST STATUS (2016): Least Concern
POPULATION TREND: Increasing

Lazuli Bunting is named for the blue gemstone lapis lazuli, due to the male's stunning blue color. Each male sings a unique song by the time they are 2 years old. When first-year males arrive on the breeding grounds, they have no song of their own and copy other males' songs to develop the elements into their own song. This can lead to a sort of sonic neighborhood, or local accents where neighboring males may all have similar songs. **LENGTH:** 5.5 inches. **WINGSPAN:** 8.75 inches. **ADULT BREEDING:** (Apr.–Sept.) Male has blue head with dark lores. Breast is rufous. Belly and undertail coverts are white. Back and wings are blue. Back is dark-streaked. Wings have two white wingbars. Tail is dark blue. Female is gray-brown overall. Two narrow whitish wingbars. **NONBREEDING:** (Sept.–Apr.) Male is like breeding males but back and head are mottled blue and rufous. Nonbreeding and immature females are like breeding females. **VOICE:** Song is a short, high-pitched warble. Call is a dry *stip stip*. **BEHAVIORS:** Hops on the ground and in vegetation. **HABITAT:** Shrubby habitat and riparian areas. **STATUS:** Uncommon migrant through the Trans-Pecos, Panhandle, and South Plains. Rare migrant east to the Oaks and Prairies. Spring migration is early April– late May, fall early August–early October.

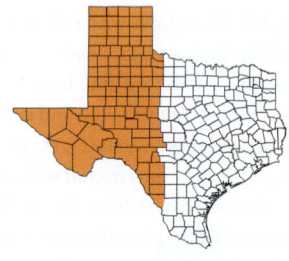

Breeding male

Immature

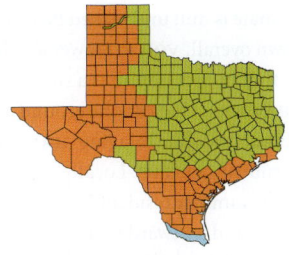

Female

Indigo Bunting

(Passerina cyanea)

IUCN RED LIST STATUS (2018): Least Concern
POPULATION TREND: Decreasing

Indigo Bunting is one of the most familiar and spectacular songbirds in eastern North America. One of the most common migrants on the Texas coast, it is not uncommon to have dozens at a time in coastal migrant traps. **LENGTH:** 5.5 inches. **WINGSPAN:** 8 inches. **ADULT BREEDING:** (Apr.–Sept.) Male is indigo blue overall. Flight feathers and tail are darker. Female is gray-brown overall. Throat is whitish. Fine streaking on breast. Tail is bluish. **NONBREEDING:** (Sept.–Apr.) Male is mottled blue and tan overall. Nonbreeding and immature females are like breeding females. **VOICE:** High-pitched warble with repeated phrases. Call is a sharp *spik spik*. **BEHAVIORS:** While foraging, hops on the ground and branches. Flights are direct. **HABITAT:** Shrubby and weedy habitats between woods, utility rights-of-way, and generating woodlands. **STATUS:** Common to abundant summer resident in the eastern half of Texas south to the coastal prairies and South Texas Brush Country, and north to the eastern edge of the Panhandle. Uncommon winter resident in the Lower Rio Grande Valley. Common to rare migrant statewide. Spring migration is late March–late May, fall mid-August–mid-October.

Male

Female

Varied Bunting

(Passerina versicolor)

IUCN RED LIST STATUS (2018): Least Concern
POPULATION TREND: Stable

Varied Bunting is a tropical species that bare-ly makes it into North America in Texas and Arizona along the Mexico border. Because the bird prefers inhospitable thornbush habitat, the intrepid birder who seeks this species will have to work for it.
LENGTH: 5.5 inches. **WINGSPAN:** 7.75 inches.
ADULT: Male has black face with red orbit-al ring. Head is violet with red nape. Body is purplish with blue undertail coverts and rump. Female is dull unstreaked brown to gray-brown overall. **VOICE:** Slower and harsh-er than other buntings. Call is a sharp *spik spik.* **BEHAVIORS:** Hops on the ground and among branches while foraging. **HABITAT:** Arid thornbush. **STATUS:** Locally uncom-mon to rare summer resident in the southern Trans-Pecos and eastward to the southwest-ern Edwards Plateau. Rare and local summer resident along the Rio Grande River south to the Lower Rio Grande Valley. Spring mi-gration is mid-April–mid-May, fall early August–mid-September.

Male

Female

Painted Bunting

(Passerina ciris)

IUCN RED LIST STATUS (2018): Least Concern
POPULATION TREND: Stable

Painted Bunting is the gaudiest of Texas songbirds, a riot of colors seemingly designed by
a child with a new box of crayons. Birders
often speak of their "hook" or "spark" bird—
the bird that made them want to know more
about birds and ignited the passion to see
more of them. No bird is cited more often as a
hook bird than the Painted Bunting.
LENGTH: 5.5 inches. **WINGSPAN:** 8.75 inches.
ADULT: Male has blue head, green back, and
red rump. Throat and underparts are bright
red. Wings and tail are dark. Female is bright
lime-green overall with pale eye ring. Underparts are lighter. **IMMATURE:** (Aug.–Oct.)
Like females but duller with less yellow tone.
VOICE: Sweet continuous warble. Call is *spep
spep*. **BEHAVIORS:** Hops when foraging. Males
sing from a high perch, often on top of a tree.
HABITAT: Semi-open country with tall patches
of weeds and grasses. **STATUS:** Uncommon
to common summer resident and migrant
statewide. Spring migration is late March–
mid-May, fall mid-July–mid-September, with
most breeding birds having departed by the
end of August.

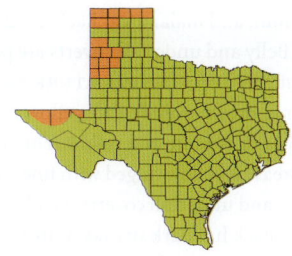

Adult

Immature

Dickcissel

(Spiza americana)

IUCN RED LIST STATUS (2018): Least Concern
POPULATION TREND: Stable

Dickcissel is a specialist of America's prairie grasslands.
LENGTH: 6.25 inches. **WINGSPAN:** 9.75 inches.
ADULT: Male has gray head with yellow supercilium and malar area. Black V on throat with white just below bill. Yellow breast and light gray below. Undertail coverts are white. Back is brownish-gray with dark streaks. Rufous shoulder patches on wings. Female is gray-brown with white throat. Yellow breast, supercilium, and malar area. Back has dark streaks. Belly and undertail coverts are pale. Chestnut patch on wing. **IMMATURE:** (Aug.–Mar.) Gray-brown overall. Pale yellow supercilium. Malar is white and yellow. Throat is white. Breast is yellow tinged with fine streaks. Pale belly and undertail coverts. Flanks are streaked. Back has dark streaks. **VOICE:** Song is a descending *dick dick dick ciss ciss cell.*

Flight call is a distinctive, raspberry-like *fppt.*
BEHAVIORS: Walks and hops on the ground. Males sing from a tall stalk. **HABITAT:** A grassland specialist, it uses a variety of grasslands between 1 and 4 feet tall. A high proportion of forbs are needed for song perches and nesting. **STATUS:** Uncommon to abundant summer resident and migrant east of the Pecos River. Spring migration is early April–late May, fall late August–late October.

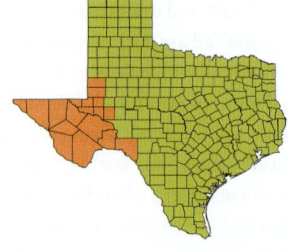

THRAUPIDAE
(TANAGERS AND ALLIES)

The tanagers are a large family of mostly neotropical songbirds. They are medium-size but have diversified into a dazzling array of color combinations. Many are some of the most colorful and dramatic birds known. Within the family they have diversified into a huge array of forms, from the thin-hooked bills of the flowerpiercers to the huge seed-crushing bills of the Darwin's finches on the Galapagos Islands.

The tanager family is now thought to be sister to the cardinal family. Recently, the North American tanagers in the genus *Piranga* have been placed in the cardinal family. Adding to the confusion for birders is the fact that there are several genera of bird in the tanager family with misleading names. Birds in the genus *Paroaria* are known as cardinals but are tanagers. Several genera are known as finches but are not part of the finch family. Other genera are known as seedeaters, flowerpiercers, dacnis, saltators, and hemispingus.

The are 384 species of tanagers worldwide. Habitat loss and introduced predators are the primary threats to the 61 species of conservation concern. Two species of Darwin's finches on the Galapagos Islands are critically endangered. The Gough Bunting of Tristan da Cunha Island is critically endangered by introduced predators. The other critically endangered species are Cherry-throated Tanager and Cone-billed Tanager. Because of the move of the *Piranga* genus to the cardinals, Texas has only one species of tanager, the Morelet's Seedeater.

Morelet's Seedeater

(Sporophila morelleti)

IUCN RED LIST STATUS (2021): Least Concern
POPULATION TREND: Increasing

Morelet's Seedeater was formerly more common in Texas than it is now. It was once relatively common in Hidalgo County but was extirpated from Texas in the 1950s and 1960s. It has been suggested that pesticides like DDT were the cause. In the 1980s, it recolonized Texas in Starr County and slowly expanded north along the Rio Grande River to Val Verde County.

LENGTH: 4.5 inches. **WINGSPAN:** 6.25 inches. **ADULT:** Male has black crown and face, including auriculars. Thick white arc under eye. Short stubby bill. Pale buffy to white collar. Breast is pale buffy to white. Dark breast band. Wings are dark with two wingbars. Female is pale buffy-olive overall. Bill is short and stubby. Two thin pale wingbars. **NONBREEDING:** Male is variable; head can be dark to completely buffy-olive. Partial or completely absent breast band. **VOICE:** Song is a sweet *tew tew tew tew sweet sweet sweet*. Call is a squeaky *stee-oop*. **BEHAVIORS:** Rarely on the ground; often remains in trees or shrubs. Climbs grass and bends seed heads down horizontal to forage. **HABITAT:** Open grassy places, often near water and canals, roadsides, and weedy fields. Often found in association with carrizo, or cane, along the Rio Grande River. **STATUS:** Uncommon to rare resident along the Rio Grande River from northern Starr County to southern Val Verde County.

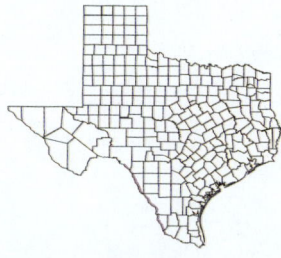

CHECKLIST OF TEXAS BIRDS

I = Introduced
RI = Reintroduction in progress - not established
E = Extinct
e = extirpated
u = uncertain origin (stable to increasing populations of introduced/native origin)
Special treatment = Review species, birds that have occurred four or fewer times per year anywhere in Texas over a ten-year average, the TBRC requests documentation for review for any occurrence of these species.
* = no species account in *Birds of Texas*

ANATIDAE (DUCKS, GEESE, AND SWANS)

- Black-bellied Whistling-Duck, *Dendrocygna autumnalis*
- Fulvous Whistling-Duck, *Dendrocygna bicolor*
- Snow Goose, *Anser caerulescens*
- Ross's Goose, *Anser rossii*
- Greater White-fronted Goose, *Anser albifrons*
- Brant, *Branta bernicla**
- Cackling Goose, *Branta hutchinsii*
- Canada Goose, *Branta canadensis*
- Trumpeter Swan, *Cygnus buccinator**
- Tundra Swan, *Cygnus columbianus*
- Muscovy Duck, *Cairina moschata*
- Wood Duck, *Aix sponsa*
- Garganey, *Spatula querquedula**
- Blue-winged Teal, *Spatula discors*
- Cinnamon Teal, *Spatula cyanoptera*
- Northern Shoveler, *Spatula clypeata*
- Gadwall, *Mareca strepera*
- Eurasian Wigeon, *Mareca penelope**
- American Wigeon, *Mareca americana*
- Mallard, *Anas platyrhynchos*
- Mexican Duck, *Anas diazi*
- American Black Duck, *Anas rubripes**
- Mottled Duck, *Anas fulvigula*
- White-cheeked Pintail, *Anas bahamensis**
- Northern Pintail, *Anas acuta*
- Green-winged Teal, *Anas crecca*
- Canvasback, *Aythya valisineria*
- Redhead, *Aythya americana*
- Ring-necked Duck, *Aythya collaris*
- Greater Scaup, *Aythya marila*
- Lesser Scaup, *Aythya affinis*
- King Eider, *Somateria spectabilis**
- Common Eider, *Somateria mollissima**
- Harlequin Duck, *Histrionicus histrionicus**
- Surf Scoter, *Melanitta perspicillata*
- White-winged Scoter, *Melanitta deglandi*
- Black Scoter, *Melanitta americana*
- Long-tailed Duck, *Clangula hyemalis*
- Bufflehead, *Bucephala albeola*
- Common Goldeneye, *Bucephala clangula*
- Barrow's Goldeneye, *Bucephala islandica**
- Hooded Merganser, *Lophodytes cucullatus*
- Common Merganser, *Mergus merganser*
- Red-breasted Merganser, *Mergus serrator*
- Masked Duck, *Nomonyx dominicus**
- Ruddy Duck, *Oxyura jamaicensis*

CRACIDAE (GUANS, CHACHALACAS, AND CURASSOWS)

- Plain Chachalaca, *Ortalis vetula*

ODONTOPHORIDAE (NEW WORLD QUAIL)

- Northern Bobwhite, *Colinus virginianus*
- Scaled Quail, *Callipepla squamata*
- Gambel's Quail, *Callipepla gambelii*
- Montezuma Quail, *Cyrtonyx montezumae*

PHASIANIDAE
(PHEASANTS, GROUSE, AND ALLIES)
- ☐ Wild Turkey, *Meleagris gallopavo*
- ☐ Greater Prairie-Chicken, *Tympanuchus cupido* (RI)*
- ☐ Lesser Prairie-Chicken, *Tympanuchus pallidicinctus*
- ☐ Ring-necked Pheasant, *Phasianus colchicus* (I)

PHOENICOPTERIDAE (FLAMINGOS)
- ☐ American Flamingo, *Phoenicopterus ruber**

PODICIPEDIDAE (GREBES)
- ☐ Least Grebe, *Tachybaptus dominicus*
- ☐ Pied-billed Grebe, *Podilymbus podiceps*
- ☐ Horned Grebe, *Podiceps auritus*
- ☐ Red-necked Grebe, *Podiceps grisegena**
- ☐ Eared Grebe, *Podiceps nigricollis*
- ☐ Western Grebe, *Aechmophorus occidentalis*
- ☐ Clark's Grebe, *Aechmophorus clarkii*

COLUMBIDAE (PIGEONS AND DOVES)
- ☐ Rock Pigeon, *Columba livia* (I)
- ☐ White-crowned Pigeon, *Patagioenas leucocephala*
- ☐ Red-billed Pigeon, *Patagioenas flavirostris*
- ☐ Band-tailed Pigeon, *Patagioenas fasciata*
- ☐ Eurasian Collared-Dove, *Streptopelia decaocto* (I)
- ☐ Passenger Pigeon, *Ectopistes migratorius (E)*
- ☐ Inca Dove, *Columbina inca*
- ☐ Common Ground Dove, *Columbina passerina*
- ☐ Ruddy Ground Dove, *Columbina talpacoti**
- ☐ Ruddy Quail-Dove, *Geotrygon montana**
- ☐ White-tipped Dove, *Leptotila verreauxi*
- ☐ White-winged Dove, *Zenaida asiatica*
- ☐ Mourning Dove, *Zenaida macroura*

CUCULIDAE (CUCKOOS)
- ☐ Smooth-billed Ani, *Crotophaga ani**
- ☐ Groove-billed Ani, *Crotophaga sulcirostris*
- ☐ Greater Roadrunner, *Geococcyx californianus*
- ☐ Dark-billed Cuckoo, *Coccyzus melacoryphus**

- ☐ Yellow-billed Cuckoo, *Coccyzus americanus*
- ☐ Mangrove Cuckoo, *Coccyzus minor**
- ☐ Black-billed Cuckoo, *Coccyzus erythropthalmus*

CAPRIMULGIDAE
(NIGHTJARS AND ALLIES)
- ☐ Lesser Nighthawk, *Chordeiles acutipennis*
- ☐ Common Nighthawk, *Chordeiles minor*
- ☐ Common Pauraque, *Nyctidromus albicollis*
- ☐ Common Poorwill, *Phalaenoptilus nuttallii*
- ☐ Chuck-will's-widow, *Antrostomus carolinensis*
- ☐ Eastern Whip-poor-will, *Antrostomus vociferus*
- ☐ Mexican Whip-poor-will, *Antrostomus arizonae*

APODIDAE (SWIFTS)
- ☐ Black Swift, *Cypseloides niger**
- ☐ White-collared Swift, *Streptoprocne zonaris**
- ☐ Chimney Swift, *Chaetura pelagica*
- ☐ White-throated Swift, *Aeronautes saxatalis*

TROCHILIDAE (HUMMINGBIRDS)
- ☐ Mexican Violetear, *Colibri thalassinus**
- ☐ Green-breasted Mango, *Anthracothorax prevostii**
- ☐ Rivoli's Hummingbird, *Eugenes fulgens*
- ☐ Amethyst-throated Mountain-gem, *Lampornis amethystinus**
- ☐ Blue-throated Mountain-gem, *Lampornis clemenciae*
- ☐ Lucifer Hummingbird, *Calothorax lucifer*
- ☐ Ruby-throated Hummingbird, *Archilochus colubris*
- ☐ Black-chinned Hummingbird, *Archilochus alexandri*
- ☐ Anna's Hummingbird, *Calypte anna*
- ☐ Costa's Hummingbird, *Calypte costae**
- ☐ Calliope Hummingbird, *Selasphorus calliope*
- ☐ Rufous Hummingbird, *Selasphorus rufus*
- ☐ Allen's Hummingbird, *Selasphorus sasin*
- ☐ Broad-tailed Hummingbird, *Selasphorus platycercus*

- Broad-billed Hummingbird, *Cynanthus latirostris*
- White-eared Hummingbird, *Basilinna leucotis**
- Violet-crowned Hummingbird, *Ramosomyia violiceps**
- Berylline Hummingbird, *Saucerottia beryllina**
- Buff-bellied Hummingbird, *Amazilia yucatanensis*

**RALLIDAE
(RAILS, GALLINULES, AND COOTS)**
- Spotted Rail, *Pardirallus maculatus**
- Paint-billed Crake, *Mustelirallus erythrops**
- Clapper Rail, *Rallus crepitans*
- King Rail, *Rallus elegans*
- Virginia Rail, *Rallus limicola*
- Sora, *Porzana carolina*
- Common Gallinule, *Gallinula galeata*
- American Coot, *Fulica americana*
- Purple Gallinule, *Porphyrio martinicus*
- Yellow Rail, *Coturnicops noveboracensis*
- Black Rail, *Laterallus jamaicensis*

ARAMIDAE (LIMPKINS)
- Limpkin, *Aramus guarauna*

GRUIDAE (CRANES)
- Sandhill Crane, *Antigone canadensis*
- Common Crane, *Grus grus**
- Whooping Crane, *Grus americana*

BURHINIDAE (THICK-KNEES)
- Double-striped Thick-knee, *Hesperoburhinus bistriatus**

RECURVIROSTRIDAE (STILTS AND AVOCETS)
- Black-necked Stilt, *Himantopus mexicanus*
- American Avocet, *Recurvirostra americana*

HAEMATOPODIDAE (OYSTERCATCHERS)
- American Oystercatcher, *Haematopus palliatus*

CHARADRIIDAE (PLOVERS AND LAPWINGS)
- Black-bellied Plover, *Pluvialis squatarola*
- American Golden-Plover, *Pluvialis dominica*
- Pacific Golden-Plover, *Pluvialis fulva**
- Killdeer, *Charadrius vociferus*
- Semipalmated Plover, *Charadrius semipalmatus*
- Piping Plover, *Charadrius melodus*
- Southern Lapwing, *Vanellus chilensis**
- Wilson's Plover, *Anarhynchus wilsonia*
- Collared Plover, *Anarhynchus collaris**
- Mountain Plover, *Anarhynchus montanus*
- Snowy Plover, *Anarhynchus nivosus*

JACANIDAE (JACANAS)
- Northern Jacana, *Jacana spinosa**

SCOLOPACIDAE (SANDPIPERS AND ALLIES)
- Upland Sandpiper, *Bartramia longicauda*
- Whimbrel, *Numenius phaeopus*
- Eskimo Curlew, *Numenius borealis**
- Long-billed Curlew, *Numenius americanus*
- Bar-tailed Godwit, *Limosa lapponica**
- Black-tailed Godwit, *Limosa limosa**
- Hudsonian Godwit, *Limosa haemastica*
- Marbled Godwit, *Limosa fedoa*
- Ruddy Turnstone, *Arenaria interpres*
- Black Turnstone, *Arenaria melanocephala**
- Red Knot, *Calidris canutus*
- Surfbird, *Calidris virgata**
- Ruff, *Calidris pugnax**
- Sharp-tailed Sandpiper, *Calidris acuminata**
- Stilt Sandpiper, *Calidris himantopus*
- Curlew Sandpiper, *Calidris ferruginea**
- Red-necked Stint, *Calidris ruficollis**
- Sanderling, *Calidris alba*
- Dunlin, *Calidris alpina*
- Purple Sandpiper, *Calidris maritima**
- Baird's Sandpiper, *Calidris bairdii*
- Least Sandpiper, *Calidris minutilla*
- White-rumped Sandpiper, *Calidris fuscicollis*
- Buff-breasted Sandpiper, *Calidris subruficollis*

- Pectoral Sandpiper, *Calidris melanotos*
- Semipalmated Sandpiper, *Calidris pusilla*
- Western Sandpiper, *Calidris mauri*
- Short-billed Dowitcher, *Limnodromus griseus*
- Long-billed Dowitcher, *Limnodromus scolopaceus*
- American Woodcock, *Scolopax minor*
- Wilson's Snipe, *Gallinago delicata*
- Spotted Sandpiper, *Actitis macularius*
- Solitary Sandpiper, *Tringa solitaria*
- Wandering Tattler, *Tringa incana**
- Lesser Yellowlegs, *Tringa flavipes*
- Willet, *Tringa semipalmata*
- Spotted Redshank, *Tringa erythropus**
- Greater Yellowlegs, *Tringa melanoleuca*
- Wilson's Phalarope, *Phalaropus tricolor*
- Red-necked Phalarope, *Phalaropus lobatus*
- Red Phalarope, *Phalaropus fulicarius**

STERCORARIIDAE (SKUAS AND JAEGERS)

- South Polar Skua, *Stercorarius maccormicki**
- Pomarine Jaeger, *Stercorarius pomarinus*
- Parasitic Jaeger, *Stercorarius parasiticus*
- Long-tailed Jaeger, *Stercorarius longicaudus**

LARIDAE (GULLS, TERNS, AND SKIMMERS)

- Black-legged Kittiwake, *Rissa tridactyla*
- Sabine's Gull, *Xema sabini*
- Bonaparte's Gull, *Chroicocephalus philadelphia*
- Black-headed Gull, *Chroicocephalus ridibundus**
- Little Gull, *Hydrocoloeus minutus*
- Laughing Gull, *Leucophaeus atricilla*
- Franklin's Gull, *Leucophaeus pipixcan*
- Black-tailed Gull, *Larus crassirostris**
- Heermann's Gull, *Larus heermanni**
- Short-billed Gull, *Larus brachyrhynchus**
- Ring-billed Gull, *Larus delawarensis*
- Western Gull, *Larus occidentalis**
- California Gull, *Larus californicus*
- Herring Gull, *Larus argentatus*
- Yellow-legged Gull, *Larus michahellis**
- Iceland Gull, *Larus glaucoides*

- Lesser Black-backed Gull, *Larus fuscus*
- Slaty-backed Gull, *Larus schistisagus**
- Glaucous-winged Gull, *Larus glaucescens**
- Glaucous Gull, *Larus hyperboreus*
- Great Black-backed Gull, *Larus marinus**
- Kelp Gull, *Larus dominicanus**
- Brown Noddy, *Anous stolidus**
- Black Noddy, *Anous minutus**
- Sooty Tern, *Onychoprion fuscatus*
- Bridled Tern, *Onychoprion anaethetus**
- Least Tern, *Sternula antillarum*
- Gull-billed Tern, *Gelochelidon nilotica*
- Caspian Tern, *Hydroprogne caspia*
- Black Tern, *Chlidonias niger*
- Roseate Tern, *Sterna dougallii**
- Common Tern, *Sterna hirundo*
- Arctic Tern, *Sterna paradisaea**
- Forster's Tern, *Sterna forsteri*
- Royal Tern, *Thalasseus maximus*
- Sandwich Tern, *Thalasseus sandvicensis*
- Elegant Tern, *Thalasseus elegans**
- Black Skimmer, *Rynchops niger*

PHAETHONTIDAE (TROPICBIRDS)

- White-tailed Tropicbird, *Phaethon lepturus**
- Red-billed Tropicbird, *Phaethon aethereus**

GAVIIDAE (LOONS)

- Red-throated Loon, *Gavia stellata*
- Pacific Loon, *Gavia pacifica*
- Common Loon, *Gavia immer*
- Yellow-billed Loon, *Gavia adamsii**

DIOMEDEIDAE (ALBATROSSES)

- Yellow-nosed Albatross, *Thalassarche chlororhynchos**

OCEANITIDAE (SOUTHERN STORM-PETRELS)

- Wilson's Storm-Petrel, *Oceanites oceanicus**

HYDROBATIDAE (NORTHERN STORM-PETRELS)

- Leach's Storm-Petrel, *Hydrobates leucorhous**
- Band-rumped Storm-Petrel, *Hydrobates castro**

PROCELLARIIDAE
(SHEARWATERS AND PETRELS)
- Trindade Petrel, *Pterodroma arminjoniana**
- Black-capped Petrel, *Pterodroma hasitata**
- Stejneger's Petrel, *Pterodroma longirostris**
- White-chinned Petrel, *Procellaria aequinoctialis**
- Cory's Shearwater, *Calonectris borealis**
- Scopoli's Shearwater, *Calonectris diomedea*
- Wedge-tailed Shearwater, *Ardenna pacifica**
- Sooty Shearwater, *Ardenna grisea**
- Great Shearwater, *Ardenna gravis**
- Manx Shearwater, *Puffinus puffinus**
- Sargasso Shearwater, *Puffinus lherminieri**

CICONIIDAE (STORKS)
- Jabiru, *Jabiru mycteria**
- Wood Stork, *Mycteria americana**

FREGATIDAE (FRIGATEBIRDS)
- Magnificent Frigatebird, *Fregata magnificens*

SULIDAE (BOOBIES AND GANNETS)
- Masked Booby, *Sula dactylatra*
- Blue-footed Booby, *Sula nebouxii**
- Brown Booby, *Sula leucogaster*
- Red-footed Booby, *Sula sula**
- Northern Gannet, *Morus bassanus*

ANHINGIDAE (ANHINGAS)
- Anhinga, *Anhinga anhinga*

PHALACROCORACIDAE
(CORMORANTS AND SHAGS)
- Double-crested Cormorant, *Nannopterum auritum*
- Neotropic Cormorant, *Nannopterum brasilianum*

PELECANIDAE (PELICANS)
- American White Pelican, *Pelecanus erythrorhynchos*
- Brown Pelican, *Pelecanus occidentalis*

ARDEIDAE
(HERONS, EGRETS, AND BITTERNS)
- Bare-throated Tiger-Heron, *Tigrisoma mexicanum**
- Least Bittern, *Botaurus exilis*
- American Bittern, *Botaurus lentiginosus*
- Little Blue Heron, *Egretta caerulea*
- Tricolored Heron, *Egretta tricolor*
- Reddish Egret, *Egretta rufescens*
- Snowy Egret, *Egretta thula*
- Yellow-crowned Night Heron, *Nyctanassa violacea*
- Black-crowned Night Heron, *Nycticorax nycticorax*
- Green Heron, *Butorides virescens*
- Great Egret, *Ardea alba*
- Western Cattle Egret, *Ardea ibis*
- Great Blue Heron, *Ardea herodias*

THRESKIORNITHIDAE (IBISES AND
SPOONBILLS)
- White Ibis, *Eudocimus albus*
- Glossy Ibis, *Plegadis falcinellus*
- White-faced Ibis, *Plegadis chihi*
- Roseate Spoonbill, *Platalea ajaja*

CATHARTIDAE (NEW WORLD VULTURES)
- Black Vulture, *Coragyps atratus*
- Turkey Vulture, *Cathartes aura*

PANDIONIDAE (OSPREY)
- Osprey, *Pandion haliaetus*

ACCIPITRIDAE
(HAWKS, EAGLES, AND KITES)
- White-tailed Kite, *Elanus leucurus*
- Hook-billed Kite, *Chondrohierax uncinatus*
- Swallow-tailed Kite, *Elanoides forficatus*
- Golden Eagle, *Aquila chrysaetos*
- Double-toothed Kite, *Harpagus bidentatus**
- Northern Harrier, *Circus hudsonius*
- Sharp-shinned Hawk, *Accipiter striatus*
- Cooper's Hawk, *Accipiter cooperii*
- American Goshawk, *Accipiter atricapillus**

- Bald Eagle, *Haliaeetus leucocephalus*
- Steller's Sea-Eagle, *Haliaeetus pelagicus**
- Mississippi Kite, *Ictinia mississippiensis*
- Crane Hawk, *Geranospiza caerulescens**
- Snail Kite, *Rostrhamus sociabilis**
- Common Black Hawk, *Buteogallus anthracinus*
- Great Black Hawk, *Buteogallus urubitinga**
- Roadside Hawk, *Rupornis magnirostris**
- Harris's Hawk, *Parabuteo unicinctus*
- White-tailed Hawk, *Geranoaetus albicaudatus*
- Gray Hawk, *Buteo plagiatus*
- Red-shouldered Hawk, *Buteo lineatus*
- Broad-winged Hawk, *Buteo platypterus*
- Short-tailed Hawk, *Buteo brachyurus**
- Swainson's Hawk, *Buteo swainsoni*
- Zone-tailed Hawk, *Buteo albonotatus*
- Red-tailed Hawk, *Buteo jamaicensis*
- Rough-legged Hawk, *Buteo lagopus*
- Ferruginous Hawk, *Buteo regalis*

TYTONIDAE (BARN OWLS)

- American Barn Owl, *Tyto furcata*

STRIGIDAE (OWLS)

- Flammulated Owl, *Psiloscops flammeolus*
- Western Screech-Owl, *Megascops kennicottii*
- Eastern Screech-Owl, *Megascops asio*
- Great Horned Owl, *Bubo virginianus*
- Snowy Owl, *Bubo scandiacus**
- Northern Pygmy-Owl, *Glaucidium gnoma**
- Ferruginous Pygmy-Owl, *Glaucidium brasilianum*
- Elf Owl, *Micrathene whitneyi*
- Burrowing Owl, *Athene cunicularia*
- Mottled Owl, *Strix virgata**
- Spotted Owl, *Strix occidentalis**
- Barred Owl, *Strix varia*
- Long-eared Owl, *Asio otus*
- Stygian Owl, *Asio stygius**
- Short-eared Owl, *Asio flammeus*
- Northern Saw-whet Owl, *Aegolius acadicus**

TROGONIDAE (TROGONS)

- Elegant Trogon, *Trogon elegans**

ALCEDINIDAE (KINGFISHERS)

- Ringed Kingfisher, *Megaceryle torquata*
- Belted Kingfisher, *Megaceryle alcyon*
- Amazon Kingfisher, *Chloroceryle amazona**
- Green Kingfisher, *Chloroceryle americana*

PICIDAE (WOODPECKERS)

- Lewis's Woodpecker, *Melanerpes lewis*
- Red-headed Woodpecker, *Melanerpes erythrocephalus*
- Acorn Woodpecker, *Melanerpes formicivorus*
- Golden-fronted Woodpecker, *Melanerpes aurifrons*
- Red-bellied Woodpecker, *Melanerpes carolinus*
- Williamson's Sapsucker, *Sphyrapicus thyroideus*
- Yellow-bellied Sapsucker, *Sphyrapicus varius*
- Red-naped Sapsucker, *Sphyrapicus nuchalis*
- Red-breasted Sapsucker, *Sphyrapicus ruber**
- Downy Woodpecker, *Dryobates pubescens*
- Ladder-backed Woodpecker, *Dryobates scalaris*
- Red-cockaded Woodpecker, *Dryobates borealis*
- Hairy Woodpecker, *Dryobates villosus*
- Northern Flicker, *Colaptes auratus*
- Pileated Woodpecker, *Dryocopus pileatus*
- Ivory-billed Woodpecker, *Campephilus principalis (E)**

FALCONIDAE (FALCONS AND CARACARAS)

- Collared Forest-Falcon, *Micrastur semitorquatus**
- Crested Caracara, *Caracara plancus*
- American Kestrel, *Falco sparverius*
- Merlin, *Falco columbarius*
- Aplomado Falcon, *Falco femoralis* (RI)
- Bat Falcon, *Falco rufigularis**
- Gyrfalcon, *Falco rusticolus**

- Peregrine Falcon, *Falco peregrinus*
- Prairie Falcon, *Falco mexicanus*

**PSITTACIDAE
(NEW WORLD AND AFRICAN PARROTS)**
- Monk Parakeet, *Myiopsitta monachus* (I)
- Carolina Parakeet, *Conuropsis carolinensis (E)**
- Green Parakeet, *Psittacara holochlorus* (u)
- Red-crowned Parrot, *Amazona viridigenalis* (u)

TITYRIDAE (TITYRAS AND ALLIES)
- Masked Tityra, *Tityra semifasciata**
- Rose-throated Becard, *Pachyramphus aglaiae**

TYRANNIDAE (TYRANT FLYCATCHERS)
- Northern Beardless-Tyrannulet, *Camptostoma imberbe*
- Greenish Elaenia, *Myiopagis viridicata**
- Small-billed Elaenia, *Elaenia parvirostris**
- White-crested Elaenia, *Elaenia albiceps**
- Dusky-capped Flycatcher, *Myiarchus tuberculifer*
- Ash-throated Flycatcher, *Myiarchus cinerascens*
- Nutting's Flycatcher, *Myiarchus nuttingi**
- Great Crested Flycatcher, *Myiarchus crinitus*
- Brown-crested Flycatcher, *Myiarchus tyrannulus*
- Great Kiskadee, *Pitangus sulphuratus*
- Social Flycatcher, *Myiozetetes similis**
- Sulphur-bellied Flycatcher, *Myiodynastes luteiventris**
- Piratic Flycatcher, *Legatus leucophaius**
- Variegated Flycatcher, *Empidonomus varius**
- Tropical Kingbird, *Tyrannus melancholicus*
- Couch's Kingbird, *Tyrannus couchii*
- Cassin's Kingbird, *Tyrannus vociferans*
- Thick-billed Kingbird, *Tyrannus crassirostris**
- Western Kingbird, *Tyrannus verticalis*
- Eastern Kingbird, *Tyrannus tyrannus*
- Gray Kingbird, *Tyrannus dominicensis**
- Scissor-tailed Flycatcher, *Tyrannus forficatus*

- Fork-tailed Flycatcher, *Tyrannus savana**
- Tufted Flycatcher, *Mitrephanes phaeocercus**
- Olive-sided Flycatcher, *Contopus cooperi*
- Greater Pewee, *Contopus pertinax**
- Western Wood-Pewee, *Contopus sordidulus*
- Eastern Wood-Pewee, *Contopus virens*
- Yellow-bellied Flycatcher, *Empidonax flaviventris*
- Acadian Flycatcher, *Empidonax virescens*
- Alder Flycatcher, *Empidonax alnorum*
- Willow Flycatcher, *Empidonax traillii*
- Least Flycatcher, *Empidonax minimus*
- Hammond's Flycatcher, *Empidonax hammondii*
- Gray Flycatcher, *Empidonax wrightii*
- Dusky Flycatcher, *Empidonax oberholseri*
- Western Flycatcher, *Empidonax difficilis*
- Buff-breasted Flycatcher, *Empidonax fulvifrons**
- Black Phoebe, *Sayornis nigricans*
- Eastern Phoebe, *Sayornis phoebe*
- Say's Phoebe, *Sayornis saya*
- Vermilion Flycatcher, *Pyrocephalus rubinus*

THAMNOPHILIDAE (TYPICAL ANTBIRDS)
- Barred Antshrike, *Thamnophilus doliatus**

VIREONIDAE (VIREOS, SHRIKE-BABBLERS, AND ERPORNIS)
- Black-capped Vireo, *Vireo atricapilla*
- White-eyed Vireo, *Vireo griseus*
- Bell's Vireo, *Vireo bellii*
- Gray Vireo, *Vireo vicinior*
- Hutton's Vireo, *Vireo huttoni*
- Yellow-throated Vireo, *Vireo flavifrons*
- Cassin's Vireo, *Vireo cassinii*
- Blue-headed Vireo, *Vireo solitarius*
- Plumbeous Vireo, *Vireo plumbeus*
- Philadelphia Vireo, *Vireo philadelphicus*
- Warbling Vireo, *Vireo gilvus*
- Red-eyed Vireo, *Vireo olivaceus*
- Yellow-green Vireo, *Vireo flavoviridis**
- Black-whiskered Vireo, *Vireo altiloquus**
- Yucatan Vireo, *Vireo magister**

LANIIDAE (SHRIKES)
- ☐ Loggerhead Shrike, *Lanius ludovicianus*
- ☐ Northern Shrike, *Lanius borealis*

CORVIDAE (CROWS, JAYS, AND MAGPIES)
- ☐ Brown Jay, *Cyanocorax morio**
- ☐ Green Jay, *Cyanocorax yncas*
- ☐ Pinyon Jay, *Gymnorhinus cyanocephalus**
- ☐ Steller's Jay, *Cyanocitta stelleri*
- ☐ Blue Jay, *Cyanocitta cristata*
- ☐ Woodhouse's Scrub-Jay, *Aphelocoma woodhouseii*
- ☐ Mexican Jay, *Aphelocoma wollweberi*
- ☐ Clark's Nutcracker, *Nucifraga columbiana**
- ☐ Black-billed Magpie, *Pica hudsonia**
- ☐ American Crow, *Corvus brachyrhynchos*
- ☐ Tamaulipas Crow, *Corvus imparatus**
- ☐ Fish Crow, *Corvus ossifragus*
- ☐ Chihuahuan Raven, *Corvus cryptoleucus*
- ☐ Common Raven, *Corvus corax*

REMIZIDAE (PENDULINE-TITS)
- ☐ Verdin, *Auriparus flaviceps*

PARIDAE
(TITS, CHICKADEES, AND TITMICE)
- ☐ Carolina Chickadee, *Poecile carolinensis*
- ☐ Black-capped Chickadee, *Poecile atricapillus**
- ☐ Mountain Chickadee, *Poecile gambeli*
- ☐ Juniper Titmouse, *Baeolophus ridgwayi*
- ☐ Tufted Titmouse, *Baeolophus bicolor*
- ☐ Black-crested Titmouse, *Baeolophus atricristatus*

ALAUDIDAE (LARKS)
- ☐ Horned Lark, *Eremophila alpestris*

HIRUNDINIDAE (SWALLOWS)
- ☐ Blue-and-white Swallow, *Pygochelidon cyanoleuca**
- ☐ Bank Swallow, *Riparia riparia*
- ☐ Tree Swallow, *Tachycineta bicolor*
- ☐ Violet-green Swallow, *Tachycineta thalassina*
- ☐ Northern Rough-winged Swallow, *Stelgidopteryx serripennis*

- ☐ Purple Martin, *Progne subis*
- ☐ Gray-breasted Martin, *Progne chalybea**
- ☐ Barn Swallow, *Hirundo rustica*
- ☐ Cliff Swallow, *Petrochelidon pyrrhonota*
- ☐ Cave Swallow, *Petrochelidon fulva*

AEGITHALIDAE (LONG-TAILED TITS)
- ☐ Bushtit, *Psaltriparus minimus*

PYCNONOTIDAE (BULBULS)
- ☐ Red-vented Bulbul, *Pycnonotus cafer* (I)

REGULIDAE (KINGLETS)
- ☐ Ruby-crowned Kinglet, *Corthylio calendula*
- ☐ Golden-crowned Kinglet, *Regulus satrapa*

BOMBYCILLIDAE (WAXWINGS)
- ☐ Bohemian Waxwing, *Bombycilla garrulus**
- ☐ Cedar Waxwing, *Bombycilla cedrorum*

PTILIOGONATIDAE (SILKY-FLYCATCHERS)
- ☐ Gray Silky-flycatcher, *Ptiliogonys cinereus**
- ☐ Phainopepla, *Phainopepla nitens*

SITTIDAE (NUTHATCHES)
- ☐ Red-breasted Nuthatch, *Sitta canadensis*
- ☐ White-breasted Nuthatch, *Sitta carolinensis*
- ☐ Pygmy Nuthatch, *Sitta pygmaea*
- ☐ Brown-headed Nuthatch, *Sitta pusilla*

CERTHIIDAE (TREECREEPERS)
- ☐ Brown Creeper, *Certhia americana*

POLIOPTILIDAE (GNATCATCHERS)
- ☐ Blue-gray Gnatcatcher, *Polioptila caerulea*
- ☐ Black-tailed Gnatcatcher, *Polioptila melanura*

TROGLODYTIDAE (WRENS)
- ☐ Rock Wren, *Salpinctes obsoletus*
- ☐ Canyon Wren, *Catherpes mexicanus*
- ☐ Cactus Wren, *Campylorhynchus brunneicapillus*
- ☐ Bewick's Wren, *Thryomanes bewickii*
- ☐ Carolina Wren, *Thryothorus ludovicianus*

- Northern House Wren, *Troglodytes aedon*
- Pacific Wren, *Troglodytes pacificus**
- Winter Wren, *Troglodytes hiemalis*
- Sedge Wren, *Cistothorus stellaris*
- Marsh Wren, *Cistothorus palustris*

MIMIDAE
(MOCKINGBIRDS AND THRASHERS)

- Blue Mockingbird, *Melanotis caerulescens**
- Black Catbird, *Melanoptila glabrirostris**
- Gray Catbird, *Dumetella carolinensis*
- Curve-billed Thrasher, *Toxostoma curvirostre*
- Brown Thrasher, *Toxostoma rufum*
- Long-billed Thrasher, *Toxostoma longirostre*
- Crissal Thrasher, *Toxostoma crissale*
- Sage Thrasher, *Oreoscoptes montanus*
- Northern Mockingbird, *Mimus polyglottos*

STURNIDAE (STARLINGS)
- European Starling, *Sturnus vulgaris* (I)

CINCLIDAE (DIPPERS)
- American Dipper, *Cinclus mexicanus**

TURDIDAE (THRUSHES AND ALLIES)
- Eastern Bluebird, *Sialia sialis*
- Western Bluebird, *Sialia Mexicana*
- Mountain Bluebird, *Sialia currucoides*
- Townsend's Solitaire, *Myadestes townsendi*
- Orange-billed Nightingale-Thrush, *Catharus aurantiirostris**
- Black-headed Nightingale-Thrush, *Catharus mexicanus**
- Veery, *Catharus fuscescens*
- Gray-cheeked Thrush, *Catharus minimus*
- Swainson's Thrush, *Catharus ustulatus*
- Hermit Thrush, *Catharus guttatus*
- Wood Thrush, *Hylocichla mustelina*
- Clay-colored Thrush, *Turdus grayi*
- White-throated Thrush, *Turdus assimilis**
- Rufous-backed Robin, *Turdus rufopalliatus**
- American Robin, *Turdus migratorius*
- Varied Thrush, *Ixoreus naevius**
- Aztec Thrush, *Ridgwayia pinicola**

MUSCICAPIDAE
(OLD WORLD FLYCATCHERS)
- Northern Wheatear, *Oenanthe oenanthe**

PEUCEDRAMIDAE (OLIVE WARBLER)
- Olive Warbler, *Peucedramus taeniatus**

ESTRILDIDAE (WAXBILLS)
- Scaly-breasted Munia, *Lonchura punctulata* (I)

PASSERIDAE (OLD WORLD SPARROWS)
- House Sparrow, *Passer domesticus* (I)

MOTACILLIDAE (WAGTAILS AND PIPITS)
- White Wagtail, *Motacilla alba**
- American Pipit, *Anthus rubescens*
- Sprague's Pipit, *Anthus spragueii*

FRINGILLIDAE
(FINCHES, EUPHONIAS, AND ALLIES)
- Evening Grosbeak, *Coccothraustes vespertinus**
- Pine Grosbeak, *Pinicola enucleator**
- Gray-crowned Rosy-Finch, *Leucosticte tephrocotis**
- House Finch, *Haemorhous mexicanus*
- Purple Finch, *Haemorhous purpureus*
- Cassin's Finch, *Haemorhous cassinii*
- Redpoll, *Acanthis flammea**
- Red Crossbill, *Loxia curvirostra*
- White-winged Crossbill, *Loxia leucoptera**
- Pine Siskin, *Spinus pinus*
- Lesser Goldfinch, *Spinus psaltria*
- Lawrence's Goldfinch, *Spinus lawrencei**
- American Goldfinch, *Spinus tristis*

CALCARIIDAE
(LONGSPURS AND SNOW BUNTINGS)
- Lapland Longspur, *Calcarius lapponicus**
- Chestnut-collared Longspur, *Calcarius ornatus**
- Smith's Longspur, *Calcarius pictus**
- Thick-billed Longspur, *Rhynchophanes mccownii**
- Snow Bunting, *Plectrophenax nivalis**

PASSERELLIDAE (NEW WORLD SPARROWS)

- Botteri's Sparrow, *Peucaea botterii*
- Cassin's Sparrow, *Peucaea cassinii*
- Bachman's Sparrow, *Peucaea aestivalis*
- Grasshopper Sparrow, *Ammodramus savannarum*
- Olive Sparrow, *Arremonops rufivirgatus*
- Black-throated Sparrow, *Amphispiza bilineata*
- Lark Sparrow, *Chondestes grammacus*
- Lark Bunting, *Calamospiza melanocorys*
- Chipping Sparrow, *Spizella passerina*
- Clay-colored Sparrow, *Spizella pallida*
- Black-chinned Sparrow, *Spizella atrogularis*
- Field Sparrow, *Spizella pusilla*
- Brewer's Sparrow, *Spizella breweri*
- Fox Sparrow, *Passerella iliaca*
- American Tree Sparrow, *Spizelloides arborea*
- Dark-eyed Junco, *Junco hyemalis*
- Yellow-eyed Junco, *Junco phaeonotus**
- White-crowned Sparrow, *Zonotrichia leucophrys*
- Golden-crowned Sparrow, *Zonotrichia atricapilla**
- Harris's Sparrow, *Zonotrichia querula*
- White-throated Sparrow, *Zonotrichia albicollis*
- Sagebrush Sparrow, *Artemisiospiza nevadensis*
- Vesper Sparrow, *Pooecetes gramineus*
- LeConte's Sparrow, *Ammospiza leconteii*
- Seaside Sparrow, *Ammospiza maritima*
- Nelson's Sparrow, *Ammospiza nelsoni*
- Baird's Sparrow, *Centronyx bairdii**
- Henslow's Sparrow, *Centronyx henslowii**
- Savannah Sparrow, *Passerculus sandwichensis*
- Song Sparrow, *Melospiza melodia*
- Lincoln's Sparrow, *Melospiza lincolnii*
- Swamp Sparrow, *Melospiza georgiana*
- Canyon Towhee, *Melozone fusca*
- Rufous-crowned Sparrow, *Aimophila ruficeps*
- Green-tailed Towhee, *Pipilo chlorurus*
- Spotted Towhee, *Pipilo maculatus*
- Eastern Towhee, *Pipilo erythrophthalmus*

ICTERIIDAE (YELLOW-BREASTED CHAT)

- Yellow-breasted Chat, *Icteria virens*

ICTERIDAE (ORIOLES AND BLACKBIRDS)

- Yellow-headed Blackbird, *Xanthocephalus xanthocephalus*
- Bobolink, *Dolichonyx oryzivorus*
- Chihuahuan Meadowlark, *Sturnella lilianae*
- Eastern Meadowlark, *Sturnella magna*
- Western Meadowlark, *Sturnella neglecta*
- Black-vented Oriole, *Icterus wagleri**
- Orchard Oriole, *Icterus spurius*
- Hooded Oriole, *Icterus cucullatus*
- Streak-backed Oriole, *Icterus pustulatus**
- Bullock's Oriole, *Icterus bullockii*
- Altamira Oriole, *Icterus gularis*
- Audubon's Oriole, *Icterus graduacauda*
- Baltimore Oriole, *Icterus galbula*
- Scott's Oriole, *Icterus parisorum*
- Red-winged Blackbird, *Agelaius phoeniceus*
- Shiny Cowbird, *Molothrus bonariensis**
- Bronzed Cowbird, *Molothrus aeneus*
- Brown-headed Cowbird, *Molothrus ater*
- Rusty Blackbird, *Euphagus carolinus*
- Brewer's Blackbird, *Euphagus cyanocephalus*
- Common Grackle, *Quiscalus quiscula*
- Boat-tailed Grackle, *Quiscalus major*
- Great-tailed Grackle, *Quiscalus mexicanus*

PARULIDAE (NEW WORLD WARBLERS)

- Ovenbird, *Seiurus aurocapilla*
- Worm-eating Warbler, *Helmitheros vermivorum*
- Louisiana Waterthrush, *Parkesia motacilla*
- Northern Waterthrush, *Parkesia noveboracensis*
- Golden-winged Warbler, *Vermivora chrysoptera*
- Blue-winged Warbler, *Vermivora cyanoptera*
- Black-and-white Warbler, *Mniotilta varia*
- Prothonotary Warbler, *Protonotaria citrea*

- Swainson's Warbler, *Limnothlypis swainsonii*
- Crescent-chested Warbler, *Oreothlypis superciliosa**
- Tennessee Warbler, *Leiothlypis peregrina*
- Orange-crowned Warbler, *Leiothlypis celata*
- Colima Warbler, *Leiothlypis crissalis*
- Lucy's Warbler, *Leiothlypis luciae*
- Nashville Warbler, *Leiothlypis ruficapilla*
- Virginia's Warbler, *Leiothlypis virginiae*
- Connecticut Warbler, *Oporornis agilis**
- Gray-crowned Yellowthroat, *Geothlypis poliocephala**
- MacGillivray's Warbler, *Geothlypis tolmiei*
- Mourning Warbler, *Geothlypis philadelphia*
- Kentucky Warbler, *Geothlypis formosa*
- Common Yellowthroat, *Geothlypis trichas*
- Hooded Warbler, *Setophaga citrina*
- American Redstart, *Setophaga ruticilla*
- Cape May Warbler, *Setophaga tigrina*
- Cerulean Warbler, *Setophaga cerulea*
- Northern Parula, *Setophaga americana*
- Tropical Parula, *Setophaga pitiayumi*
- Magnolia Warbler, *Setophaga magnolia*
- Bay-breasted Warbler, *Setophaga castanea*
- Blackburnian Warbler, *Setophaga fusca*
- Yellow Warbler, *Setophaga petechia*
- Chestnut-sided Warbler, *Setophaga pensylvanica*
- Blackpoll Warbler, *Setophaga striata*
- Black-throated Blue Warbler, *Setophaga caerulescens*
- Palm Warbler, *Setophaga palmarum*
- Pine Warbler, *Setophaga pinus*
- Yellow-rumped Warbler, *Setophaga coronata*
- Yellow-throated Warbler, *Setophaga dominica*
- Prairie Warbler, *Setophaga discolor*
- Grace's Warbler, *Setophaga graciae*
- Black-throated Gray Warbler, *Setophaga nigrescens*
- Townsend's Warbler, *Setophaga townsendi*
- Hermit Warbler, *Setophaga occidentalis*
- Golden-cheeked Warbler, *Setophaga chrysoparia*

- Black-throated Green Warbler, *Setophaga virens*
- Fan-tailed Warbler, *Basileuterus lachrymosus**
- Rufous-capped Warbler, *Basileuterus rufifrons**
- Golden-crowned Warbler, *Basileuterus culicivorus**
- Canada Warbler, *Cardellina canadensis*
- Wilson's Warbler, *Cardellina pusilla*
- Red-faced Warbler, *Cardellina rubrifrons**
- Painted Redstart, *Myioborus pictus*
- Slate-throated Redstart, *Myioborus miniatus**

CARDINALIDAE (CARDINALS AND ALLIES)

- Hepatic Tanager, *Piranga flava*
- Summer Tanager, *Piranga rubra*
- Scarlet Tanager, *Piranga olivacea*
- Western Tanager, *Piranga ludoviciana*
- Flame-colored Tanager, *Piranga bidentata**
- Crimson-collared Grosbeak, *Periporphyrus celaeno**
- Northern Cardinal, *Cardinalis cardinalis*
- Pyrrhuloxia, *Cardinalis sinuatus*
- Yellow Grosbeak, *Pheucticus chrysopeplus**
- Rose-breasted Grosbeak, *Pheucticus ludovicianus*
- Black-headed Grosbeak, *Pheucticus melanocephalus*
- Blue Bunting, *Cyanocompsa parellina**
- Blue Grosbeak, *Passerina caerulea*
- Lazuli Bunting, *Passerina amoena*
- Indigo Bunting, *Passerina cyanea*
- Varied Bunting, *Passerina versicolor*
- Painted Bunting, *Passerina ciris*
- Dickcissel, *Spiza americana*

THRAUPIDAE (TANAGERS AND ALLIES)

- Red-legged Honeycreeper, *Cyanerpes cyaneu**
- Yellow-faced Grassquit, *Tiaris olivaceus**
- Morelet's Seedeater, *Sporophila morelleti*

RECOMMENDED RESOURCES

SCIENCE, CONSERVATION, AND RECREATION ORGANIZATIONS

United States Fish and Wildlife Service
Southwest Region
fws.gov/about/region/southwest

Texas Parks and Wildlife Department
tpwd.texas.gov

North American Breeding Bird Survey
www.pwrc.usgs.gov/bbs

eBird
ebird.org/home

Cornell Lab of Ornithology
birds.cornell.edu

Great Texas Wildlife Trails
tpwd.texas.gov/huntwild/wildlife/wild-
life-trails

TEXBIRDS
freelists.org/list/texbirds

Birding Hotspots
birdinghotspots.org

BIRDING AND CONSERVATION ORGANIZATIONS

Audubon Dallas
audubondallas.org

Audubon Outdoor Club of Corpus Christi
audubonoutdoorclub.com

Bastrop County Audubon Society
bastropcountyaudubon.org

Bexar Audubon Society
bexaraudubon.org

Big Country Audubon Society
bigcountryaudubon.org

El Paso/Trans-Pecos Audubon Society
trans-pecos-audubon.com

Fort Worth Audubon Society
fwas.org

Frontera Audubon
fronteraaudubon.org

Golden Triangle Audubon Society
goldentriangleaudubon.org

Gulf Coast Bird Observatory
gcbo.org

Houston Audubon
houstonaudubon.org

Huntsville Audubon Society
huntsvilleaudubon.org

Monte Mucho Audubon Society
montemucho.com

Prairies and Timbers Audubon Society
prairieandtimbers.org

Rio Brazos Audubon Society
riobrazosaudubon.org

San Antonio Audubon Society
saaudubon.org

Texas Master Naturalist Program
txmn.tamu.edu

Texas Ornithological Society
texasbirds.org

Texas Panhandle Audubon Society
txpas.org

Texoma Audubon Society
texomaaudubon.org

Travis Audubon
travisaudubon.org

Twin Lakes Audubon Society
twinlakesaudubon.org

Tyler Audubon Society
tyleraudubon.org

REFERENCES

All About Birds
allaboutbirds.org

Birds of the World
birdsoftheworld.org

Xeno-canto
xeno-canto.org

OPPOSITE Short-eared Owl

ACKNOWLEDGMENTS

WITHOUT THE SUPPORT of my wife Donna this field guide might not have been possible. She was there to listen to me bemoan the task ahead many times, then gently and purposefully steered me back on the path to success.

I would like to thank the photographers who contributed to this work and whose excellent images make this field guide better: Fay Ratta, Norman Welsh, Bob Friedrichs, Joseph H. Hood, and Quentin Thigpen.

Special thanks to my all my birding friends who I peppered with questions throughout this process, but especially John Berner and John O'Brien, who had to endure miles and miles of my prattling on about this volume while we chased rare birds across Texas.

And lastly, I must acknowledge the role of my close friend David Dauphin, who we lost in 2015. He was mentor extraordinaire and taught me much about birds and the natural world, the amazing birding of Texas, the amazing places of Texas, about being a leader in organizations, and about people in general. His influence was ever present as I worked these months on this field guide.

OPPOSITE Lazuli Bunting

BIBLIOGRAPHY

Behrstock, Robert A., Ted L. Eubanks Jr., and Ron J. Weeks. 2006. *Birdlife of Houston, Galveston, and the Upper Texas Coast*. College Station, TX: Texas A&M University Press.

Birds of the World. Cornell Lab of Ornithology. BirdsoftheWorld.org. Cornell University, NY.

Choate, Ernest A. 1985. *Dictionary of American Bird Names, Revised Edition*. Boston: The Harvard Common Press.

Crossley, Richard. 2011. *The Crossley ID Guide: Eastern Birds*. Princeton, NJ: Princeton University Press.

Crossley, Richard, Kevin Karlson, and Michael O'Brien. 2006. *The Shorebird Guide*. New York: Houghton Mifflin Company.

Dunn, Jon, and Kimball Garrett. 1997. *A Peterson Field Guide to Warblers of North America*. New York: Houghton Mifflin Company.

Freeman, Brush, and Mark W. Lockwood. *The TOS Handbook of Texas Birds*. College Station, TX: Texas A&M University Press, 2004

Holloway, Joel Ellis. 2003. *Dictionary of Birds of the United States*. Portland, OR: Timber Press.

Peterson, Roger Tory. 1980. *A Field Guide to the Birds East of the Rockies*. 4th ed. New York: Houghton Mifflin Company.

Sibley, David Allen. 2001. *The Sibley Guide to Bird Life and Behavior*. New York: Alfred A. Knopf.

Sibley, David Allen. 2014. *The Sibley Guide to Birds*. 2nd ed. New York: Alfred A. Knopf.

Tveten, John L. 1993. *The Birds of Texas*. Fredericksburg, TX: Shearer Publishing.

Vuilleumier, François. 2020. *American Museum of Natural History Birds of North America*. 3rd ed. New York: DK Publishing.

PHOTO AND ILLUSTRATION CREDITS

All photographs are by the author except the following:

Ben Horstmann, 34 (bottom)
Bob Friedrichs, 12, 53, 100, 116, 122, 132, 158 (main), 196, 215 (left), 289, 294, 311 (left), 356, 359, 360, 396, 466–469, 476, 508, 512 (inset), 523, 525, 531, 534 (right), 535–537, 539, 545, 547, 549 (right), 560–562, 564, 569
Fay Ratta, 71 (left), 89, 124, 125, 130 (right), 278 (left), 308 (left), 317 (right), 391, 420, 483, 511, 580, 600
Javier De Leon, 30–31
Jesse Huth, 145
Joseph H. Hood, 54 (nset), 71 (right), 72, 95, 99, 103, 113, 159 (inset), 163, 169, 172 (left), 184 (right), 185, 193, 199, 314, 330, 376 (left), 383, 519 (right)
Michael Good, 582 (right)
Michael Gray, 178
Norman Welsh, 21 (top), 29, 43, 48, 52, 58, 61, 63 (main), 64 (right), 65, 66, 67 (main), 74 (main), 75 (main), 76, 78 (right), 79, 83, 101, 117, 128 (left), 139, 149, 157, 173, 176, 187 (main), 189 (inset), 212, 216 (right), 233 (left), 255 (left), 268, 269 (left, right top), 270, 272 (left, right bottom), 276 (right), 278 (right), 290, 298, 308 (right), 324, 338, 341, 342, 351 (right top), 358, 367, 389, 392, 411, 427, 433, 434, 437, 453 (left), 459 (right), 478, 486, 488, 498, 502, 505, 507 (left), 526, 544, 571, 575 (inset), 576 (inset), 577, 581 (right), 584 (inset), 599
Quentin Thigpen, 129 (left), 285, 355, 400, 444, 451, 503, 554
William_H_Powell, 4–5, 36–37

FLICKR

Jason Crotty, 175
Jeff Davis, 119

CREATIVE

Andy Reago & Chrissy McClarren, 118
Ann Harkness, 393 (right)
Brian Ralphs, 311 (right)
Dominic Sherony, 274 (right), 504
Don Faulkner, 375
Eric Ellingson, 221
Greg Schecter, 222
Joyce Cory, 205
Shawn Taylor, 121

PUBLIC DOMAIN 1.0

Alan Schmierer, 332

INDEX

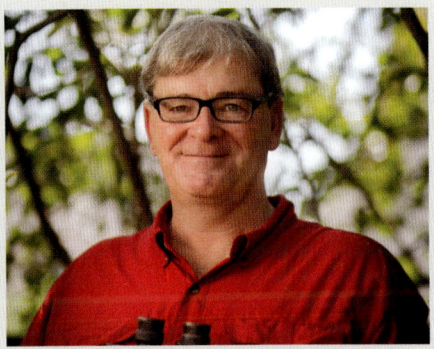

David Sarkozi, while still in middle school, read about Kenn Kaufman's record-breaking 1973 Big Year and thought that sounded like a fun idea. Starting out with his father's binoculars and the family encyclopedia, he began birding. Soon he spent his lawn-mowing money on a Golden Guide and spent many mornings birding at the local park. At the University of Houston, birding was put on the back burner. Soon after college, though, while at the beach, a bird he didn't recognize flew by, rekindling his curiosity, and he retrieved his long-unused Golden Guide. The new bird was a Black Tern, it turns out (pun intended!). He discovered he liked sharing what he learned about birds and started leading field trips for the Houston Ornithology Group. He became its chairman for two terms, as well as one of the founders of the Friends of Anahuac Refuge and president of the Texas Ornithological Society. After retiring from the University of Houston as manager of public safety systems, he returned to the idea of a Big Year of birding and in 2017 completed a 509-species big year in Texas. In 2019 he became the first person to bird all counties in Texas in a single year. He has a special interest in the birds of Belize and has led more than 30 birding trips there.